Managing Forest Ecosystems

Volume 35

Aims & Scope

Well-managed forests and woodlands are a renewable resource, producing essential raw material with minimum waste and energy use. Rich in habitat and species diversity, forests may contribute to increased ecosystem stability. They can absorb the effects of unwanted deposition and other disturbances and protect neighbouring ecosystems by maintaining stable nutrient and energy cycles and by preventing soil degradation and erosion. They provide much-needed recreation and their continued existence contributes to stabilizing rural communities.

Forests are managed for timber production and species, habitat and process conservation. A subtle shift from multiple-use management to ecosystems management is being observed and the new ecological perspective of multi-functional forest management is based on the principles of ecosystem diversity, stability and elasticity, and the dynamic equilibrium of primary and secondary production.

Making full use of new technology is one of the challenges facing forest management today. Resource information must be obtained with a limited budget. This requires better timing of resource assessment activities and improved use of multiple data sources. Sound ecosystems management, like any other management activity, relies on effective forecasting and operational control.

The aim of the book series *Managing Forest Ecosystems* is to present state-of-the-art research results relating to the practice of forest management. Contributions are solicited from prominent authors. Each reference book, monograph or proceedings volume will be focused to deal with a specific context. Typical issues of the series are: resource assessment techniques, evaluating sustainability for even-aged and uneven-aged forests, multi-objective management, predicting forest development, optimizing forest management, biodiversity management and monitoring, risk assessment and economic analysis.

More information about this series at http://www.springer.com/series/6247

Christoph Fischer • Berthold Traub
Editors

Swiss National Forest Inventory – Methods and Models of the Fourth Assessment

Springer

Editors
Christoph Fischer
Swiss Federal Institute for Forest, Snow
and Landscape Research WSL
Birmensdorf, Switzerland

Berthold Traub
Swiss Federal Institute for Forest, Snow
and Landscape Research WSL
Birmensdorf, Switzerland

ISSN 1568-1319 ISSN 2352-3956 (electronic)
Managing Forest Ecosystems
ISBN 978-3-030-19292-1 ISBN 978-3-030-19293-8 (eBook)
https://doi.org/10.1007/978-3-030-19293-8

This Springer imprint is published by the registered company Springer Nature Switzerland AG.
The registered company address is: Gewerbestrasse 11, 6330 Cham, Switzerland

Preface

The first National Forest Inventory of Switzerland (NFI1) was conducted between 1983 and 1985. Since then the goals and methods of the NFI have been continuously developed to meet changing and emerging needs. Conducting a national forest inventory involves addressing a large number of issues. To cover the full thematic extent, a wide range of methods are required and must be coordinated. Consequently, changes in methods must be conducted with care in order not to jeopardise the time series in the NFI's longitudinal data.

This book was written to provide a complete synopsis of all methods used in the fourth assessment cycle of the NFI (NFI4), conducted from 2009 to 2017, and the interconnections between them. The book is divided into parts and chapters that reflect the broad range of topics covered in NFI4, with a particular emphasis on data handling and evaluation. This book can also be understood as an update to the last complete description of NFI methods published in 2001, where the focus was on the methods used in the second NFI (NFI2) carried out between 1993 and 1995.

We hope this book may serve as a guideline not only for all who are involved in the further development of already established national forest inventories, but also for those who have to develop such an inventory from scratch. The procedures and methods presented here have been confirmed to be consistent and efficient, and found to meet high-quality standards. They form a reliable basis for developing any long-term monitoring system for future forest inventories.

Birmensdorf, Switzerland
February 2019

Christoph Hegg
Christoph Fischer
Berthold Traub

Acknowledgements

Many people have contributed to this book. These include not only the authors of the individual chapters, but also colleagues from other National Forest Inventories, as well as renowned researchers in the field, in particular:

Radim Adolt
Susana Barreiro
Antoine Colin
Dominik Cullmann
Karsten Dunger
Jiří Fejfar
Patrizia Gasparini
Thomas Gschwantner
Gerald Kändler
Christoph Kleinn
Michael Köhl
Miloš Kučera
Paul Magdon
Steen Magnussen
Jiří Nechvíle
Heino Polley
Thomas Riedel
Klemens Schadauer
Göran Ståhl

Without their help, we would not have been able to complete the book. We would also like to thank the reviewers for helping us improve our work through their critical feedback. For any remaining errors, the authors and editors take full responsibility.

We also owe special thanks to Silvia Dingwall and Melissa Dawes, who have improved the English in this book with patience, knowledge and good sense. In addition, we are grateful to Valeria Rinaudo, Ineke Ravesloot and Jill Ritchie from Spinger for their support along the way.

We would also like to thank the staff of the Swiss National Forest Inventory for their support. The book was created within the framework of the Swiss National Forest Inventory, a joined research program of the Swiss Federal Office for the Environment (FOEN) and the Swiss Federal Institute for Forest, Snow and Landscape Research (WSL).

Last, but certainly not least, we thank our families for their constant patience and moral support.

Contents

Part I Swiss NFI at a Glance

1 **Swiss NFI at a Glance** 3
Urs-Beat Brändli and Martin Hägeli

Part II Sampling Design and Estimation Procedures

2 **Sampling Design and Estimation Procedures** 39
Adrian Lanz, Christoph Fischer, and Meinrad Abegg

Part III Remote Sensing

3 **Remote Sensing Data Sources** 95
Christian Ginzler

4 **Remote Sensing Data Management** 101
Rolf Meile

5 **Stereo GIS Application for Aerial-Image Interpretation** 105
Barbara Schneider

6 **Variables on the Sample Plot Captured by the Stereo GIS Application** 111
Christian Ginzler

7 **Area-Wide Products** 125
Christian Ginzler, Bronwyn Price, Ruedi Bösch, Christoph Fischer, Martina L. Hobi, Achilleas Psomas, Natalia Rehush, Zuyuan Wang, and Lars T. Waser

Part IV Terrestrial Inventory and Interview Survey

8 Planning and Organisation of the Terrestrial Inventory and Interview Survey 145
Fabrizio Cioldi and Markus Keller

9 Field Assessment 159
Christoph Düggelin, Markus Keller, and Fabrizio Cioldi

10 Interview Survey with the Local Forest Service 187
Christoph Fischer and Marielle Fraefel

11 Time Study of the Terrestrial Inventory and Interview Survey 197
Fabrizio Cioldi

Part V Modelling of Forest Resources

12 State and Change of Forest Resources 205
Anne Herold, Jürgen Zell, Brigitte Rohner, Markus Didion, Esther Thürig, and Erik Rösler

13 Assortments 231
Brigitte Rohner, Christoph Fischer, Anne Herold, and Erik Rösler

14 Whole Tree Biomass and Carbon Stock 243
Markus Didion, Anne Herold, and Esther Thürig

15 Other Models 249
Christoph Fischer, Urs-Beat Brändli, and Markus O. Huber

16 Calculation of Potential Timber Harvesting Costs (HeProMo) 257
Christoph Fischer and Golo Stadelmann

17 Forest Development Model MASSIMO 265
Golo Stadelmann, Christian Temperli, Brigitte Rohner, Markus Didion, Anne Herold, Erik Rösler, and Esther Thürig

18 Model of Carbon Cycling in Dead Organic Matter and Soil (Yasso07) 281
Markus Didion and Jürgen Zell

19 Scenario Simulations 285
Golo Stadelmann, Markus Didion, and Esther Thürig

Part VI The Swiss National Forest Inventory Data Analysis System

20 The Swiss National Forest Inventory Data Analysis System 299
Berthold Traub, Rolf Meile, Simon Speich, and Erik Rösler

Part VII Data Quality

21 Quality in Aerial-Image Interpretation and Field Survey 339
Berthold Traub, Fabrizio Cioldi, Christoph Düggelin, Markus Keller, and Christian Ginzler

22 Compilation of a High-Quality Forest Road Data Set 367
Marielle Fraefel and Christoph Fischer

Part VIII Raw Data Collection Software in the Swiss NFI

23 Raw Data Collection Software in the Swiss NFI 377
Sandro Bischof, Enikő Stüdeli-Fey, and Rolf Meile

Part IX Coordination of the Swiss National Forest Inventory Data Analysis System

24 Coordination of the Swiss National Forest Inventory Data Analysis System 405
Meinrad Abegg

Glossary 417

Index 427

Contributors

Meinrad Abegg Scientific Service NFI, Swiss Federal Institute for Forest, Snow and Landscape Research WSL, Birmensdorf, Switzerland

Sandro Bischof GIS, Swiss Federal Research Institute WSL, Birmensdorf, Switzerland

Ruedi Bösch Remote Sensing, Swiss Federal Research Institute WSL, Birmensdorf, Switzerland

Urs-Beat Brändli Scientific Service NFI, Swiss Federal Research Institute WSL, Birmensdorf, Switzerland

Fabrizio Cioldi Scientific Service NFI, Swiss Federal Research Institute WSL, Birmensdorf, Switzerland

Markus Didion Resource Analysis, Swiss Federal Research Institute WSL, Birmensdorf, Switzerland

Christoph Düggelin Scientific Service NFI, Swiss Federal Research Institute WSL, Birmensdorf, Switzerland

Christoph Fischer Scientific Service NFI, Swiss Federal Research Institute WSL, Birmensdorf, Switzerland

Marielle Fraefel GIS, Swiss Federal Research Institute WSL, Birmensdorf, Switzerland

Christian Ginzler Remote Sensing, Swiss Federal Research Institute WSL, Birmensdorf, Switzerland

Martin Hägeli GIS, Swiss Federal Research Institute WSL, Birmensdorf, Switzerland

Anne Herold Resource Analysis, Swiss Federal Research Institute WSL, Birmensdorf, Switzerland

Martina L. Hobi Stand Dynamics and Silviculture, Swiss Federal Research Institute WSL, Birmensdorf, Switzerland

Markus O. Huber Scientific Service NFI, Swiss Federal Research Institute WSL, Birmensdorf, Switzerland

Markus Keller Scientific Service NFI, Swiss Federal Research Institute WSL, Birmensdorf, Switzerland

Adrian Lanz Statistics Lab, Swiss Federal Research Institute WSL, Birmensdorf, Switzerland

Rolf Meile GIS, Swiss Federal Research Institute WSL, Birmensdorf, Switzerland

Bronwyn Price Remote Sensing, Swiss Federal Research Institute WSL, Birmensdorf, Switzerland

Achilleas Psomas Remote Sensing, Swiss Federal Research Institute WSL, Birmensdorf, Switzerland

Natalia Rehush Scientific Service NFI, Swiss Federal Research Institute WSL, Birmensdorf, Switzerland

Brigitte Rohner Resource Analysis, Swiss Federal Research Institute WSL, Birmensdorf, Switzerland

Erik Rösler Resource Analysis, Swiss Federal Research Institute WSL, Birmensdorf, Switzerland

Barbara Schneider Remote Sensing, Swiss Federal Research Institute WSL, Birmensdorf, Switzerland

Simon Speich Scientific Service NFI, Swiss Federal Research Institute WSL, Birmensdorf, Switzerland

Golo Stadelmann Resource Analysis, Swiss Federal Research Institute WSL, Birmensdorf, Switzerland

Enikő Stüdeli-Fey GIS, Swiss Federal Research Institute WSL, Birmensdorf, Switzerland

Christian Temperli Resource Analysis, Swiss Federal Research Institute WSL, Birmensdorf, Switzerland

Esther Thürig Resource Analysis, Swiss Federal Research Institute WSL, Birmensdorf, Switzerland

Berthold Traub Scientific Service NFI, Swiss Federal Research Institute WSL, Birmensdorf, Switzerland

Zuyuan Wang Remote Sensing, Swiss Federal Research Institute WSL, Birmensdorf, Switzerland

Lars T. Waser Remote Sensing, Swiss Federal Research Institute WSL, Birmensdorf, Switzerland

Jürgen Zell Resource Analysis, Swiss Federal Research Institute WSL, Birmensdorf, Switzerland

Part I
Swiss NFI at a Glance

Chapter 1
Swiss NFI at a Glance

Urs-Beat Brändli and Martin Hägeli

Abstract This chapter provides a general overview of the Swiss National Forest Inventory (NFI), as well as of its tasks, organisation, contents, methods and applications. Cross-references are given to those chapters in the book where the particular topic is dealt with in more detail.

1.1 Introduction

This chapter provides a general overview of the Swiss National Forest Inventory (Swiss NFI, syn. NFI), as well as of its tasks, organisation, contents, methods and applications. In the following sections cross-references are given to those chapters in the book where the particular topic is dealt with in more detail.

1.2 History and Objectives

This section is based on a recent publication by Lanz et al. (2016), which begins by describing how forests and wood resources in Switzerland have been used intensively since the eighteenth century. The overexploitation and clear-cutting of forests in the alpine and mountainous parts of the country during the first half of the nineteenth century are widely believed to have led to debris flows and floods that caused extensive damage in the lower-elevation parts of the country.

Shortly after Switzerland was founded as a modern federal state in 1848, a series of important decisions on forestry were taken nationally. In 1855 the then new Polytech, today's Swiss Federal Institute of Technology (ETH), was established with

U.-B. Brändli (✉)
Scientific Service NFI, Swiss Federal Research Institute WSL, Birmensdorf, Switzerland
e-mail: urs-beat.braendli@wsl.ch

M. Hägeli
GIS, Swiss Federal Research Institute WSL, Birmensdorf, Switzerland

C. Fischer, B. Traub (eds.), *Swiss National Forest Inventory – Methods and Models of the Fourth Assessment*, Managing Forest Ecosystems 35,
https://doi.org/10.1007/978-3-030-19293-8_1

a School (or Department) of Forestry. In 1874 the federal administration set up a forestry inspectorate, which later developed into the Federal Office for the Environment (FOEN). Finally, in 1876 a national Forest Act specifying legislation for the mountain forests came into force. Its primary aim was to prevent natural hazards, which is why clear-cutting regimes were banned, and the restoration of forests mostly through reforestation was promoted. In 1915 the Act was extended to include all Swiss forests. As a result, forest cover in Switzerland increased from around 19% in 1850 to nearly 32% in the year 2010.

The information on forests available remained limited and mainly only on public forests, which cover 69% of Switzerland's forest area. Estimates for e.g. timber volumes at the national level were unreliable because the cantons, the member states of the Swiss Confederation, had different systems for collecting data. In the late 1950s, a professor of forestry at ETH in Zurich, Alfred Kurt, initiated a political discussion about the need for a forest survey (Kurt 1957), which led, 25 years later, to the first National Forest Inventory (NFI1) in Switzerland being started in 1983 (Mahrer and Vollenweider 1983). Kurt maintained that a national forest inventory should provide enough information to: (a) establish a national forest policy to ensure that Switzerland's forestry remains internationally competitive, (b) support forest managers in maintaining social, protective and recreational services and raise public awareness about forests, and (c) take forests into account in planning regional and national land use (Kurt 1967). His original vision was, therefore, to carry out a national inventory in the form of a thematically broad survey that covered the many different products and services of forests.

By the 1960s, it had become clear that reliable data about forest conditions in Switzerland was urgently needed to improve silvicultural planning and its implementation. The first small NFI was carried out by what was then called the *Eidgenössische Anstalt für das forstliche Versuchswesen* (EAFV), which later became the Swiss Federal Institute for Forest, Snow and Landscape Research WSL. The survey involved interpreting not only aerial images, but also the responses to questionnaires sent out to the cantonal forest administrations (Ott 1972).

Work then began on preparing a more extensive national forest inventory, where the focus was mainly on assessing the productive functions of forests rather than on conducting a thematically broad survey as originally planned. This change in focus was probably largely due to the methodological difficulties involved in assessing and validating other forest goods and services, but it was also influenced by the 1973 oil crisis. Politically, it was seen as essential to have more information on the immediate and potential availability of wood resources, the expected timber harvesting costs, and the requirements for investment in forest roads and other infrastructure (BAR 1981). Monitoring social and protective functions was still, however, considered important.

It was not until new legislation was introduced in 1991 in the Federal Act on Forest that it became mandatory to consider all forest functions in sustainable forest management. The Act also required the national authorities to inform the public periodically about the state of the forests. By then the NFI had already become well established and was recognised as the main instrument for fulfilling the requirement to monitor and report on the state of Switzerland's forest.

Field data was collected on permanent sample plots established for the first National Forest Inventory (NFI1 1983/1985) in 1983–1985. The inventory was repeated 10 years later (NFI2 1993/1995). One of the main findings from these surveys was that it would be possible to increase the timber yield (SAEFL 1999). Since the third inventory conducted 11 years later (NFI3 2004/2006), the NFI plot-level scenario model MASSIMO (Chap. 17) has been used to analyse the role and capacity of Swiss forests in mitigating greenhouse gas effects and fulfilling the demand for wood.

With the fourth NFI (NFI4 2009/2017), it was decided to change to a 9-year 'rotating system' whereby field data on the sample of permanent plots are no longer collected periodically but continuously, using nine sub-grids, each of which is selected from the full grid and covers the entire country. A different sub-grid is used each year until the full 9-year cycle has been completed. This continuous inventory design allows representative reports on the state of the Swiss forest to be produced more frequently. The plan is for the NFI methods and/or content to be re-evaluated every 9 years, by when the whole sample has been measured. The inventory is carried out by WSL in collaboration with FOEN on a 4-year contract basis. The current mandate includes the main inventory module and specific modules on: (a) the greenhouse gas inventory (GHGI) (Chap. 14 and Sect. 19.3) and (b) scenario modelling (Sect. 19.2). In addition, (c) the forests' protective functions (Sect. 15.3) and (d) biodiversity functions (Sect. 15.2) are analysed in more detail. A further requirement is for (e) international reporting and cooperation to take place. The NFI additionally supports several cantons in the planning and implementation of regional inventories.

1.3 Legal Status, Obligations and Organisations

The first National Forest Inventory (NFI1 1983/85) was carried out after the Federal Council decided a survey was necessary in August 1981. The initial legislative basis for the NFI was first created in 1991 with the Federal Act on Forest (Forest Act), which has been in force since 1993. Today the NFI is mentioned in two federal acts and is subject to their regulations. In Art. 33 of the Forest Act (SR 921.0), the regulations for the data collections and the disclosure obligations of the authorities and private individuals are specified.

Since 2008 the Federal Office for the Environment (FOEN) has been responsible for collecting data on forests, according to Art. 37a of the Forest Ordinance (SR 921.01). It collects the basic data on the sites, the functions and the condition of the forest in the NFI in collaboration with the Swiss Federal Institute for Forest, Snow and Landscape Research WSL (Fig. 1.1). Art. 34 of the Forest Act specifies informing the authorities and the public as a requirement, which means also publishing the findings. The cantons are responsible for forest planning and management (SR 921.0, Art. 20), which are thus not part of NFI's tasks.

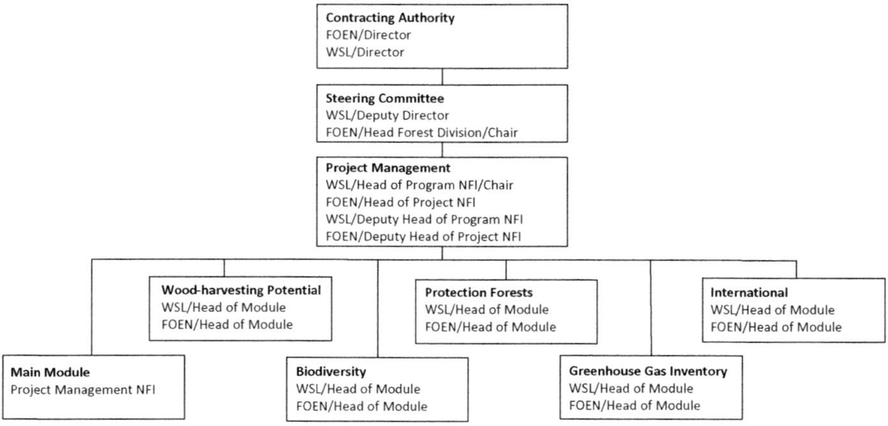

Fig. 1.1 NFI Organisational chart 2017

The data storage, availability, access and use of the NFI data as basic geo-data for Switzerland have been regulated in the Federal Act on Geoinformation GeoIA (SR 510.62) and in the Ordinance on Geoinformation (SR 510.620) since 2008, as have the responsibilities of WSL and FOEN. Access of the survey teams to the permanent sample plots for the data collection is guaranteed in Art. 20 and Art. 21 of GeoIA, as are the support of the local authorities and the marking of the sample-plot centres. This ensures NFI has the right to access private property to collect data and to install markings. The ETH Law (SR 414.110) grants, in Art. 5, NFI the freedom to do research and independence from political influence in the scientific interpretation of the data and in publishing findings.

With these legal bases, NFI has been organised as a Project that is jointly managed by FOEN and WSL (Fig. 1.1). The two Directors act as the joint Contracting Authorities on behalf of the Swiss Confederation. The aims and financial terms are regulated in a contract between FOEN and WSL for 4 years, which is revised at the end of each 4-year period. The Steering Committee of NFI directs the Project through the Project Management, which manages NFI operationally. A Master Plan serves as the basis for planning and management, which essentially operationalises the agreed services. It can be revised every 6 months. All the management staff are on the same levels as FOEN and WSL employees, with equal representation. They make decisions by consensus.

The Project itself is currently organised as a basic module plus five additional modules: biodiversity, wood-harvesting potential, protection forests, greenhouse gas inventory and international. The heads of each module are responsible for the module and for reporting to the Project Management (Fig. 1.1). WSL is in charge of the planning (Sects. 1.4, 1.5 and Chap. 8), development of methods (Part II, III and V), implementation (Part III, IV and VII), analysis (Part VI), scientific reporting and scientific services (Sect. 1.6). FOEN is responsible for the political interpretation of the results and for political support.

At WSL, NFI is organised as a programme with a matrix structure. It receives support from WSL's different Research Units, as well as the Service and Support Units, with the largest contributions coming from the Research Unit "Forest

Resources and Management". Expertise and information are consistently shared between at least two people (following the deputising principle) to ensure that the operation can be maintained reliably. The field surveyors are employed on annual contracts and paid an hourly wage. Piece-working is excluded to ensure the data is of high quality (Part VII).

About 30% of the funding is spent collecting data in the field surveys, conducting remote sensing, and interviewing the forestry services. Around 50% goes towards modelling, analyses, publications and reporting. About 15% is spent on the design, survey method and inventory statistics, and approx. 5% on data storage and providing access to the data on the Internet. All these proportions include the costs for research and development in the corresponding categories.

1.4 Assessing Information Needs and NFI Impacts

The federal government has been legally bound, since 1991, to carry out periodic surveys (Sect. 1.3) of forest sites, the forest's functions and condition, and the production and use of wood, as well as of the structures and economic situation in forestry (SR 921.0, Art. 33). To this end, FOEN and WSL should collect basic data about the sites, functions and condition of the forest in the National Forest Inventory (SR 921.01, Art. 37a). FOEN is responsible for informing the authorities and the public about the surveys.

The legislation contains no specifications about the NFI's contents, methods, products and services. There is considerable freedom to act – constrained, of course, by the limited resources available. This means the focus must be on the most important social issues. The information needs must also be clarified with the target groups. The relevance of a national inventory can be increased through periodic discussion (needs assessments), a good provision of products and services (output), a widespread use of the NFI (impact) and a retrospective evaluation. Satisfied users are potential advocates for long-term data series even in times when funding is tight. This is the conviction the NFI has followed for the past three decades.

What should be recorded in an NFI and at what spatial resolution? Even before the first National Forest Inventory (**NFI1**), the 'father' of the NFI (Kurt 1967) wrote: "It should not be underestimated that procedural questions are much easier to answer than those referring to the specific aims." A national work group with representatives from the present-day institutions FOEN and WSL, from the Association of Forest Owners (WaldSchweiz), from forestry and the wood industry and from the head foresters of the cantons (heads of the cantonal forest administrations) was directly or indirectly involved up to 1981 in designing the first NFI (Mahrer and Vollenweider 1983). During the long preparatory design process, the original idea of conducting an intensive, fully comprehensive forest inventory (Sect. 1.2) with the canton as the recording unit was reduced to a programme that ensured that the most important goals could be achieved at the level of Switzerland, with the creation of a sound basis for the forest policy of the federal government. As collection methods and models for the forest functions were lacking, the main focus was on forest resources and their availability (network of forest roads, costs of timber harvesting).

To define the aims for the second NFI (**NFI2**), a ten-member advisory committee was set up with representatives from FOEN, the cantons, WSL and other experts. In addition to recording the developments since the first NFI, the thematic areas forest functions, forest soil and ground vegetation were discussed as possible extensions (Zierhofer 2000). Models of the forest functions and ways to assess the forest services were developed and introduced in NFI2. The funding was not, however, sufficient to include soil and vegetation. In 1991, 276 national stakeholders were asked for the first time in a written questionnaire about what they needed the NFI data for and how they used it. The results served as input for the data catalogue NFI2 and for future reports on results and services (Brändli 1992). The responses (response rate of 47%) were almost all only positive, and showed that the NFI is well known among forest experts and widely applied.

In the run up to the third NFI (**NFI3**) after the completion of the second NFI, the effects of NFI1 and NFI2 were analysed and the needs for a third survey were ascertained. The formal bases (programme agreements), the NFI services (output) and the satisfaction and use of NFI users (impacts) were evaluated, and corresponding recommendations were made for implementing the NFI in the future (e.g. new products and user groups). The NFI's core business and its spatial and temporal resolution were defined and questions to be answered in NFI3 were formulated on the basis of structural interviews with 30 individual experts, subsequent extensive discussions in four workshops with 60 experts, and a final internet survey with 300 stakeholders (users). These expected responses were grouped in accordance with the Helsinki criteria (Ministerial Conference 1994), and priorities were defined for the Swiss government, cantons and research. NFI3 was designed in the form of eight modules, which makes it easier to assess its implementation capacity, decide on priorities and plan budgeting (Bättig et al. 2002). Bättig et al.'s study was also decisive for **NFI4** and for the planning of NFI5.

The output, impact and outcomes of NFI3 were evaluated in 2012 in a similar way to the previous NFI (Bernath et al. 2013) and assessed, with an independent consultant again performing the evaluation with WSL support. The feedback was once again largely very positive, and helpful in deciding on the priorities and outputs for future surveys. One of the recommendations implemented was to run periodic workshops for NFI users. These NFI-user workshops with 20–40 participants have been of great value in improving NFI products and services (Brändli and Bernasconi 2014, Abegg and Brändli 2016).

1.5 Content, Data Collection and Methods

Which survey method is used in the NFI depends on the information and degree of accuracy needed (Mahrer and Vollenweider 1983). With the first NFI, the goals to be taken into account were: to determine the forest area and the total wood volume in Switzerland with an accuracy of $\pm 0.5\%$ (relative standard error); to provide a basis for estimating the increment and use; and to develop a sampling grid, which could later be made denser for cantonal inventories. With these requirements, a nationwide uniform quadratic sampling grid (1 km × 1 km) was established with permanent sample plots at the nodes in the sampling grid (Part II). To minimise the amount of work in the field needed for the forest/non-forest decision and for determining the location of the sample plots, the aerial images from swisstopo (Federal Office of Topography) were interpreted before the field data was collected (Part III). The information required for NFI1 already included ownership, forest access roads, the methods and distances for transporting (hauling) wood to the roads, and the amount of potential wood for harvesting and its availability (harvesting costs). It was therefore decided to obtain this information for each sample plot, through interviews with the local foresters and cantonal forest administrations (Mahrer and Vollenweider 1983, Chap. 10).

1.5.1 Content

The content of the NFI (data catalogue) has been continuously developed further to take into account the aims (Sect. 1.2), the legal requirements (Sect. 1.3), the current information needs (Sect. 1.4), the methodological state of the art, as well as the technical and financial feasibility.

The data catalogue today contains around 350 attributes, the most important of which are summarised in Table 1.1 (Appendix). The development of the NFI content can be presented quantitatively and thematically by referring to the manuals for data collection (Zingg and Bachofen 1988; Stierlin et al. 1994; Ginzler et al. 2005; Keller 2011; Keller 2013; Düggelin and Keller 2017). In NFI1, the focus was not only on the forest resources, but also on the health of the forest and the use of the wood. In NFI2 many features concerning the forest as a habitat (biodiversity) were added, and in NFI3 the introduction of forest functions led to the inclusion of all the criteria for sustainable forest management. The level of detail of the information also increased considerably and the number of features nearly doubled. When the continuous inventory approach of NFI4 started in 2009, a few additions to the NFI's central themes of wood production and biodiversity were made. Since 2018 various attributes in NFI5 have been augmented, and three new groups of features related to biodiversity and resources have been introduced, namely: microhabitats on sample trees, problematic herbaceous neophytes and stump inventories (Appendix). Including this last feature means that the time needed just to record the data has reached a mean of almost 120 min per sample forest plot, which is the maximum a team of two can take to cover, on average, the target of two sample plots per day.

In relation to the international indicators, NFI today obtains about 50% of the required information on sustainable forest management (Brändli and Duc 2009). Important sources of national data for the other indicators include, for example, the federal Land-Use Statistics, the Sanasilva Inventory, the Forest Statistics and the Biodiversity Monitoring (Brändli et al. 2010a).

Of the information obtained with the different data-collection methods, 6% of the features are derived from interpreting aerial images, 74% from field surveys and 20% from the interview survey and forest road survey. Thus, even though only a small proportion of the NFI features are recorded through remote sensing, this approach is becoming increasingly important. To supplement the results obtained with the statistical samples, the following remote sensing products (maps) covering the whole country have been developed for the NFI since the start of the continuous survey: a vegetation-height model, a forest cover map, a tree type map (coniferous/broadleaved mixture) and a map of the total wood volume. The field-survey data is generally used in developing and checking these models (Chap. 7).

1.5.2 Design, Sampling Methods and Periodicity

Sections 1.5.2 and 1.5.3 are largely based on the recent publication by Lanz et al. (2016).

In developing the sample-plot network of NFI1 1983/1985, the results of a pilot inventory in Canton Nidwalden were used, the forest area was assumed to be 1.3 million ha, and the target accuracy for assessing the growing stock in Switzerland was set at around ±0.5% (p-value = 68%). The result was a systematic 1 km × 1 km grid covering the whole country, i.e. one sample plot per square kilometre of forest (Mahrer and Vollenweider 1983). As the forest area was smaller than originally assumed, the number of sample plots was 11,863 (including shrub forest). The field surveys in accessible forest excluding shrub forest (10,975 sample plots) recorded the growing stock as 330 m^3 ha^{-1} ± 0.7%. The corresponding estimation error for the most important survey units, the five production regions Jura, Central Plateau, Pre-Alps, Alps and Southern Alps (Southern Slopes of the Alps), varied between ±1.3% and ± 2.6%.

Although all the sample plots and trees had been marked for re-measurement in NFI1, budget constraints meant the number of plots had to be reduced by half in NFI2 1993/1995. Since then, field data has been recorded on a 1.41-km grid. To obtain, nevertheless, the target level of precision, a dense 500 m × 500 m grid of systematically distributed plots was introduced for stereo-image interpretation. The auxiliary variables gained from photo interpretation were combined with the field data using an estimation procedure known as two-phase (double) sampling for post-stratification (Köhl 1994; Mandallaz 2008).

The two main design changes introduced in NFI3 2004/2006 were to extend: (a) the field-data collection to include 'shrub forest', and (b) the aerial-photo interpretation to include tree resources outside forest and the FAO's 'other wooded land' category (OWL). 'Shrub forest' is a subcategory of OWL, and refers to a low productivity type of forest found mainly in high mountain areas.

The NFI switched from a periodic to a continuous survey in 2009, when NFI4 2009/2017 began, to ensure balanced budgets, continuity in specialist knowhow and regular up-to-date information. It was decided to collect the data over 9 years and to use, each year, a different one of the nine independent sub-grids selected from the original grid

(Sect. 2.4.4). Each sub-sample of re-measured plots on the annual sub-grid is fully representative of the entire country and each region. This rotating system makes it possible to rapidly introduce new attributes as needed and to obtain corresponding results within a year. The main content, general methods, models and statistical algorithms should not, however, be changed within the 9-year period, but only after the whole sample has been measured, if necessary. For NFI4, the photo interpretation has been reduced to the 1.41 × 1.41 km grid of the terrestrial survey, and the new forest cover map (Sect. 7.2) was used to derive first-phase sampling information (Sect. 2.5.2).

Since the new continuous survey system was introduced, the results for NFI4 2009/2011 and NFI4 2009/2013 have been compiled and published (www.lfi.ch). The final report for NFI4 2009/2017 will be available in 2020. The aerial-photo interpretation for all of NFI4 was again restricted to the grid on which the field-data collection takes place. Thus the main focus was on detecting new forest plots and measuring tree resources outside the forest and OWL.

Depending on the data collection method used (aerial-photo interpretation, field survey or interview survey) and survey feature, the information obtained from surveying the sample plots will refer to different parts of the plot: the sample-plot centre, the interpretation area, the raster points, the (concentric) sample circles, the transects, or a taxation stretch on the forest boundary line (Fig. 1.2).

In the field surveys (Chap. 9), standing and lying stems are measured and described in detail (e.g. any damage is recorded) on two nested circular sample plots. The smaller plot has a horizontal radius of 7.98 m and is 200 m^2 in area, while the larger one has a horizontal radius of 12.62 m and is 500 m^2 in area. On the smaller plot, trees and shrubs with a diameter at breast height (dbh) $\geq$12 cm and <36 cm are measured, whereas all trees with a dbh $\geq$36 cm are measured on the larger plot (Keller 2011). Tree height and upper stem diameter at a height of 7 m are measured only on a randomly selected sub-sample of the living and standing tally trees. These trees are used in the NFI two-stage

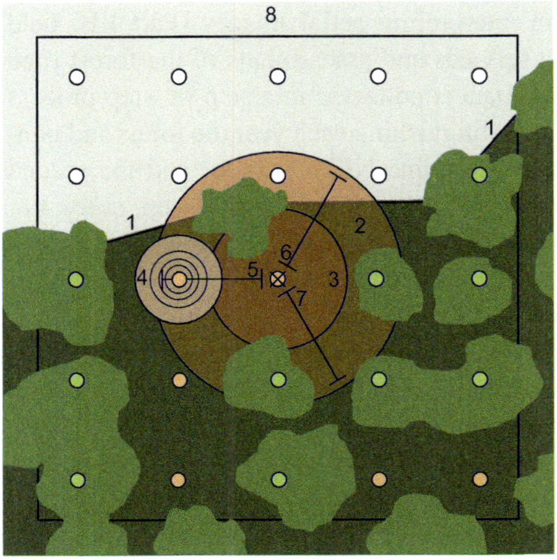

Field survey

x	Sample-plot center
1	Forest-boundary-line (FBL)
2	Circle for trees with DBH ≥ 36 cm
3	Circle for trees with DHB ≥ 12 cm
4	Circles for survey of young forest
5,6,7	Transect for survey of lying deadwood
8	Interpretation area (50 x 50 m)

Aerial photo interpretation

○	Raster-point outside FBL
● (green)	Raster-point on stocking element
● (orange)	Raster-point on other ground cover

Fig. 1.2 Design of an NFI4 forest plot

procedure to estimate the tree volume (and biomass) with the aim to avoid potential bias in these estimates (Mandallaz 2008). On average, 11.9 tally trees are measured per plot, and a sub-sample of 2.3 tariff trees are selected from them.

Young trees and the main shrub species with a minimum height of 10 cm and a dbh of less than 12 cm are assessed on a separate set of four circular sample plots with radii between 0.9 and 4 m. The centre of this set of nested plots is displaced from the plot centre by 10 m.

For the assessment of lying deadwood, a line-intersect sampling technique is used (Böhl and Brändli 2007), involving three transect lines 10 m in length. The transects start 1 m from the plot centre and go in three different directions (35 gon, 170 gon and 300 gon). The diameter threshold for the lying deadwood is 7 cm at the point of intersection with the transect line.

If the plot centre is less than 25 m from a forest edge, the woody plant species and several structural and biodiversity-related features of the stocking at the forest edge are measured along a taxation stretch of 50 m.

For the assessment of several site and stand features in the field and in the aerial images, a quadratic interpretation area of 50 × 50 m is used. The land cover is then assessed and vegetation height measured at 25 systematically distributed raster points on the interpretation area in the photos (Ginzler et al. 2005).

Finally, the NFI also interviews the local foresters and cantonal forest administration about each sample plot and the hauling route to the next road (Sect. 10.3). From the forest-road survey with the local foresters, a complete overview of the forest-road network, including several NFI features, can then be obtained for the whole country (Keller 2013; Müller et al. 2016).

1.5.3 Data Collection and Analyses

The data used in the NFI comes from interpreting aerial images (Part III), field surveys, oral interviews with the forest services and assessments of the forest-road network on topographic maps (Part IV). Data is collected in a step-by-step process (Lanz et al. 2010). The first step involves distinguishing each year the forest and non-forest plots in the aerial images on one of the nine NFI sub-grids, and the second collecting field data on the (potential) forest plots identified. In a third step, every 4 or 5 years, interviews referring to about half of all plots are conducted. In the fourth step every 9 years, the GIS layers of the forest-road network are updated. In addition, information from various external sources is used, such as digital maps of the administrative regions, soil characteristics or forest reserves. Further external input sources are specific models, e.g. for single-tree volume and biomass, the extent of the potential vegetation or the prediction of site quality (Brassel and Lischke 2001; Part V).

The **interpretation of aerial images** (Part III) is intended to reduce the cost of NFI by excluding plots that are clearly non-forest from the field survey, and by providing reference points to measure the sample-plot centre of new forest plots in the field. The main variables assessed in the manual photo interpretation are land cover (11 classes) and vegetation height on 25 systematically distributed raster points in the interpretation area (Ginzler et al. 2005). NFI data from the photo interpretation have been used for: assessing tree resources outside forest areas (Ginzler et al. 2011), analysing landscape patterns and diversity (Mathys et al. 2006), estimating small areas (Steinmann et al. 2013) and calculating forest road lengths (Brändli et al. 2016).

In the **terrestrial survey** (Part IV) a forest engineer and a forester form one field team. The season for field data collection lasts from April to October. About 820 plots are visited in the field per year, of which about 700 are forest plots and 40 are shrub-forest plots (OWL). Since 2009, three teams have normally been in action at any one point in time. Fieldwork includes periodic training courses (three per year) and the re-assessment of about 8% of plots by a second team (blind checks). All variables assessed in the field (Appendix Table 1.1) are described in the field manual (Düggelin and Keller 2017).

Every 4–5 years, the foresters from the local **forest service** are interviewed by the field team to obtain a range of information for each forest or shrub-forest sample plot, relating to forest functions, planning, silviculture, harvesting techniques and forest history (Chap. 10). Every 9 years, the entire network of forest roads is updated in collaboration with the local foresters (Keller 2013; Müller et al. 2016; Sect. 10.3). The cantonal forest service is periodically asked in written form for an updated digital map of the forest districts and for information on the status of the regional forest planning (forest development plans) for each plot.

Field teams are equipped with a tablet computer and specially designed software developed at WSL for data collection (Sect. 23.2). Plausibility tests based on current input and previous measurements are implemented. The mobile phone network is used to transfer field data to the raw database at WSL.

All variables needed for producing the result tables are transferred to a separate database (Sect. 20.3) of derived variables. The derivation may be a direct copy from the raw database, but it is often the result of a complex combination (model) of various raw data and external data sources (Part V). All scripts are stored in a structured way so that re-calculations with the original raw data can be made at any time. The current NFI database runs on Oracle and most of the scripts for the derivations of the variables are written in Structured Query Language (SQL), but some complex derivations are written in SAS©.

The production of result tables is governed through a web interface software written in PHP (Sect. 20.2.1). Users can select the inventories, target variables, domains of interest and various technical parameters they want. The result tables are produced in real time with the statistical estimation procedures of a data analysis application (software). In the background, the steering parameters for the production of a specific table are stored in the database and can be reused (Traub et al. 2017). All scripts are currently being rewritten in Python in order to make NFI and WSL independent of user licenses and allow them to give third parties free access to the source code and to the NFI Data Analysis System NAFIDAS (Chap. 20).

1.6 Reporting, Products and Services

How valuable a forest inventory is depends on how the data is used. The more intensive, widespread and relevant the uses of the data are, the more valuable the inventory is as the source of data. To ensure that stakeholders use and trust the data and products, these aspects of the inventory should cover a wide range and be relevant and reliable. This is why the outreach activities of the NFI, which includes reports, products, services and publicity work, has—since the first NFI—been considered as strategically very important and resources have, correspondingly, been made available. The results of the impact assessments (Sect. 1.4) show that the outreach activities of the NFI have, to a great extent, met its goals for the primary stakeholders, partly thanks to the periodic needs analyses carried out (Bättig et al. 2002; Bernath et al. 2013).

Initially it was necessary, given Switzerland's federal system, to overcome the considerable scepticism of the forest services in some cantons. Today. practically all the federal and cantonal forest services are in favour of the NFI, and all of them actively support the data collection (in the interviews and forest-road survey). The NFI has become an important source of data for the federal and cantonal reporting on the forest, and thus also for forest policy. It is moreover increasingly appreciated in environmental science research, thanks to its characteristic, systematic, representative and unbiased design. Researchers constitute today the largest group of users interested in the data supplied by the NFI (Fig. 1.3). The impact assessments for NFI2 already revealed that 44% of the people involved in forest research who were asked said they could not do without the NFI in their personal field of activity (Bättig et al. 2002). The corresponding proportion for all stakeholders from forestry circles was around 60%.

WSL is responsible for the NFI data analyses, the scientific interpretation of the results and the reporting, as well as for the products and services, while the Federal Office for the Environment (FOEN) is responsible for the political evaluation of the results.

1.6.1 Reporting

Since the introduction of a continuous survey in 2009, the NFI has been able to produce up-to-date results every year. This is required for the Greenhouse Gas Inventory and is currently provided in the form of annual reports (Sect. 19.3). While FOEN generally needs information that is relevant for political issues and is as up to date as possible, most other stakeholders still prefer to receive results that relate to the whole sample, with a higher spatial resolution and very low estimation errors (statistically significant changes), only once a decade (Abegg and Brändli 2016).

WSL publishes the main NFI results in the form of carefully edited books with illustrations and tables every 9–10 years (EAFV/BFL 1988; Brassel and Brändli 1999; Brändli 2010). Since NFI3, the content has been structured according to the Pan-European criteria and indicators (Forest Europe 2011), with scientific interpretations of the results. In addition, thousands of output tables (Fig. 1.4) and maps

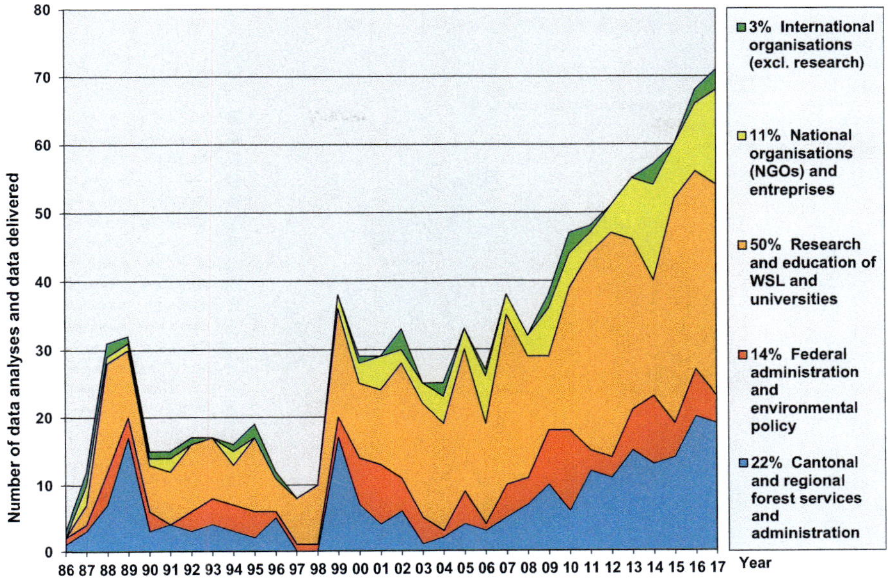

Fig. 1.3 Annual requests for data and data analyses from 1986 to 2017, classified according to user group

NFI4b

forest area: altitude (400 m classes) · production region
unit: 1000 ha
unit of evaluation: forest
grid:
state 2009/13

	production region											
	Jura		Plateau		Pre-Alps		Alps		Southern Alps		Switzerland	
altitude (400 m classes)	1000 ha	± %	1000 ha	± %	1000 ha	± %	1000 ha	± %	1000 ha	± %	1000 ha	± %
above 1800 m	.	.	.	.	2.5	38	103.7	5	27.1	10	133.3	5
1401-1800 m	4.4	29	1.8	45	42.5	9	171.4	4	52.0	7	272.0	3
1001-1400 m	61.3	7	4.9	27	95.1	5	107.0	5	47.6	7	316.0	3
601-1000 m	85.4	5	94.9	5	85.1	5	61.6	7	34.6	9	361.7	3
to 600 m	52.6	7	132.2	4	6.8	23	9.7	19	23.7	11	225.0	3
total	203.7	1	233.8	1	232.0	1	453.4	1	185.1	1	1308.0	1

calculated by unit of reference: production region

© WSL, Swiss National Forest Inventory, 03.11.2014 #161305/147684

definitions (in German) methods (in German) download display map

Table citation

Abegg, M.; Brändli, U.-B.; Cioldi, F.; Fischer, C.; Herold-Bonardi, A.; Huber M.; Keller, M.; Meile, R.; Rösler, E.; Speich, S.; Traub, B.; Vidondo, B., 2014: Swiss national forest inventory - Result table No. 161305: forest area
Birmensdorf, Swiss Federal Research Institute WSL
https://doi.org/10.21258/1332151

Fig. 1.4 Example of a result table from www.lfi.ch showing forest area (including shrub forest) according to altitude and production region

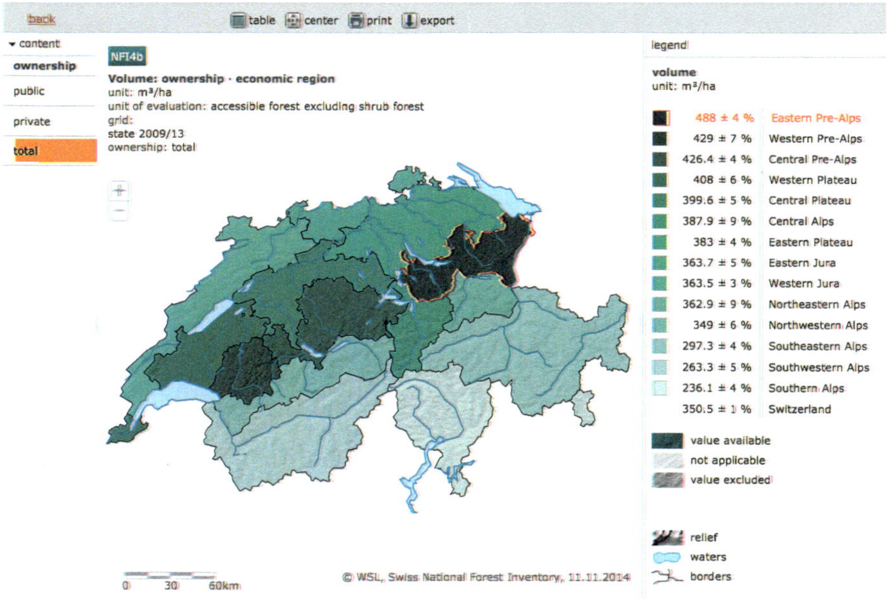

Fig. 1.5 Volume per hectare by ownership and economic region. Example of a result map available on the Internet (www.lfi.ch)

(Fig. 1.5) have been put on the Internet in four languages, German, French, Italian and English, and can be downloaded (http://www.lfi.ch/resultate, Speich et al. 2010; Speich and Meile 2013; Abegg et al. 2014). These tables, such as the scenario simulations for future supplies of wood (Sect. 19.2), are updated every 4 to 5 years.

The NFI supplies about 50% of the data for the national reporting on sustainable forest management, e.g. the Forest Report 2005 (SAEFL, WSL 2005) and the Forest Report 2015 (Rigling and Schaffer 2015). The conclusions from such reports provided the basis for the development of the Action Programme 2004–2015 for the Swiss National Forest Programme (SAEFL 2004) and the Forest Policy 2020 (FOEN 2013).

The NFI also provides information for international reporting. Examples are: The Global Forest Resources Assessment (FRA) of the Food and Agriculture Organization of the United Nations; the FOREST EUROPE reports on the State of Forests and Sustainable Forest Management in Europe; and Switzerland's Greenhouse Gas Inventory (Sect. 19.3), which is submitted under the United Nations Framework Convention on Climate Change and the Kyoto Protocol.

1.6.2 Other Products

To supplement the reporting (Sect. 1.6.1), scientific staff at NFI are also responsible for further analyses of the NFI data as part of their own research, for example on forest growth and (future) forest development. In addition, they cooperate with researchers from other institutions on projects and papers if possible, rather than just supplying raw data. This requires specialisation in topics such as resources, biodiversity, protection forests and socioeconomics, but ensures at the same time better professional networking and a greater output of scientific articles and special reports, such as scenarios for the wood supply. Publications on methods make the know-how of NFI specialists accessible to third parties.

As interest and demand have grown, the NFI management team decided to use remote sensing to produce area-wide records and maps in NFI4, in addition to the digital forest-road network, and to update it every year. The first four products are the vegetation-height model, the forest cover map, the coniferous/broadleaved tree map and the wood-volume map (Chap. 7). These data can be obtained under the same conditions as those from the sample surveys.

The online portal www.lfi.ch (Brändli and Heller 2000; Brändli and Speich 2005) should be of use to all stakeholder groups who are interested in information on the forest. This is why you will find there, in addition to countless tables and maps, other products, some of which are explicitly aimed at PR, such as a virtual walk in the forest (Brändli 2000), which introduces schoolchildren to findings of interest for laypeople. Another highly valued section is that on Forest trees of Switzerland (Brändli 1998), which now contains 150 distribution maps of trees and shrubs. Those who would like to have a general overview of forest development over the past 30 years will find what they are looking for in the NFI-Cockpit with its interactive graphics (Brändli and Brändli 2015). The vivid NFI-Film (Brändli et al. 2010b) explains the goals and methods of the NFI in four different languages.

In order to inform forest managers and forest owners about NFI findings, the most important results are summarised in an easily comprehensible form and published in the specialist media. As part of the public relations work of different institutions and the forest services, a travelling exhibition (Brändli 1993) was produced, together with leaflets (Brändli and Speich 2015) and a series of weatherproof posters for foresters (Brändli et al. 2015) in all the national languages.

The wider public receives information about the NFI programme and its findings through periodic media releases. From time to time reporters accompany the field team or interview members of the NFI management team about the findings in newspapers or on the radio or television.

1.6.3 Services

NFI data is made available to third parties, Swiss institutions and their representatives as soon as the specific official results have been published by WSL. Data extracts are provided for specific research and statistical purposes under contractual agreements (http://www.lfi.ch/dienstleist/daten-en.php). In the case of research projects, a collaboration with NFI researchers is usually sought. Since the first NFI, hundreds of responses to data requests have been sent. Special data analyses are also

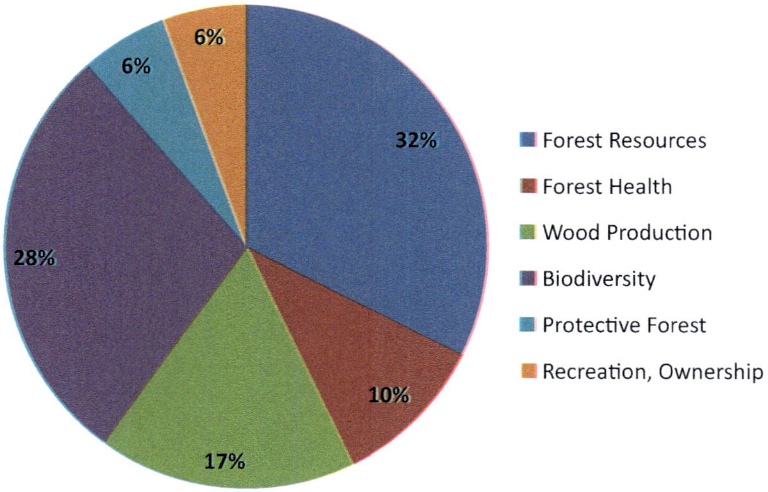

Fig. 1.6 Topics of NFI data extracts for third parties and analyses on request

made on request, amounting currently, together with data extracts, to around 65 per year (Fig. 1.3). The forest road network layer and the national area-wide remote sensing products have led to a steadily growing demand for data. The main consumer groups are research and education institutions (50%), cantonal and regional forest administrations (22%), the federal administration (14%), non-governmental national organisations and enterprises (11%) and international organisations (3%). The main topics of data requests concern forest resources, biodiversity and wood production (Fig. 1.6).

NFI know-how is in demand and appreciated by monitoring specialists in Switzerland and abroad. As far as NFI's personnel resources allow, NFI staff may also become involved in inventory consulting and cooperation. The NFI provides support in planning, material, training of field teams, data hosting (database) and data analyses of cantonal inventories according to NFI methods. To date, the Principality of Liechtenstein and the cantons Aargau, Appenzell, Berne, Grisons, Lucerne, Obwalden and Zurich have conducted such regional inventories using NFI methods on denser sampling grids (http://www.lfi.ch/dienstleist/inventur-en.php). International consulting to advise on inventory design and statistics has so far benefited Kirgizstan, Romania, France, Czech Republic, Slovakia and Bulgaria, and is currently being carried out in Belize. The NFI participates in, and sometimes coordinates, projects and activities of the European Network of Forest Inventories (ENFIN, http://enfin.info/). In the Carpathian Biosphere Reserve CBR (Ukraine), first and second inventories of Europe's largest primeval beech forest were planned and carried out with NFI support (Commarmot et al. 2013).

Appendix

Table 1.1 Development of the NFI data catalogue during the four inventories, NFI1, NFI2, NFI3 and NFI4, and during NFI5 since 2018 according to data-collection method and the criteria for sustainable forest management (SFM)

		Inventory					SFM criterion (topic) to which the feature contibutes					
Data collection method	Feature	NFI1	NFI2	NFI3	NFI4	NFI5	Forest resources	Health and vitality	Forest use	Biodiversity	Protection forest	Social economy
Aerial photo	Pre-determined data for 'forest/non-forest' decision	X	X	X	X	X	X					
	Reference points for the field plot location	X	X	X	X	X	X					
	Altitude	X	X				X					
	Forest boundary line			X	X	X	X					
	Degree of crown cover		X	X	X	X	X					
	Forest gaps			X	X	X	X					
	Deciduous/evergreen tree cover at 25 raster points		X	X	X	X	X					
	Land cover class at 25 raster points		X	X	X	X	X					
	Vegetation height/ Altitude at 25 raster points		X	X	X	X	X					
	Stage of development		X	X	X	X	X					
	Degree of closure		X	X	X	X	X					
	Relief		X				X					

(continued)

Table 1.1 (continued)

Data collection method	Feature	Inventory					SFM criterion (topic) to which the feature contibutes					
		NFI1	NFI2	NFI3	NFI4	NFI5	Forest resources	Health and vitality	Forest use	Biodiversity	Protection forest	Social economy
	Stocking element outside forest			X	X	X	X					
	Single trees outside forest			X	X		X					
	Potential forest use				X	X	X					
Terrestrial area	Forest/non-forest decision	X	X	X	X	X	X					
	Reason for forest decrease			X	X	X	X					
	Reason for forest increase			X	X	X	X					
	Previous use before forest increase			X	X	X	X					
	Photos of the sample plot				X	X	X					
	Accessiblity and walkability	X	X	X	X	X	X					
	Soil samples from topsoil	X		(x)			X					
	Thickness of soil horizon in topsoil				(x)		X					
	Slope	X	X	X	X	X	X					
	Aspect	X	X	X	X	X	X					
	Relief	X	X	X	X	X	X					
	Landslide	X	X	X	X	X					X	
	Erosion	X	X	X	X	X					X	
	Rockfall	X	X	X	X	X					X	

	Boulder-axis lengths			X	X	X					X	
	Boulder shape			X	X	X					X	
	Snow movement, Avalanches	X	X	X	X	X					X	
	Traces of fire	X	X	X	X	X		X				
	Pasturing (type, intensity)	X	X	X	X	X		X				
	Ecological pressure, disturbances	X	X	X	X	X		X				
	Tracks from skidding during wood harvesting			X	X	X		X				
	Vehicle tracks from wood harvesting (length, depth, width, degree of cover, stand or skid track, age, puddles)			X	X	X		X				
	Type of damage to soil			X	X	X		X				
	Obstacles to and constraints on logging	X	X	X	X	X			X			
	Presence of stumps	X	X	X	X	X			X			
	Category of use	X	X	X	X	X	X					
	Special sites and water bodies		X							X		
	Water bodies			X	X	X				X		
	Azonal site types			X	X	X				X		
	Vegetation type			X						X		
	Heaps of branches		X	X	X	X				X		
	Snags		X	X	X	X				X		
	Places without vegetation		X	X	X	X				X		

(continued)

Table 1.1 (continued)

Data collection method	Feature	Inventory					SFM criterion (topic) to which the feature contibutes					
		NFI1	NFI2	NFI3	NFI4	NFI5	Forest resources	Health and vitality	Forest use	Biodiversity	Protection forest	Social economy
	Dry stonewalls, heaps of stones		X	X	X	X				X		
	Geomorphological objects		X	X	X	X				X		
	Recreational facilities		X	X	X	X						X
	Root plates			X	X	X				X		
	Ants (presence, species)				X					X		
	Ant-heaps (number, dimensions)				X	X				X		
	Number of tree cavities with detritus				X					X		
	Problematic herbaceous neophytes					X				X		
Terrestrial stand	Forest type	X	X	X	X	X	X					
	Forest origin and management type	X	X	X	X	X	X					
	Stage of development	X	X	X	X	X	X					
	Stand age	X	X	X	X	X	X					
	Degree of mixture	X	X	X	X	X	X					
	Degree of cover of tree species in the upper layer			X	X	X	X					
	Degree of crown closure	X	X	X	X	X	X					
	Type of gap		X	X	X	X					X	
	Stand structure	X	X	X	X	X	X					

	Degree of cover of stand layers			X	X	X	X					
	Type of cover of stand layers			X	X	X	X					
	Stand's top height				X	X	X					
	Stand stability (loads, resilience)	X	X	X	X	X		X				
	Type of areal forest damage (areal damage)			X	X	X		X				
	Extent of areal damage			X	X	X		X				
	State of clearing			X	X	X		X				
	Stocking border, forest edge	X	X	X	X	X	X					
	Stand border	X	X	X	X	X	X					
	Inner edges			X	X	X				X		
	Stand size			X	X	X	X					
	Type of regeneration	X	X	X	X	X	X					
	Protection of regeneration	X	X	X	X	X	X					
	Distribution of regeneration	X					X					
	Degree of cover of regeneration		X	X	X	X	X					
	Degree of cover of established regeneration			X	X	X	X					
	Main tree species in established regeneration				X	X	X					
	Degree of cover of shrub layer		X	X	X	X	X					

(continued)

Table 1.1 (continued)

Data collection method	Feature	Inventory					SFM criterion (topic) to which the feature contibutes					
		NFI1	NFI2	NFI3	NFI4	NFI5	Forest resources	Health and vitality	Forest use	Biodiversity	Protection forest	Social economy
	Degree of cover of ground vegetation		X	X	X	X	X					
	Degree of cover of berry bushes		X	X	X	X				X		
	Main species of berry bush		X	X	X	X				X		
	Urgency of silvicultural treatment to be done next	X	X	X	X	X			X			
	Type of silvicultural treatment to be done next	X	X	X	X	X			X			
	Presence of woody plant species (200 m^2 plot)			X	X	X				X		
	Proportion of woody plant species in shrub forest			X						X		
Terrestrial tree (dbh $\geq$ 12 cm)	Tree species	X	X	X	X	X	X			X		
	dbh up to 60 cm, circumference if diameter > 60 cm	X	X	X	X	X	X					
	Circumference		X				X					
	Diameter at 7 m tree height (d7), tree height	X	X	X	X	X	X					
	Distance, azimuth from centre	X	X	X	X	X	X					
	Inclination towards centre				X	X	X					

	Living / dead / standing / lying	X	X	X	X	X	X					
	Date of death				X	X	X					
	Tree age				X	X	X					
	Stem height of broken snag				X	X	X					
	Treetop used				X	X			X			
	Growth form			X	X	X	X					
	Tree status		X	X	X	X	X		X			
	Tree-rings on stumps of used trees		X	X	X	X	X					
	Height of stumps of used trees			X	X	X	X					
	Diameter of stumps of used trees				X	X				X		
	Inventory of stumps					X	X			X		
	Layer	X	X	X	X	X	X					
	Crown class (length)	X	X	X	X	X	X					
	Crown class (shape)	X	X	X			X					
	Social position		X				X					
	Height where crown starts			X	X	X	X					
	Crown dimensions			X								
	Damages (type, location, dimensions)	X	X	X	X	X		X				
	Mistletoe			X				X				
	Stem quality	X		(x)					X			
	Lopsided or bent tree		X	X	X	X		X				

(continued)

Table 1.1 (continued)

Data collection method	Feature	Inventory					SFM criterion (topic) to which the feature contibutes					
		NFI1	NFI2	NFI3	NFI4	NFI5	Forest resources	Health and vitality	Forest use	Biodiversity	Protection forest	Social economy
	Fork	X	X	X	X	X			X			
	Length of sawable round wood					X			X			
	Stump shoots		X	X	X	X	X					
	Seed trees in regeneration, pioneer planting		X	X	X	X	X					
	Proportion of dry parts > 20%		X	X	X	X				X		
	Coppicing above 2 m height		X	X	X	X	X					
	Reaction to bark marking of dbh position		X					X				
	Stem heights of broken snags			X	X	X				X		
	Woodpecker cavity			X	X	X				X		
	Fungus species (perennial fruiting bodies)			X	X	X				X		
	Lichen species			X						X		
	Length of lying tree			X	X	X				X		
	Twigs on dead tree/ deadwood twig			X	X	X				X		
	Bark cover on dead tree			X	X	X				X		
	Hardness of dead tree			X	X	X				X		
	Ground contact of dead tree			X	X	X				X		

	Degree of moss cover on dead tree			X	X	X				X		
	Degree of lichen cover on dead tree			X	X	X				X		
	Number of shrubs on dead tree			X	X	X				X		
	Number of saplings on dead tree			X	X	X				X		
	Microhabitats (cavities, holes, cracks, bark pockets, fungal fruit bodies, cover with ivy, moss and lichens)					X				X		
Terrestrial regeneration (dbh < 12 cm)	Tree species	X	X	X	X	X	X					
	Height class, dbh	X	X	X	X	X	X					
	Damage, browsing	X	X	X	X	X	X					
	Length of leading shoot (next plant)			X	X		X					
	Regneration type (next plant)			X	X	X	X					
	Natural or artificial regeneration (next plant)				X	X	X					
	Measures to protect against game (next plant)			X	X	X		X				
	Substrate (next plant)			X	X	X	X					
	Silvicultural assessment (next plant)			X	X	X	X					
Terrestrial subplot regeneration	Degree of shading			X	X		X					
	Proportion naturally poor regeneration conditions			X	X	X	X					

(continued)

Table 1.1 (continued)

Data collection method	Feature	Inventory					SFM criterion (topic) to which the feature contibutes					
		NFI1	NFI2	NFI3	NFI4	NFI5	Forest resources	Health and vitality	Forest use	Biodiversity	Protection forest	Social economy
	Proportion anthropogenically poor regeneration conditions			X	X	X	X					
	Proportional area with good regeneration conditions			X	X	X	X					
	Proportional area with competing vegetation			X	X	X	X					
	Competing plant species			X	X	X	X					
	Main type of topsoil			X	X		X					
Lying dead-wood (Line intersect sampling)	Broadleaved/coniferous wood				X	X				X		
	Cross diameter			X	X	X				X		
	Length </≥ 1 m				X	X				X		
	Slope			X	X	X				X		
	Degree of decomposition				X	X				X		
	Length of intersect with heap of branches			X	X	X				X		
	Piece of deadwood in heap of branches				X	X				X		
Terrestrial forest edge	Exposition of forest edge		X	X	X	X				X		
	Composition of forest edge		X	X	X	X				X		
	Width of shelterbelt		X	X	X	X				X		

	Width of shrub belt		X	X	X	X				X		
	Width of herbaceous fringe		X	X	X	X				X		
	Shape of forest edge		X	X	X	X				X		
	Density of forest edge		X	X	X	X				X		
	Condition of forest edge		X	X	X	X				X		
	Barrier delimiting forest edge		X	X	X	X				X		
	Immediate neighbourhood of forest edge		X	X	X	X				X		
	Woody species along the forest edge		X	X	X	X				X		
Interview survey and forest road assessment	Ownership	X	X	X	X	X						X
	Forest road network (condition, type of surface, newly built, renaturalised)	X	X	X	X	X			X			
	Width of forest road				X	X			X			
	Vehicle type				X	X			X			
	Obstacles on forest road				X	X			X			
	Connection of forest roads to main road network				X	X			X			
	Concept for opening-up forest land			X	X	X			X			
	Intentions concerning concept for opening-up forest land			X	X	X			X			
	Pre-hauling distance	X	X	X	X	X			X			

(continued)

Table 1.1 (continued)

Data collection method	Feature	Inventory					SFM criterion (topic) to which the feature contibutes					
		NFI1	NFI2	NFI3	NFI4	NFI5	Forest resources	Health and vitality	Forest use	Biodiversity	Protection forest	Social economy
	Hauling distance	X	X	X	X	X			X			
	Means of hauling	X	X	X	X	X			X			
	Limitations in choice of means of hauling	X	X	X	X	X			X			
	Marking of skid roads				X	X		X				
	Hauling destination		X						X			
	Hauling direction		X	X	X	X			X			
	Hauling route			X	X	X			X			
	Altitude of helicopter timber depot			X	X	X			X			
	Length of cable line			X	X	X			X			
	Location of mobile tower yarder			X	X	X			X			
	Timber sorting (short, long)		X	X	X	X			X			
	Year of last use	X	X	X	X	X			X			
	Type of last use		X	X	X	X			X			
	Year and type of silvicultural treatment since the previous inventory			X	X	X			X			
	Salvage cuts (proportion, reason)		X	X	X	X		X				
	Date of next silvicultural treatment			X	X	X			X			

	Type of next silvicultural treatment			X	X	X			X			
	Principle goal of plentering in mountain forests			X	X	X			X			
	Conditions for next silvicultural treatment			X	X	X			X			
	Timber harvesting (type, in-/outsourcing)		X	X	X	X			X			
	Basis for local planning (type, year of establishment)		X	X	X	X			X			
	Regional Forest Development Plan (status)			X	X	X			X			
	Certification (type, label)			X	X	X			X			
	Size of managed unit			X	X	X			X			
	Forest functions			X	X	X	X					
	Primary functions			X	X	X	X					
	Location in catchment area of drinking-water source			X	X	X						X
	Intensity of recreational use (visitor frequency)			X	X	X						X
	Seasonality of recreational use			X	X	X						X
	Type of recreational use			X	X	X						X
	Type of forest origin			X	X	X	X					
	Year of afforestation			X	X	X	X					
	Type of stand origin			X	X	X	X					
	Year of last pasturing			X	X	X		X				

References

Abegg M, Brändli UB (2016) Auswertung LFI4 – Ergebnisse des zweiten LFI User-Workshops. Infobl Arbeitsgr Waldplan -Manage 16(1):10–12

Abegg M, Brändli UB, Cioldi F, Fischer C, Herold-Bonardi A, Huber M, Keller M, Meile R, Rösler E, Speich S, Traub B (2014) Fourth national forest inventory – result tables and maps on the Internet for the NFI 2009–2013 (NFI4b). [Published online 06.11.2014] Available from World Wide Web http://www.lfi.ch/resultate/. Swiss Federal Research Institute WSL, Birmensdorf

BAR (Schweizerisches Bundesarchiv) (1981) Bericht über die Geschäftsführung des Bundesrates, des Bundesgerichts und des Eidgenössischen Versicherungsgerichts im Jahr 1981. Geschäftsberichte des Bundesrates, Schweizerisches Bundesarchiv, Bern

Bättig C, Bächtiger C, Bernasconi A, Brändli U-B, Brassel P (2002) Landesforstinventar. Wirkungsanalyse zu LFI1 und 2 und Bedarfsanalyse für das LFI3. Schlussbericht. ARGE Interface, Pan, WSL. Umwelt-Materialien Nr. 143, Wald. Bern, Bundesamt für Umwelt, Wald und Landschaft. 89 S

Bernath K, Kissling I, von Felten N, Salathé M, Hallenbarter D (2013) Evaluation landesforstinventar 3. Schlussbericht 21. März 2013. Im Auftrag von WSL und BAFU. Ernst Basler + Partner AG, Zollikon. 62 pages. Unpublished

Böhl J, Brändli UB (2007) Deadwood volume assessment in the third Swiss National Forest Inventory: methods and first results. Eur J For Res 126:449–457

Brändli U-B (1992) Resultate der Umfrage zum Bericht 305, Schweizerisches Landesforstinventar, Ergebnisse der Erstaufnahme 1982–1986, Eidg. Forschungsanstalt WSL, Birmensdorf. 12 pages. Unpublished

Brändli U-B (1993) The national forest inventory: a window on the swiss forest. Verified Knowledge thanks to Systematic Observation. Birmensdorf, Federal Institute of Forest, Snow and Landscape Research. 16 p

Brändli U-B (1998) Die häufigsten Waldbäume der Schweiz. Ergebnisse aus dem Landesforstinventar 1983–85: Verbreitung, Standort und Häufigkeit von 30 Baumarten. 2. Auflage. Ber. Eidgenöss. Forsch.anst. Wald Schnee Landsch. 342: 279 S

Brändli U-B (2000) Waldspaziergang im Internet. Schweizerisches Landesforstinventar LFI allgemein zugänglich. Inf.bl. Forsch.bereich Landsch. 48: 4

Brändli U-B (Red.) (2010) Schweizerisches Landesforstinventar. Ergebnisse der dritten Erhebung 2004–2006. Eidgenössische Forschungsanstalt für Wald, Schnee und Landschaft WSL, Birmensdorf; Bundesamt für Umwelt, BAFU, Bern, 312 S

Brändli U-B, Bernasconi A (2014) Ergebnisse des ersten LFI User-workshop. Infobl. Arbeitsgr. Waldplan. -manage. 14/1: 9–11

Brändli UB, Brändli R (2015) Swiss NFI cockpit – forest indicator trends. Poster. [Published online 12.11.2015] Available on the Internet http://www.lfi.ch/publikationen/publ/NFIcockpit-en.php. Birmensdorf, Swiss Federal Research Institute WSL

Brändli U-B, Duc P (2009) Indikatoren und Ergebnisse zur nachhaltigen Waldnutzung im Landesforstinventar LFI. Langzeitforschung für eine nachhaltige Waldnutzung. Forum für Wissen 2009:59–66

Brändli, U.-B.; Heller-Kellenberger, I., 2000: Schweizerisches Landesforstinventar LFI. [Published online 4.9.2000] Available from World Wide Web http://www.lfi.ch. Birmensdorf, Eidg. Forschungsanstalt WSL

Brändli U-B, Speich S, (2005) Swiss national forest inventory NFI. [Published online 13.12.2005] Available from World Wide Web <http://www.lfi.ch/index-en.php>. Birmensdorf, Swiss Federal Research Institute WSL

Brändli U-B, Speich S (2015) The Swiss National Forest Inventory. 2nd revised edition. [Flyer] Birmensdorf, Swiss Federal Institute for Forest, Snow and Landscape Research WSL. 12 p

Brändli U-B, Brang P, Lanz A, Abegg M (2010a) LFI-Bilanz zur Nachhaltigkeit. In: Brändli, U.-B. (Red.) Schweizerisches Landesforstinventar. Ergebnisse der dritten Erhebung 2004–2006.

Birmensdorf, Eidgenössische Forschungsanstalt für Wald, Schnee und Landschaft WSL. Bern, Bundesamt für Umwelt, BAFU. 265–288
Brändli U-B, Moll M, Egli J (2010b) Dem Geheimnis Wald auf der Spur. Das Schweizerische Landesforstinventar. Sur les traces du mystère de la forêt. L'Inventaire Forestier National suisse. Alla scoperta dei misteri del bosco. L'Inventario Forestale Nazionale svizzero. Uncovering the forest's secrets. The Swiss National Forest Inventory. Birmensdorf, Eidg. Forschungsanstalt WSL. DVD
Brändli UB, Denzler L, Dingwall S (2015) Results of the third NFI 2004–06. Poster series in 15 parts. [Published online 24.08.2015] Available from World Wide Web http://www.lfi.ch/publikationen/publ/posterLFI3-en.php. Birmensdorf, Swiss Federal Research Institute WSL
Brändli U-B, Fischer C, Camin P (2016) Stand der Walderschliessung mit Lastwagenstrassen in der Schweiz. Schweiz Z Forstwes 167(3):143–151. https://doi.org/10.3188/szf.2016.0143
Brassel P, Brändli UB (Red.) (1999) Schweizerisches Landesforstinventar. Ergebnisse der Zweitaufnahme 1993–1995. Eidgenössische Forschungsanstalt für Wald, Schnee und Landschaft WSL, Birmensdorf; Bundesamt für Umwelt, Wald und Landschaft BUWAL, Bern. Haupt, Bern, Stuttgart, Wien
Brassel P, Lischke H (eds) (2001) Swiss national forest inventory: methods and models of the second assessment. Swiss Federal Research Institute WSL, Birmensdorf
Commarmot B, Brändli U-B, Hamor F, Lavnyy V (2013) Inventory of the largest primeval beech forest in Europe. A Swiss-Ukrainian Scientific Adventure. Birmensdorf, Swiss Federal Research Institute WSL; L'viv, Ukrainian National Forestry University; Rakhiv, Carpathian Biosphere Reserve. 69 pp
Düggelin C, Keller M (2017) Schweizerisches Landesforstinventar. Feldaufnahme-Anleitung 2017. Birmensdorf, Eidgenössische Forschungsanstalt für Wald, Schnee und Landschaft WSL. 220 S
EAFV/BFL (Eidgenössische Anstalt für das forstliche Versuchswesen; Bundesamt für Forstwesen und Landschaftsschutz) (Red.) (1988) Schweizerisches landesforstinventar. Ergebnisse der Erstaufnahme 1982–1986. Bericht 305, Eidgenössische Anstalt für das forstliche Versuchswesen EAFV, Birmensdorf
FOEN, Federal Office for the Environment (2013) Forest Policy 2020. Visions, objectives and measures for the sustainable management of forests in Switzerland. Federal Office for the Environment, Bern: 66 pp
FOREST EUROPE, UNECE, FAO (2011) State of Europe's forests 2011. Status and trends in sustainable Forest Management in Europe. Ministerial conference on the protection of forests in Europe. FOREST EUROPE Liaison Unit Oslo, Norway. 337 p
Ginzler C, Bärtschi H, Bedolla A, Brassel P, Hägeli M, Hauser M, Kamphues M, Laranjeiro L, Mathys L, Uebersax D, Weber E, Wicki P, Zulliger D (2005) Luftbildinterpretation LFI3 – Interpretationsanleitung zum dritten Landesforstinventar. Eidgenössische Forschungsanstalt für Wald, Schnee und Landschaft WSL, Birmensdorf
Ginzler C, Mathys L, Thürig E (2011) Die Baumbedeckung in der Schweiz. Schweiz Z Forstwes 162(9):344–349
Keller M (2011) Swiss national forest inventory. Manual of the Field Survey 2004—2007. Swiss Federal Research Institute WSL Birmensdorf, 269 pp
Keller M. (2013) Schweizerisches Landesforstinventar. Anleitung für die Umfrage- und Erschliessungserhebung 2013. Birmensdorf, Eidgenössische Forschungsanstalt für Wald, Schnee und Landschaft WSL. 40 S
Köhl M (1994) Statistisches Design für das zweite Schweizerische Landesforstinventar: Ein Folgeinventurkonzept unter Verwendung von Luftbildern und terrestrischen Aufnahmen. Mitt. Eidgenöss. Forsch.anst. Wald Schnee Landsch 69(1):1–141
Kurt A (1957) Die Produktionsmöglichkeiten im Schweizer Wald. In: Schweizerische Wald- und Holztagung 1956 in Bern. Beiheft der Schweizerischen Zeitschrift für Forstwesen 28:31–39
Kurt A (1967) Ein forstliches Landesinventar als Grundlage schweizerischer Forstpolitik. Wald und Holz 49:94–99
Lanz A, Brändli UB, Brassel P, Ginzler C, Kaufmann E, Thürig E (2010) Switzerland. In: Tomppo E, Gschwantner T, Lawrence M, McRoberts RE (eds) National forest inventories – pathways for common reporting. Springer, Heidelberg

Lanz A, Abegg M, Brändli UB, Camin P, Cioldi F, Ginzler C, Fischer C (2016) Switzerland. [Chap. 43]. In: Vidal C, Alberdi I, Hernández L, Redmond J (eds) National forest inventories. Assessment of wood availability and use. Springer, Cham, pp 783–805

Mahrer F, Vollenweider C (1983) Das Landesforstinventar der Schweiz. Bericht 247, Eidgenössische Anstalt für das forstliche Versuchswesen EAFV, Birmensdorf

Mandallaz D (2008) Sampling techniques for forest inventories, Applied Environmental Statistics. Chapman & Hall/CRC, Boca Raton

Mathys L, Ginzler C, Zimmermann NE, Brassel P, Wildi O (2006) Sensitivity assessment on continuous landscape variables to classify a discrete forest area. For Ecol Manag 229 (1–3):111–119

Ministerial Conference (1994) European criteria and most suitable quantitative indicators for sustainable management. Adopted by the first Expert Level Follow-Up Meeting of the Helsinki Conference. Geneva, June 24, 1994. Helsinki, Finland, Ministry of Agriculture and Forestry

Müller K, Fraefel M, Cioldi F, Camin P, Fischer C (2016) Der Datensatz «Walderschliessungsstrassen 2013» des Schweizerischen Landesforstinventars. Schweiz Z Forstwes 167(3):136–142. https://doi.org/10.3188/szf.2016.0136

Ott E (1972) Erhebungen über den gegenwärtigen Zustand des Schweizer Waldes als Grundlage waldbaulicher Zielsetzungen. Mitteilung Nr. 48, Eidgenössische Anstalt für das Forstliche Versuchswesen EAFV, Birmensdorf

Rigling A, Schaffer HP (eds) (2015) Forest report 2015. Condition and use of Swiss Forests. Federal Office for the Environment, Bern, Swiss Federal Institute WSL, Birmensdorf, 144p

SAEFL (Swiss Agency for the Environment, Forests and Landscape) (1999) The Swiss forest – taking stock. Interpretation of the Second National Forest Inventory in terms of forest policy. Swiss Agency for the Environment, Forests and Landscape, Bern

SAEFL (Swiss Agency for the Environment, Forests and Landscape) (2004) Swiss National Forest Programme (Swiss NFP), Environmental documentation 363. Swiss Agency for the Environment, Forests and Landscape, Bern. 150 pages

SAEFL, WSL (2005) Forest Report 2005. Facts and Figures about the Condition of Swiss Forests. Berne, Swiss Agency for the Environment, Forest and Landscape; Birmensdorf, Swiss Federal Institute WSL

Speich S, Meile R (2013) Interactive maps of results of the National Forest Inventory NFI. [Published online 02.05.2013] Available from World Wide Web http://www.lfi.ch/resultate/aktuell-en.php. Birmensdorf, Swiss Federal Research Institute WSL

Speich S, Abegg M, Brändli U-B, Cioldi F, Duc P, Keller M, Meile R, Rösler E, Schwyzer A, Ulmer U, Zandt H (2010) Third national forest inventory – result tables on the internet. [Published online 16.3.2010] Available from World Wide Web http://www.lfi.resultate/. Birmensdorf, Eidg. Forschungsanstalt WSL

SR 414.110 (1991) Bundesgesetz über die Eidgenössischen Technischen Hochschulen (ETH-Gesetz) vom 4. Oktober 1991. AS 1993 210

SR 510.62 (2007) Bundesgesetz über Geoinformation (Geoinformationsgesetz, GeoIG) vom 5. Oktober 2007. AS 2008 2793

SR 510.620 (2008) Verordnung über Geoinformation (Geoinformationsverordnung, GeoIV) vom 21. Mai 2008. AS 2008 2809

SR 921.0 (1991) Bundesgesetz über den Wald (Waldgesetz, WaG) vom 4. Oktober 1991. AS 1992 2521

SR 921.01 (1992) Verordnung über den Wald (Waldverordnung, WaV) vom 30. November 1992. AS 1992 2538

Steinmann K, Mandallaz D, Ginzler C, Lanz A (2013) Small area estimations of proportion of forest and timber volume combining Lidar data and stereo aerial images with terrestrial data. Scand J For Res 28(4):373–385

Stierlin HR, Brändli UB, Herold A, Zinggeler J (1994) Schweizerisches landesforstinventar. Anleitung für die Feldaufnahmen der Erhebung 1993–1995. Birmensdorf, Eidgenössische Forschungsanstalt für Wald, Schnee und Landschaft. 204 S

Traub B, Meile R, Speich S, Rösler E (2017) The data storage and analysis system of the Swiss National Forest Inventory. Comput Electron Agric 132:97–107. https://doi.org/10.1016/j.compag.2016.11.016

Zierhofer W (2000) Wirkungsanalyse LFI1 und LFI2, Bedarfsanalyse LFI3 – Ergebnisse der Pilotstudie. Im Auftrag der Eidg. Forstdirektion, BUWAL. Nur für internen Gebrauch. Wettingen. 36 pages (unpublished)

Zingg A, Bachofen H (1988) Schweizerisches Landesforstinventar. Anleitung für die Erstaufnahmen 1982–1986. Ber. Eidgenöss. Forsch.anst. Wald Schnee Landsch. 304: 134 S

Part II
Sampling Design and Estimation Procedures

Chapter 2
Sampling Design and Estimation Procedures

Adrian Lanz, Christoph Fischer, and Meinrad Abegg

Abstract The topic of this chapter is the statistical sampling approach of the Swiss National Forest Inventory (NFI). We start with the main populations and population parameters of interest, and how these are – under the infinite population or Monte-Carlo approach to forest inventory – re-shaped into and understood as continuous populations in space. The spatial arrangement of field data collection on a grid of permanently installed sample plots recently changed from a periodic into a continuous (annual) system, which are presented in the following sections, together with estimation procedures which make use of remote sensing and other spatial data to increase the precision of the estimates. We then emphasise on the estimation of change and change components, and depict specific solutions such as for the estimation of the average annual change and the change per unit area. Finally, the NFI system of two-stage subsampling of tally trees for stem volume estimation and the respective estimation procedures are described in this chapter.

2.1 Introduction

Swiss forests cover more than 30% of the country's territory and are stocked with about 500 million trees (with a stem diameter at breast height ≥12 cm). The sheer size of this study object inhibits full observation and enumeration of Switzerland's forest and trees. An additional challenge is that the information in demand extends beyond the current state of the forest and the stock of living trees as well as recent changes. Indeed, it includes many other aspects of forest ecosystems, such as young

A. Lanz (✉)
Statistics Lab, Swiss Federal Research Institute WSL, Birmensdorf, Switzerland
e-mail: adrian.lanz@wsl.ch

C. Fischer
Scientific Service NFI, Swiss Federal Research Institute WSL, Birmensdorf, Switzerland

M. Abegg
Scientific Service NFI, Swiss Federal Institute for Forest, Snow and Landscape Research WSL, Birmensdorf, Switzerland

C. Fischer, B. Traub (eds.), *Swiss National Forest Inventory – Methods and Models of the Fourth Assessment*, Managing Forest Ecosystems 35,
https://doi.org/10.1007/978-3-030-19293-8_2

(regeneration) trees, standing and lying deadwood, and the occurrence of shrubs, plants and animals, as well as various site and stand characteristics, including the recreational use of forests and forest management activities.

Therefore, only a small part of the total forest area can be observed in a national survey. In fact, with the current amount of field data collected and forest area covered annually by the Swiss NFI (NFI), it would take around 40,000 years to observe Switzerland's entire forest.

2.1.1 Survey Sampling

In this situation, *statistical sampling* comes into play because randomly selected samples eliminate selection bias. The resulting samples are objective and acceptable to the public (Särndal et al. 2003). Survey sampling is a specialised field of statistics consisting of a theory and methods for extracting information from observational data to solve real-world problems (Madigan et al. 2014). Survey sampling techniques are widely used in public statistics at the national and regional levels, including in forest inventories. Statistical knowledge is essential in survey sampling. The statistician's role is to: (a) design the acquisition of data in a way that minimises bias and confounding factors, and maximises information content; (b) verify the quality of data after its collection; and (c) analyse data in a way that provides insights or information that support decision making.

A *sample survey* is built on three pillars: (a) a well-defined *population* and agreed *population parameters* of interest; (b) a *sampling design* for data acquisition; and (c) a set of *estimator algorithms* for estimating the population parameters of interest from the collected data. The three pillars interact and depend on each other. For instance, a change in the definition of the population usually gives rise to a change in the sampling design, which then requires an adaptation of the estimators. The role of the statistician is to ensure smooth interactions and to find optimal overall *survey designs*.

2.1.2 Chapter Contents

This chapter is divided into four sections. In the first section, we provide the definitions of the main *populations of interest* and the respective *population elements* with associated target variables in the case of the NFI. In the second section, we first outline the statistical theory of *probability sampling* in forest inventories in general, and then explain the *sampling design* of the NFI for the main populations of interest, i.e. trees, young (regeneration) trees and lying deadwood. The NFI implements a unique, two-stage sampling and measurement design with related procedures for estimating stem volumes and the biomass of living trees. The design is explained in detail in this section. Another focus of the second section is the spatial and temporal organisation of the field data collection on annual panels of *permanently installed sample plots*.

In the third section, the overall *estimation procedures* are presented, with emphasis on aspects that are specific to the NFI, such as: (a) the two-phase estimation procedures combining remote sensing and other spatial auxiliary data with field data for more precise estimation; (b) the two-stage estimation procedure for growing stock and biomass estimation; and (c) the implementation of area domain and sub-population estimation. The estimators are given in 20.8.2.

The *estimation of change and components of change* between inventories is paramount in the NFI. In the final section of this chapter, the panel survey system of the NFI, with periodic re-measurement of permanent plots, is described in detail. The historical context of estimating change is described, including the *méthode de contrôle* (Gurnaud 1886; Biolley 1901, 1920) established by Biolley in 1890 in Canton Neuchâtel. The NFI system with permanent plots is essentially a continuation and application of Biolley's inventory system with respect to the form of statistical sampling. The estimators and components of change are presented in detail, particularly regarding the treatment of changing area domains and the estimation of change components in the stock of living trees under the two-stage sampling design.

2.2 Sampling Design and Estimation Procedures

2.2.1 Population of Interest

Probability sampling methods always build upon a well-defined population of elements with associated target variables for which means and totals (sums) are then estimated. In the context of the NFI, it is therefore essential to have an exact definition of *forest*, but also of various population types, such as the *population of trees*, the *population of deadwood pieces*, and the *population of young trees in the forest regeneration*. These populations are briefly described in this section so that the sampling scheme used in the NFI can be understood by the reader.

2.2.2 Forest and Shrub Forest Area

The current definition of *forest*, as applied in the NFI, was published in 1976 (Mahrer 1976). The definition is designed to comply as much as possible with the legal definition of a forest according to Swiss federal law. The corresponding regulation, published in 1965, leaves space for a certain margin of discretion. This enables the use of clear and objective criteria for the NFI forest definition, which are still in agreement with the legal definition. Furthermore, it is important that the forest definition can be applied when using remote sensing alone, thereby making it possible to include inaccessible forest in NFI forest assessments. For the NFI, criteria related to the following aspects are used in the definition of forest:

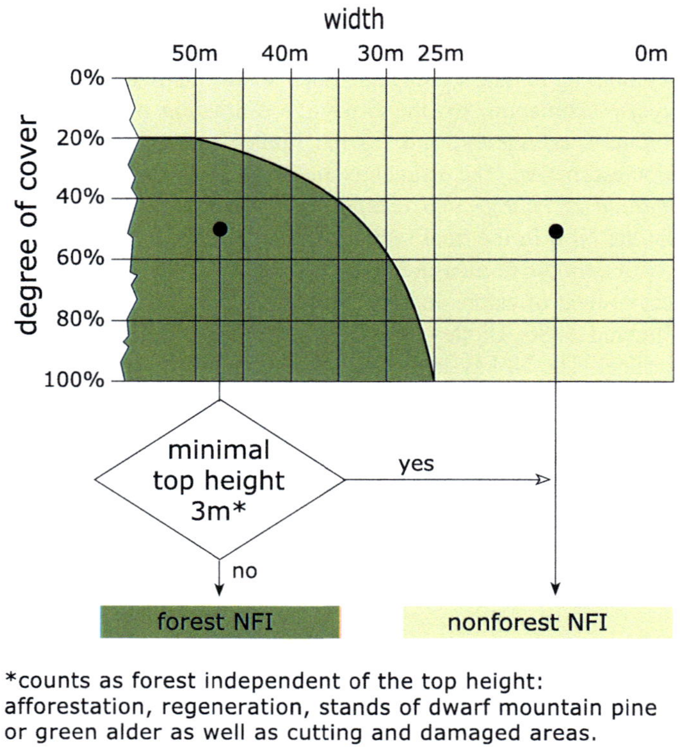

Fig. 2.1 The relationship between crown cover and width of a stocking, according to the NFI definition

- Stocking of trees
- Minimum crown cover
- Minimum height of trees
- Minimum width of stocking
- Land use

Eligible elements of a forest stocking are defined as trees of any species or members of a closed list of shrub species. Orchards are not considered because the land use is predominantly agricultural (Düggelin and Keller 2017). According to the NFI definition, a forest stocking has a minimum crown cover of 20% and a minimum width of 25 m. The required minimum width of the stocking is directly dependent on the crown cover (Fig. 2.1) and can reach up to 50 m if the crown cover is only 20%. The minimum width indirectly defines the minimum size of the stocking. The purpose of including the minimum width criterion is that linear stockings that are narrow but very long, for example bordering a brook, are not defined as forest.

The elements of the stocking must have a height of at least 3 m because this is the detectability threshold for trees in the aerial-photo interpretation. However, at the upper treeline in alpine regions there are relatively large areas covered primarily by two species, *Alnus viridis* and *Pinus mugo*, that rarely reach 3 m in height. An exception was therefore established for these two species, and they are considered stocking even if their height is <3 m. Additionally, temporarily unstocked areas, for example owing to harvesting, natural disturbances or damage, and afforested areas are counted as forest. Further, small clearings which are not clearly separable from the surrounding forested land are included as part of the forest, but only if the width of the clearing is less than 25 m. Finally, there are land-use criteria which lead to certain areas being classified as 'forest' according to the NFI forest definition, even if they are unstocked, and other areas not being considered forest even though they are stocked with trees. The following elements are not considered forest according to the NFI definition (Düggelin and Keller 2017):

- Roads and streets wider than 6 m
- Streams with a streambed wider than 6 m
- Railway tracks, cable cars, ski lifts or similar infrastructure
- Buildings with a non-forestry purpose
- Allotments or orchards
- Tree nurseries for gardening purposes
- Parks
- Rows of trees bordering a street, alleys or avenues

The following unstocked areas are considered forest:

- Forest roads up to 6 m wide
- Timber yards: small unsurfaced storage yards in permanent use, located next to a forest road and forest stand
- Recreation areas: huts, resting places, car parks and other recreation areas with a width of less than 25 m
- Tree nurseries for forestry purposes, adjacent to forest land and in small clearings up to 25 m wide
- Streams up to 6 m wide
- Erosion channels, avalanche tracks, skidding tracks or other channels with an unstocked area <25 m wide
- Small clearings of meadows, cropland or bare land

The forest area in Switzerland is divided into two subtypes of forest: *shrub forest* and *forest without shrub forest* because shrub forests are of no economic interest for the forestry industry, which was the main focus of the first NFI. Shrub forest is distinguished from the rest of the forest according to the criterion of a crown cover by shrub species of more than 3 m in height, or *Alnus viridis* or *Pinus mugo* stands, of at least two-thirds.

2.2.3 Trees

In the NFI, trees with a height $\geq$10 cm are sampled. However, the measurement procedure differs for two groups of trees: (a) *Tally trees* are living and dead, standing and lying trees and certain shrubs according to a given species list with a diameter at breast height ($d_{1.3}$) $\geq$12 cm. (b) *Young trees* consists of standing living trees, as well as certain shrubs, >10 cm in height but <12 cm in $d_{1.3}$. In NFI5, standing dead trees with a height >1.3 m will be sampled as well. The population of trees considered in the NFI only includes trees inside forest areas according to the NFI forest definition.

2.2.3.1 Tally Trees

All trees with a $d_{1.3}$ $\geq$12 cm are considered tally trees, whereas only shrubs included in a specific list (Chap. 9) are assigned to this group. This list contains all native and most non-native shrub species occurring in forests. The $d_{1.3}$ is measured on stems at 1.3 m above the highest ground level at the bottom of the tree. Tally trees can be living, dead, standing or lying. Trees and shrubs are distinguished according to a defined species list. As the NFI is based on permanent sample plots, tally trees are tracked over successive inventories – even if they leave the sample temporarily – so that their development can be assessed when they re-appear in the sample.

2.2.3.2 Young Trees

To assess regeneration, all trees, as well as shrubs of seven species (Chap. 4), with a height $\geq$10 cm but a $d_{1.3}$ <12 cm are measured. For regeneration trees <1.3 m in height, signs of game browsing is additionally assessed. Two different aspects of regeneration are evaluated using two different methods: (a) the number of individuals is estimated by counting the number of individuals in concentric circles on the regeneration plot; and (b) the relative area 'occupied' by young trees with certain characteristics of interest is estimated by assessing the young tree individual located closest to the plot centre (Schwyzer and Lanz 2010).

2.2.4 Lying Deadwood

Deadwood is important as habitat, and two different types of deadwood are thus assessed in the NFI: (a) standing and lying dead tally trees on NFI plots, and (b) pieces of lying deadwood found on line intersects. The elements of the population of lying deadwood are pieces with a minimum length of 1 m and a minimum diameter of 7 cm. Stumps are not included. The detailed measurement protocol for line-intersect sampling is described in Chap. 9 and the inference applied is described in Sect. 2.3.5.

2.2.5 *Further Elements*

2.2.5.1 Forest Edge Survey

Forest edges are considered important habitats for a wide range of species. In order to describe these habitats, the NFI includes an assessment of the characteristics of forest edges located close to NFI sample-plot centres (<25 m). For further details on these assessments, see Chap. 9. The data is collected along a transect 50 m in length. Due to the method of sampling and the fractal property of forest edges (Mandelbrot 1967), statistical inference of the length of forest edges in Switzerland, for instance, would require specific assumptions and estimation techniques, which are not implemented in the NFI. However, based on the available sample of approximately 1000 edge transects, it is possible to evaluate the average composition of the forest edges and to assess changes between inventories.

2.2.5.2 Forest Road Survey

The forest road survey is a census survey conducted by the NFI that provides the basis for reporting on the state of the forest road infrastructure in Switzerland. Forest truck roads are defined as roads on forest land or roads bordering forest land with a minimum width of 2.5 m and a carrying capacity of 10 t per axle. Various characteristics of the individual road sections are verified and assessed as part of the NFI interview survey with the local forest service (Chap. 10). The methods of data analysis and the way of combining the road survey data with the NFI field data changed during the course of NFI4. The earlier method is described by Paschedag and Zinggeler (2001). The new method makes use of spatially explicit data such as the NFI forest cover map (Sect. 7.2), and increased automation in the calculations.

2.3 Sampling Design

2.3.1 *Population Parameters of Interest*

A forest inventory, like any statistical survey, is concerned with the estimation of summary statistics, such as totals (sums) or means, and of target variables associated with the elements of the population under study. The typical elements in a forest inventory are trees located on forest land. Thus, the population bears a clear reference to space, a characteristic which plays a role not only in the definition of the population, but also in the sampling method and the mathematical notation.

The formal description of the population of interest assumes a suitable projection of the real forest landscape with trees located in the plane in $\Re^2$. Points in $\Re^2$ are denoted by ω and the area domain of interest, the forest land, is a bounded region F in $\Re^2$ with surface area λ_F.

The population P_F with N_F elements, usually trees, in F is labelled $1, \ldots, i, \ldots, N_F$ and element locations u_i are assumed to be well defined and fixed. $u_i \in F$ is true for all elements of P_F.

The element-level target variables of interest are assumed to be measurable without error. Typical target variables of interest are the stem volume of trees, the basal area of stems measured 1.3 m aboveground, and total aboveground biomass of trees. Target variables are denoted by X. If two different target variables are needed, they are referred to as X and Z. The variable $X_i \equiv 1$ for all $i \in P_F$, which is a useful, but also simple, variable for counting or estimating the number of stems in the population.

The population parameters of interest in a forest inventory are:

The total surface area of domain F, for example the forest land

$$\lambda_F = \int_{\Re^2} \iota_F(\omega) d\omega = \int_F \iota_F(\omega) d\omega \tag{2.1}$$

where the integral refers to Riemann integration in $\Re^2$ and

$$\iota_F(\omega) = \begin{cases} 1 & \text{if } \omega \in F \\ 0 & \text{else} \end{cases} \tag{2.2}$$

is a Bernoulli variable defined for all $\omega \in \Re^2$, indicating points in domain F.

The total of target variable X over the population P_F of elements in domain F

$$T_F^{(X)} = \sum_{i=1}^{N_F} X_i; \tag{2.3}$$

The mean spatial density of target variable X over the population P_F of elements in domain F

$$Y_F^{(X)} = \frac{1}{\lambda_F} T_F^{(X)}; \tag{2.4}$$

The ratio of the totals of two different target variables X and Z over the population P_F of elements in domain F

$$R_F^{(X/Z)} = \frac{T_F^{(X)}}{T_F^{(Z)}} = \frac{Y_F^{(X)}}{Y_F^{(Z)}}. \tag{2.5}$$

2.3.2 Infinite Population Approach

In a forest inventory covering a large area, we do not have a list of all trees for sample selection. For the sampling, a *sampling frame* $G \supseteq F$ is used, for a region of known surface area λ_G that is as large as the study area, the forest land F.

The points ω in G are the *sampling units* from which a random sample will be selected in the survey. The challenge is to define, prior to sampling and for all sampling units, $\omega \in G$ – a *local (per unit area) density* (Mandallaz 2008) function $y_F^{(X)}(\omega)$ with regards to target variable X in domain F, with the property

$$\int_G y_F^{(X)}(\omega)d\omega = \sum\nolimits_{i=1}^{N_F} X_i = T_F^{(X)} \tag{2.6}$$

In other words: the *local density* functions need to be defined such that the population parameter $T_F^{(X)}$ of interest, the total of X over the population P_F of N_F elements in F, can be computed through Riemann integration of $y_F^{(X)}(\omega)$ over all points ω in frame G.

Once the local density function has been defined, the random sampling of population elements (trees) that form the finite, but not available, list of elements in the population can be replaced by the random sampling units (points) from the infinite, but well-defined, number of units in the sampling frame G. This concept is known as the *infinite population approach* (Mandallaz 2008; Eriksson 1995b) or *Monte-Carlo approach* (Gregoire and Valentine 2007) to sampling in forest inventories.

Local density functions can be understood as *association rules* between the sampling units $\omega \in G$ (points) and nearby population elements $u_i \in F$ (trees). In mathematical notation, the local density can be written as

$$y_F^{(X)}(\omega) = \sum_{i \in S(\omega)} X_i f_i(\omega) = \sum_{i=1}^{N_F} \iota_{i.S(\omega)} X_i f_i(\omega), \tag{2.7}$$

i.e. a weighted sum of target variable values X of population elements associated with ω. The term $\iota_{i.S(\omega)}$ formally describes the sampling unit at point ω. The $\iota_{i.S(\omega)}$ describe the association of population elements with sampling units: $\iota_{i.S(\omega)} = 1$ if population element i is a member of the local sample $S(\omega)$ of trees associated with sampling unit (point) ω; otherwise $\iota_{i.S(\omega)} = 0$.

$f_i(\omega)$ are element-level *extrapolation factors*, and we conclude from Eqs. (2.6) to (2.7) that a local density function should be defined such, that the condition

$$\int_G \iota_{i.S(\omega)} f_i(\omega)d\omega = \int_{A_{i.G}} f_i(\omega)d\omega = 1 \tag{2.8}$$

is true for all elements i in population P_F. $A_{i,G}$ of size $\lambda_{A_{i,G}}$ is the so-called *inclusion zone* (or inclusion field) of population element i. It includes all sampling units (points) ω in G for which i is a member of the respective local sample $S(\omega)$.

Inclusion zones and related extrapolation factors are mathematical concepts that correspond with sample-selection rules applied in field operations (duality principle). This link will be illustrated for the NFI field protocol in the next section.

Before moving on, it is important to mention that we assume fixed target variable values on population elements and fixed local densities on sampling units, which implies association rules between sampling units and population elements defined prior to sampling. The randomness on which inference is based is only in the method of selecting sampling units (points) from the sampling frame (Sect. 2.4.2).

2.3.2.1 Local Density Functions in NFI

In this section, the NFI plot configurations, the association rules between sampling units (plot centres) and population elements, and the related local density functions are explained for the most important populations of interest: trees, pieces of lying deadwood and young trees.

2.3.2.2 Area Domain Estimation

In this first section, we describe the NFI 'plot configuration' and the local density function for (forest) area estimation.

The population of interest is in this case a geographic (area) domain $F \subseteq G$ of unknown size λ_F. Area domains frequently need to be estimated in the NFI, to estimate not only Switzerland's total forest area, but also the proportion and/or surface area of various area (sub)domains, such as forest areas classified according to ownership or stand-age categories.

The population of interest is formally defined as a bounded region F in $\Re^2$ and the response for sampling units in G is simply a domain membership indicator variable $\iota_F(\omega) = 1$ for units located in F, and $\iota_F(\omega) = 0$ otherwise. Consequently, the one-to-one association rule between sampling units ω in G and population elements ω in F results in a local density function $y(\omega) = \iota_F(\omega)$.

2.3.3 *Sampling of Tally Trees*

The typical association rule, or *sampling function*, between sampling units (points) and trees in a forest inventory is called *fixed radius plot sampling*, i.e. the selection of all trees in a circular plot centred at point $\omega \in G$ of radius ρ. Therefore, for a tree positioned at point $u_i \in F$, the resulting *inclusion zone*

$$A_{i.G} = \{\omega \in G | |\omega - u_i| \leq \rho\} \tag{2.9}$$

is usually circular, centred in u_i and of size $\lambda_{A_{i.G}} = \pi\rho^2$, and the extrapolation factor is $f_i(\omega) = 1/\lambda_{A_{i.G}} = 1/\pi\rho^2$. For trees near the sampling frame's boundary, the inclusion zone may extend outside of G, so that $f_i(\omega) > 1/\pi\rho^2$.

Under *concentric (nested) fixed-radius plot sampling*, predefined *subpopulations of trees* are selected from circular plots with different radii. The subpopulations may be defined according to any criteria, for example tree species. A typical approach in forest inventories is to use larger circles for trees with large stem diameters and smaller circles for trees with smaller stem diameters.

The NFI uses two concentric circular plots with fixed sizes and the same centre points. The subpopulation of trees with a $d_{1.3} \geq 12$ cm but <36 cm is associated with a plot 200 m^2 in area, and the subpopulation with a $d_{1.3}$ >36 cm is associated with a plot 500 m^2 in area.

The local densities are calculated as a per hectare density and the *tally tree extrapolation factors* in the NFI are therefore 50 ha^{-1} for the 200 m^2 plots and 20 ha^{-1} for the 500 m^2 plots.

Slope Correction Because all area-related population parameters of interest are understood as statistics after projection to the plane, circular plots become ellipses when they are projected onto sloped terrain and a locally smooth terrain surface is assumed. To facilitate fieldwork and to avoid tree individual maximum distances between the plot centre and tree locations, the NFI radius for tally tree selection, applied parallel to the possibly sloped surface of the terrain, is defined as

$$\rho(\omega) = \frac{\rho_{\text{flat}}}{\sqrt{\cos\beta(\omega)}} \tag{2.10}$$

where $\beta(\omega)$ is the locally determined slope of the terrain, and ρ_{flat} is the standard radius of 12.61 m for the subpopulation of large trees and 7.98 m for the subpopulation of small trees.

$\rho(\omega)$ is chosen such that the surface area of the back-projected ellipse is equal to the surface area of the original circular plot ρ_{flat} in radius.

NFI Sampling Frame Boundary Correction The field protocol of the NFI leads to a *specific sampling frame for tree sampling*. The field protocol is such that only sampling units ω located on forest land are used for tally tree selection and measurement. As a consequence, trees near the forest boundary line have a lower overall probability of becoming a member of the sample, and boundary correction measures are therefore required.

The mathematical explanation for this boundary effect is related to the shape and size of tree inclusion zones. Under fixed radius plot sampling, the tree inclusion zones are bounded by circles centred at tree locations u_i. However, the circular tree inclusion zone is truncated for trees near the boundary.

The following boundary correction measures are applied in the NFI:

- For sampling units qualifying for tree sampling, i.e. located on forest land, a *reduction line*, roughly corresponding to the forest boundary, is locally determined and registered as part of the field protocol.
- The reduction line is defined in such a way that (a) no eligible tree in P_F is outside this line, and (b) the forest boundary is outside and does not cross the reduction line.
- After automated analysis of field data, the derived extrapolation factors for tally trees are then $f_i(\omega) = f_i(\omega)^{(0)}/p(\omega)$, where $f_i(\omega)^{(0)}$ is the standard extrapolation factor of tally tree i and $p(\omega)$ is the proportion of the plot, either the small or the large plot depending on the i-th tally tree $d_{1.3}$, located inside the reduction line.

This procedure of correcting tally tree extrapolation factors is, as a matter of principle, not exact (Gregoire 1982), but it is assumed to have a minor impact on the overall estimates of tree resources. Based on a modified field data collection protocol in use since NFI5, there are plans to apply tree-level boundary correction in the next NFI reports.

2.3.4 *Sampling of Tariff Trees*

In NFI, a subsample of *tariff trees* is randomly selected from the sample of tally trees for the more time-consuming measurement of their upper-stem diameters and stem lengths. The overall sampling scheme has been described as two-stage sampling (Mandallaz 2008). Related *two-stage estimation procedures* have been applied in NFI since NFI2 for growing stock and biomass estimation.

2.3.4.1 Motivation

Traditionally, single stem volumes are predicted by means of *tariff functions*, which predict stem volumes as a function of the stem $d_{1.3}$. Tariff functions are calibrated on the basis of section-wise measured stems. Mean tariff functions are accurate, but they are known to over-estimate or under-estimate the true stem volume of individual stems because stem forms depend on many factors, such as tree species, local growth conditions and individual growing conditions for stems in the stand. Because of these influencing factors, so-called local tariff functions under specific growth conditions and for different tree species are used in forest inventories.

In Switzerland, the need for accurate stem volume functions gained attention in connection with the development and large-scale implementation of permanent-plot forest inventories at the forest enterprise level. The starting point was in the 1960s, when a set of tree-species-specific *stem volume functions* were established that predict the stem volume as a function of the stem $d_{1.3}$, the stem upper diameter

(7 m aboveground) and the stem length (Chap. 12). This approach was further developed when the NFI began in the 1980s.

In the NFI, these stem volume functions are used to: (a) derive the stem volumes for all second-stage tariff trees in the sample, and (b) calibrate highly localised tariff functions, in which various tree, stand and site variables are used as explanatory variables. The tariff functions are then used to predict a less accurate stem volume for all tally trees in the sample.

The details and accuracy of the different volume and tariff functions are explained in Chap. 12. Briefly, the tariff-function-based stem volumes available for the first-stage sample of tally trees are moderately accurate, while the volume-function-based stem volumes available for the second-stage subsample of tariff trees are very accurate.

2.3.4.2 Poisson Sampling of Tariff Trees

In NFI1 to NFI4 (years 2009–2014), a subsample $\ddot{S}(\omega)$ of tariff trees was randomly selected from the sample $\dot{S}(\omega)$ of tally trees at ω. Individual tariff trees were selected independently and with predefined (conditional) second-stage sample inclusion probabilities p_i.

Thus, under this *Poisson sampling* of tariff trees, the *generalised (two-stage) local density function* can be written as

$$\ddot{y}_F^{(X)}(\omega) = \sum_{i \in \dot{S}(\omega)} \hat{X}_i f_i(\omega) + \sum_{i \in \ddot{S}(\omega)} \frac{R_i f_i(\omega)}{p_i} \tag{2.11}$$

where $R_i = X_i - \hat{X}_i$ denotes the difference between the true[1] (volume function) stem volume X_i and the predicted (tariff function) stem volume $\hat{X}_i$.

With single-stage sampling, the local densities $\dot{y}_F^{(X)}(\omega)$ are predefined and fixed values for all sampling units in the frame. This is not the case for the generalised local densities under two-stage sampling. For a given sampling unit, $\ddot{y}_F^{(X)}(\omega)$ is a random variable with expectation $y_F^{(X)}(\omega)$, the local single-stage density of target variable X, the true (volume function) stem volume (Mandallaz 2008).

2.3.4.3 Fixed Association of Tariff Trees

The independent Poisson selection of tariff trees in each inventory has certain disadvantages if repeatedly applied in an inventory with permanent plots, especially for change estimation but also for the simple purpose of establishing an NFI database of re-measured tariff trees.

---

[1] It is assumed to be true in terms of the best value available.

For this reason, the approach for selecting tariff trees was changed during NFI4. Tariff trees are now associated with sampling units ω according to predefined associations rules, as done for tally trees. No randomness is involved in the selection of tariff trees, and a tariff tree remains associated with the sampling unit over successive inventories as long as the tree is not removed from the population.

In this revised approach, the local density function at point ω can formally be written as

$$\ddot{y}_F^{(X)}(\omega) = \sum_{i \in \dot{S}(\omega)} \hat{X}_i \dot{f}_i(\omega) + \sum_{i \in \ddot{S}(\omega)} R_i \ddot{f}_i(\omega) \tag{2.12}$$

where $R_i = X_i - \hat{X}_i$ denotes the difference between the true (volume function) stem volume X_i and the predicted (tariff function) stem volume $\hat{X}_i$, and $\dot{f}_i(\omega)$ and $\ddot{f}_i(\omega)$ are the respective extrapolation factors for tally trees and tariff trees.

As the sampling of tally trees is done independently from the sampling of tariff trees, the Riemann integral of the local density function $\ddot{y}_F^{(X)}(\omega)$ is

$$\int_G \ddot{y}_F^{(X)}(\omega)d\omega = \int_G y_F^{(\hat{X})}(\omega)d\omega + \int_G y_F^{(R)}(\omega)d\omega = T_F^{(\hat{X})} + T_F^{(R)} = T_F^{(X)} \tag{2.13}$$

the sum of the true (volume function) stem volume X over the population P_F of trees in F. We implicitly required R_i to be available for all tariff trees, so that $\ddot{S}(\omega) \subseteq \dot{S}(\omega)$, which, however, is a restriction easily fulfilled with appropriate allocation rules.

2.3.4.4 Optimal Sampling Schemes for Tariff Trees

The rules for optimal tally and tariff tree selection are as follows (Mandallaz 2008):

- First-stage (tally) trees should be selected with sample inclusion probabilities proportional to predictions: $\lambda_{A_{i}.G}/\lambda_G \propto \hat{X}_i$,
- Second-stage (tariff) trees should be selected with sample inclusion probabilities proportional to residuals: $\lambda_{A_{i}.G}/\lambda_G p_i \propto R_i$,
- The optimal conditional second-stage (tariff) tree sample inclusion probabilities are therefore $p_i \propto R_i/\hat{X}_i$.

Under the nested (concentric) fixed-area plot sampling of the NFI, the first-stage tally tree inclusion probabilities are constant for trees with $d_{1.3}$ >= 12 cm and $d_{1.3} < 36$ cm, and higher but again constant for tally trees with $d_{1.3} \geq 36$ cm. Because the tally tree inclusion probabilities are constant within each of the two categories of tree diameters, the optimal conditional inclusion probabilities p_i of tariff trees are proportional to the residuals R_i. The increase in the residuals is roughly proportional to the true volume: $R_i \propto X_i$.

2.3.4.5 Tariff Tree Selection in NFI1

A fixed association of tariff trees was applied, which included tally trees with $d_{1.3}$ >=12 cm and $d_{1.3}$ <60 cm if the trees are located in the plot sector 0–150 gon (151 gon out of the 400 gon of the full circle), and all tally trees in the $d_{1.3}$ class ≥60 cm. Therefore, when the inclusion probability of tariff trees is written in the form $\ddot{f}_i(\omega) = \dot{f}_i(\omega)/q_i$, we get

$$q_i = \begin{cases} 151/400 & \text{if } 12 \leq d_{1.3.i} \leq 59 \\ 1 & \text{if } d_{1.3.i} \geq 60 \end{cases} \tag{2.14}$$

Tariff Tree Selection in NF2 to NFI4 (Years 2009–2014) The optimised Poisson sampling selection included tally trees with $d_{1.3}$ >=12 cm and $d_{1.3}$ <60 cm if the trees are located in the plot sector 0–149 gon (150 gon out of the 400 gon of the full circle) with the (conditional) inclusion probabilities p_i given below. All tally trees with $d_{1.3}$ ≥60 cm were selected as tariff trees. The resulting conditional sample inclusion probabilities of tariff trees were

$$p_i = \begin{cases} 150/400 \times 0.000015 \times (d_{1.3.i})^2 \times 50 & \text{if } 12 \leq d_{1.3.i} \leq 35 \\ 150/400 \times 0.000015 \times (d_{1.3.i})^2 \times 20 & \text{if } 36 \leq d_{1.3.i} \leq 57 \\ 150/400 & \text{if } 58 \leq d_{1.3.i} \leq 59 \\ 1 & \text{if } d_{1.3.i} \geq 60 \end{cases} \tag{2.15}$$

Tariff Tree Selection NFI4 (Years 2015–2017) Two aspects of tariff tree sampling were changed. First, the proportion of tariff trees was adapted so that the originally planned inclusion probabilities could be followed better. Second, a fixed association of tally trees was introduced to maximise the number of tariff trees re-measured over successive inventories.

With $\ddot{f}_i(\omega) = \dot{f}_i(\omega)/q_i$, the new values are

$$q_i = \begin{cases} 5.625 \times 10^{-6} \times (d_{1.3.i})^{2.00} \times 50 & \text{if } 12 \leq d_{1.3.i} \leq 35 \\ 1.125 \times 10^{-8} \times (d_{1.3.i})^{3.75} \times 20 & \text{if } 36 \leq d_{1.3.i} \leq 59 \\ 1 & \text{if } d_{1.3.i} \geq 60 \end{cases} \tag{2.16}$$

where 5.625×10^{-6} is equal to 150/400 × 0.000015, the value used in NFI2 to NFI4 (years 2009–2014). 1.12510^{-8} was adopted to constantly increase inclusion probabilities for tariff tree selection in these $d_{1.3}$ classes. Figure 2.2 illustrates the conditional and relative inclusion probabilities of tariff trees in the different inventories, divided into diameter classes, and Fig. 2.3 shows the respective number of tariff trees it was expected would be measured in an annual sample.

The optimal proportion of tariff trees is roughly proportional to the basal area, with appropriate basal area factors defined accordingly. Therefore, an obvious

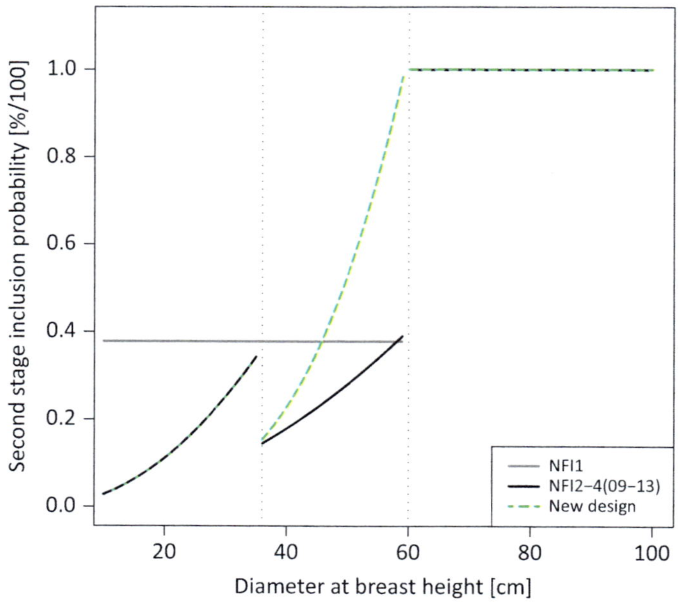

Fig. 2.2 Conditional second-stage inclusion probabilities p_i of tariff trees under Poisson sampling and relative inclusion probabilities q_i of tariff trees under a fixed association in NFI1, NFI2–NFI4 (years 2009–2014), and NFI4 (year 2015–2017)

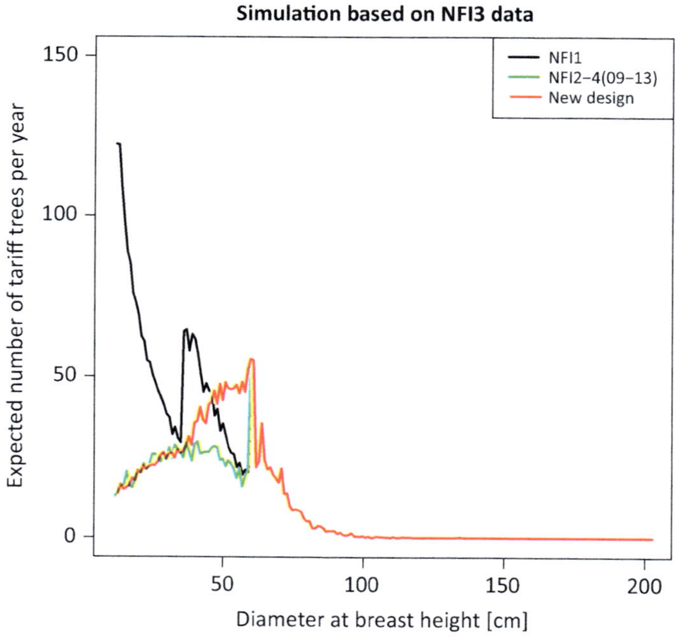

Fig. 2.3 Expected number of tariff trees in an annual panel of the NFI as a function of $d_{1.3}$. Values are given for NFI1, NFI2–NFI4 (years 2009–2014), and NFI4 (2015–2017)

choice for the combined plot configuration of tally and tariff trees would be to overlap the fixed-radius plot sampling of tally trees with an angle-count sampling of tariff trees.

In the NFI, a concept similar to the well-known angle count sampling of tally trees was adopted for the new system of tariff tree selection. Because the tariff trees are concentrated in the plot sector between 0 and 150 gon, the new, sector sampling of tariff trees continues to favour tariff tree association in these sectors of the plots. The idea is to replace the optimal horizontal angles with optimal sectors for tariff trees selection. Using two plots of different sizes for selecting the tally tree causes a slight difficulty. To give an example: assuming the planned proportion of tariff trees in $d_{1.3}$ class 30 cm is $q = 0.25$ translates into a sector sampling of tariff trees in sector 0–100 gon, i.e. 25% of the full circle.

The resulting sector openings $s_{\text{small}.\ i}$ and $s_{\text{large}.\ i}$ for the small 200 m^2 plot and the larger 500 m^2 plot for tariff tree selection are

$$s_{\text{small}.i} \leq \begin{cases} 400 \times 5.625 \times 10^{-6} \times (d_{1.3.i})^{2.00} \times 50 & \text{if } 12 \leq d_{1.3.i} \leq 35 \\ 400 \times 5.625 \times 10^{-6} \times (35)^{2.00} \times 50 & \text{if } 36 \leq d_{1.3.i} \leq 44 \end{cases} \tag{2.17}$$

And

$$s_{\text{large}.i} \leq \begin{cases} 400 \times 1.125 \times 10^{-8} \times (d_{1.3.i})^{3.75} \times 20 \times Q_i & \text{if } 36 \leq d_{1.3.i} \leq 44 \\ 400 \times 1.125 \times 10^{-8} \times (d_{1.3.i})^{3.75} \times 20 & \text{if } 45 \leq d_{1.3.i} \leq 59 \\ 400 & \text{if } d_{1.3.i} \geq 60 \end{cases} \tag{2.18}$$

where stem diameters $d_{1.3.i}$ are down-rounded to centimetres and

$$Q_i = \frac{50}{50-20} \times \left(1 - \frac{5.625 \times 10^{-6} \times (35)^{2.00}}{1.125 \times 10^{-8} \times (d_{1.3.i})^{3.75}}\right). \tag{2.19}$$

is just a numerical term for trees with $d_{1.3}$ >=36 cm and $d_{1.3}$<=44 cm class trees with different sector openings are selected in the small and large NFI plots.

The inclusion probabilities and respective sector openings are given in Table 2.1 in the appendix.

Tariff Tree Non-response The number of tariff trees selected in the NFI sample is always smaller than planned, mainly because the tariff trees selected unconditionally or randomly do not qualify for upper stem and/or stem length measurements. At the estimation stage, the nominal shares q_i and the conditional inclusion probabilities p_i are adjusted for non-response. The proportion of non-response is determined in *response homogeneity groups*, which are formed for the five production regions of Switzerland, three elevation classes, the categories coniferous and broadleaved, and seven $d_{1.3}$ classes.

Boundary Correction for Tariff Trees Because of the anisotropy in the new association of tariff trees in plot sectors, the theoretically correct extrapolation factors for tariff trees can become extremely large for tariff trees near the southern border of the sampling frame. On the other hand, they can become smaller than the first-stage tally tree extrapolation factors for tariff trees near the northern border of the sampling frame.

In this sense, an isotropic association of tariff trees would clearly be preferable. The only reason an anisotropic sector association of tariff trees is currently in use in the NFI is to maximise the number of tariff trees with repeated upper stem diameter and stem length measurements. Because anisotropy is not expected in the distribution of frame boundaries, the extrapolation factors are always $\dot{f}_i(\omega)/q_i$, where the first-stage tally tree extrapolation factor $\dot{f}_i(\omega)$ may be boundary corrected.

2.3.5 *Sampling Pieces of Lying Deadwood*

The infinite population principles can be adapted for use in the line intersect sampling of pieces of lying deadwood on forest land (Kaiser 1983; Gregoire and Valentine 2007). Compared to tree sampling, the population elements are now *objects or particles* of arbitrary form and dimension located on forest land F. Instead of circular plots, one or more lines are used for identifying population elements associated with sampling units ω in G.

The exact solution requires: (a) the definition of position, orientation and length of transect lines associated with a sampling unit ω in G, (b) rules for the association of particles with transect lines and, therefore, for the clear-cut definition of the members of the local sample at point ω, (c) a clear field protocol for the measurements conducted on these particles; and (d) the local density function itself.

The NFI line-intersect sampling method for pieces of deadwood is one of several options used in inventory practice. It involves the following rules (Düggelin and Keller 2017):

- Three transect lines of fixed orientations 35 gon, 170 gon and 300 gon are defined for each sampling unit ω in G
- Transect lines are 10 m in length and start with an offset of 1 m from the sampling unit's position at ω. The length of a transect line is the length projected to the plane; the length applied in the field, following the inclination of the terrain, may therefore be more than 10 m (cosine formula)
- Transect lines stop at the boundary of the sampling frame for tally tree sampling (Chap. 9), and the length of this reduced transect line is registered in the field protocol
- A piece of deadwood is considered a member of the local sample at ω only if at least one of the transect lines intersects with the *central axis* of the piece. If the deadwood piece is forked or otherwise branched, all parts and branches are examined separately

- The diameter of a deadwood piece is measured perpendicular to its central axis at the point of intersection between the central axis and the transect line
- The inclination (angle) of the central axis of the deadwood piece is measured and recorded at this point of intersection during fieldwork

The local density function used in the NFI for the *volume of lying deadwood pieces* in domain F at sampling unit ω is defined as

$$y_F(\omega) = \frac{1}{m(\omega)} \sum_{h=1}^{m(\omega)} \frac{\pi^2}{8\,l_h} \sum_{i \in S_h} \frac{d_{m.i}^2}{\cos \alpha_i} \tag{2.20}$$

where $m(\omega)$ is the number of transect lines installed at point ω (usually $m(\omega) = 3$), l_h is the projected length of transect line h, and $d_{m.i}$ and α_i denote the means of two perpendicular diameter measurements and the inclination, respectively, of deadwood piece i, which is part of the local sample S_h of deadwood pieces intersected by transect line h. The estimator is an approximation. Its derivation can be found in Gregoire and Valentine (2007).

2.3.6 *Sampling Young Trees*

In the NFI, the sampling scheme for *young trees*, also termed a *young forest assessment,* encompasses the population of *living trees* with a height of at least 10 cm but a $d_{1.3}$ <12 cm. The scheme was changed several times between inventory cycles (Schwyzer and Lanz 2010). The common feature of all sampling schemes is that the centre point(s) for young-tree data collection is offset from the standard sampling unit's position by 10 m, and that the trees are not identified for re-measurement in a consequent inventory. The (auxiliary) plot centres for the young tree assessment are, however, considered permanent when the net change in the population of young trees was occasionally assessed between inventories.

Since NFI4, *nested (concentric) fixed-area plot sampling* has been used for the association of living young trees with sampling units, so that the local density function follows the usual rules explained in Sect. 2.3.2.1. The only deviation from standard procedures is the *offset of the plot centre for young tree sampling from the standard sampling unit location* at point ω. To obtain exact and design-unbiased estimates, boundary correction measures would be needed for young trees near the sampling frame boundary, which are currently not implemented. A correction currently applied in the NFI is the establishment of an alternative plot centre in the direction away from the sampling frame boundary in cases where the plot centre for young tree sampling is located outside the sampling frame. Such replacements approximately compensate for young forest plot centres that are not detected and established because the centre of the standard sampling unit is located outside the sampling frame. Identical frames are used for tree and young tree sampling.

In some of the earlier inventories, *two subplot centres,* offset by 10 m from the centre of the sampling unit in opposite directions, were used *for young tree sampling*. Such a sampling unit, consisting of two or more subunits, is usually understood as a cluster, and cluster sampling gives rise to certain adaptations in the mode of sampling and in the choice of estimators. However, because the information needed on young forest does not have to be so precise, the local density for a sampling unit at point ω was simply computed as the arithmetic mean of the two densities measured on the young tree subplots.

Nearest Tree Sampling The assessment of some of the young tree attributes is complex and costly. For this reason, the field protocol restricts the assessments of some of the young tree target variables and characteristics to the young tree individual positioned closest to the subplot centre. A typical target variable of interest in this type of sampling is the presence and type of damage from game browsing observed on young trees.

For this data, the local density function is defined as

$$y_F^{(C)}(\omega) = \imath_F(\omega)\,\imath_i^{(C)}(\omega) \tag{2.21}$$

where $\imath_F(\omega)$ indicates whether ω is in the domain F of interest and the indicator variables

$$\imath_i^{(C)}(\omega) = \begin{cases} 1 & \text{if } i \text{ closest to } \omega \text{ has characteristic } C \\ 0 & \text{else} \end{cases} \tag{2.22}$$

are defined with respect to characteristic C of interest, formally for the entire population P_F of N_F young trees in F.

Under such a nearest tree sampling design and with a local density function defined as above, the infinite population is actually a *Voronoi partitioning* of the domain F of interest (and $y_F^{(C)}(\omega) = 0$, if $\omega \notin F$). The young tree positions $u_i \in F$ are the seeds (centres) of associated Voronoi cells V_i of surface area λ_{V_i}. With N_F young trees in F, then $F = \cup_{i=1}^{N_F} V_i$ and $\lambda_F = \sum_{i=1}^{N_F} \lambda_{V_i}$. For a given young tree characteristic C, the total $T_F^{(C)} = \int_G y_F^{(C)}(\omega)d\omega = \lambda_{F_c}$ is the total surface area of Voronoi cells in F occupied by young trees with characteristic C.

In the NFI, a separate young tree that is nearest to plot centre is selected for several subpopulations of young trees, which are defined according to height and diameter class (Düggelin and Keller 2017).

2.4 Sampling Design of the NFI

With the definition and introduction of a sampling frame and sampling units with associated local densities, the actual population of interest and its elements, such as trees or pieces of lying deadwood, become obscured. The randomised sampling is actually implemented on the continuous plane (or universe) of points in the frame.

2.4.1 *Basic Sampling Methods in Area Sampling*

Horvitz-Thompson Theorem for the Continuous Universe The Horvitz-Thompson estimator plays a central role in survey sampling (Horvitz and Thompson 1952), but the theory was developed for finite populations and is mainly recognised in that context. We briefly mention an extension of the theorem to sampling from an infinite population.

A common reference is Cordy (1993), who defines *inclusion densities* $\pi(\omega)$ on the frame G, which may be thought of as a local measure of the number of points to be selected per unit area. In forest inventory practices, $\pi(\omega)$ are constant over the entire sampling frame or within sampling strata. Cordy also mentions a continuous universe version of importance sampling, which results in unequal inclusion densities.

For a sample S of m points ω in G with associated inclusion densities $\pi(\omega)$ and observed responses $y_X(\omega)$, the extended Horvitz-Thompson estimator

$$\hat{T}_G^{(X)} = \sum_{\omega \in S} \frac{y_X(\omega)}{\pi(\omega)} \tag{2.23}$$

is shown to be unbiased for the total $\hat{T}_G^{(X)} = \int_G y_X(\omega) d\omega$ of the Riemann integrable function $y_X(\omega)$ defined over G. Cordy also derives expressions for the theoretical variance $\mathbb{V}\left\langle \hat{T}_G^{(X)} \right\rangle$ and a sample-based estimator of this variance for which the pairwise inclusion densities $\pi(\omega, \omega')$ have to be calculated for the sampled points ω and ω'. The inclusion densities play a similar role to *inclusion probabilities* in finite population sampling. They describe the design according to which the sample is selected, and they are needed for design-based estimation.

Under a uniform distribution of random points in G and with m independently selected points ω, the inclusion densities are constant over G and given by $\pi(\omega) = m/\lambda_G$, with joint densities $\pi(\omega, \omega') = m(m-1)/\lambda_G^2$.

The Infinite Population Approach to Forest Inventories In forest inventories sampling units are usually distributed with constant density, although sometimes densities vary among strata. The original description of the infinite population approach starts with an assumed sample of only one random point ω, uniformly distributed in G (Mandallaz 2008; Eriksson 1995b).

Then,

$$t_X(\omega) = \lambda_G y_X(\omega) = \hat{T}_G^{(X)} = \hat{T}_F^{(X)} = \sum_{i=1}^{N_F} \frac{X_i \iota_{i.S(\omega)}}{\pi_i}. \tag{2.24}$$

turns out to be the Horvitz-Thompson estimator of the total $T_F^{(X)}$ of target variable X in F, and also of target variable X in G, because $G \supseteq F$, by definition.

$y_X(\omega)$ is the local density and $t_X(\omega)$ is the local density expanded to totals, without explicit reference to the area domain F of interest, $\pi_i = 1/\lambda_G f_i(\omega)$ are sample inclusion probabilities, and $\iota_{i.\ S(\omega)}$ sample membership indicator variables for elements in P_F.

The expectation of the $t_X(\omega)$ for a random point ω in G is, by construction, the total $T_G^{(X)}$ of X in G, which is equal to the population parameter of interest, the total of X in F.

The theoretical variance of $\hat{T}_G^{(X)}$ can be given using the Horvitz-Thompson theorem. Then the joint inclusion probabilities $\pi_{i\,i'} = \lambda_{A_{i.G} \cap A_{i'.G}}$ have a physical interpretation as surface areas of overlapping inclusion zones between pairs of trees. Because most second-order inclusion probabilities are zero, the Horvitz-Thompson variance estimator is not available for samples originating from the random selection of a single point.

However, with a replicate sample of m independently and uniformly distributed random points, a design-unbiased estimator of the true variance of $\hat{T}_G^{(X)} = \lambda_G \hat{Y}_G^{(X)}$ $= \frac{\lambda_G}{m} \sum_{j=1}^{m} y_X(\omega_j)$ is immediately available and given by

$$\hat{\mathbb{V}}\left\langle \hat{T}_G^{(X)} \right\rangle = \frac{\lambda_G^2}{m} \frac{\sum_{j=1}^{m} \left(y_X(\omega_j) - \hat{Y}_G^{(X)} \right)^2}{m-1}. \tag{2.25}$$

Systematic Sampling In forest inventories, sampling units are usually selected in a systematic way and require the definition of two non-collinear vectors in $\Re^2$ which define a partitioning of G (actually $\Re^2$) into fundamental cells of known size λ_C and with the same shape. Under aligned systematic sampling, the selection of a single random point in one of the fundamental cells determines the position of the sampling units in all cells.

Systematic sampling usually leads to sample inclusion probabilities $\pi_i > 0$ for all elements in the population, but the second-order sample inclusion probabilities $\pi_{i\,i'} = 0$ for most pairs of population elements. The immediate consequence is that

$$\hat{T}_G^{(X)} = \lambda_C \sum_{\omega_j \in G} y_X(\omega_j) \tag{2.26}$$

is a design-unbiased point estimate of the true total $T_G^{(X)}$ of X over frame G, but – as a matter of mathematical principle – there is no design-unbiased estimator for the variance of $\hat{T}_G^{(X)}$ available under systematic sampling.

The advantage of systematic sampling over independent random point sampling is the even spread of the sample over the entire study area, as well as the balanced coverage of area subdomains of interest, with the number of sampling units proportional to the size of the subdomains.

The NFI approach to producing estimates of the unknown variance under systematic sampling is to assume that the sampling units are independently and uniformly distributed over G. It is often argued, and has been demonstrated in simulation studies for different populations, that the independent random point estimators tend to overestimate the true variance for systematic sampling. The implicit assumption of independent sampling units may, however, be reasonable because silvicultural forest treatments in Switzerland tend to be small scale and the spatial correlation ranges observed in geostatistical case studies are low.

Cluster Sampling and Stratification For cluster sampling, specific instructions are need for sampling units at the sampling frame boundary, and also for subplot centres off-set from the centre of sampling units as in the NFI young tree survey. For young tree sampling, plot centres are relocated to avoid unintentional manipulation of the population prior to observation and measurement in the field. In earlier NFIs, two separate subplots, i.e. true clusters, were used for assessing young trees. The three transect lines used in the assessment of lying deadwood can also be considered clusters. In both cases, the NFI estimation procedure is to simply pool together the two plots or three lines to form a single unit, partially ignoring the theoretically correct solutions for the treatment of clusters at sampling frame boundaries (Gregoire and Valentine 2007; Mandallaz 2008).

The density of sampling units (points) in NFI is constant throughout all of Switzerland. Stratification has been proposed occasionally to reduce the markedly higher costs of accessing field plots in mountainous regions (Lanz 2000). On the other hand, the information required for forest policy and management in the protective forests in these regions is particularly important. Hence, a uniform distribution of sampling units is largely accepted with the current requirements for information. Post-sampling stratification is, however, applied in the computation of the estimates (Sect. 2.5.2).

2.4.2 Sampling Frames of the NFI

The NFI sampling frame covers the entire surface area of the country and includes water in the form of lakes, rivers and streams, and large areas of unproductive land, such as rocks and glaciers in the mountains.

This sampling frame, which includes approximately 65% non-forest land, is primarily needed for estimating the forest area. Important gains in precision could be achieved mathematically if the sampling frame were reduced to include less non-forest land, for example only 10%. No such reduced sampling frame has been available for the NFI so far, mainly because the landscape in Switzerland is very fragmented.

For cost reasons, the NFI uses a *specific sampling frame for field-data collection*, which is limited to points located in forests or shrub forests. The manual aerial-photo interpretation still covers the entire country, and is used mainly to detect new forest plots and plots clearly located outside forest land. These plots are then eliminated from the (annual) list of sampling units to be visited by field crews for data collection. Protection against omission errors is built into the aerial-image interpretation to guarantee the visit of sampling units in cases of ambiguity. An average of approximately 5% of sampling units visited by field crews turn out to be located on non-forest land.

At the estimation stage of the inventory, all sampling units located outside forests are assigned a local density of $y_X(\omega) = 0$.

Inaccessible Sampling Units Sampling units ω located on forest land and not accessible for field-data collection remain in the sample, with a local density set to $y_X(\omega) = 0$ for all target variables except for forest area estimation.

A boundary between parts of the forest that are accessible for data collection and those that are inaccessible may cross a plot located on forest land, in which case the sampling unit is treated as: (a) an inaccessible field plot without field-data collection if the sampling unit centre ω is located in the inaccessible part of the plot; or (b) a sampling unit with field-data collection if the sampling unit centre is located in the accessible part of the plot. In case (b), the boundary between the accessible and inaccessible parts of the plot is considered a reduction line (Sect. 2.3.3).

The estimates in the standard result tables in NFI, such as growing stock or biomass, always refer to parts of the forest that are accessible for data collection Inaccessible sampling units could be treated as non-response, but have not so far in NFI for two main reasons: (a) imputation is unreliable because auxiliary (replacement) data may be lacking; and (b) effects of omitting these plots are considered marginal because many of these inaccessible plots are located on unproductive land and topographically very exposed sites, such as steep slopes and rocky terrain.

2.4.3 NFI Panels for Terrestrial Data Collection

The square grid used for field-data collection in the NFI has a density of one sampling unit per 2 km^2, resulting in a total of approximately 20,000 sampling units, of which around 7500 are located on forest land. The plot centres were permanently installed during the first inventory cycle, NFI1 (1983–1985), and the field-data collection was repeated in NFI2, NFI3 and NFI4. Statistically, the inventory may be understood as a *panel or longitudinal survey* with one single random event that defined the orientation and position of the original grid in 1983.

The chronology of field-data collection is illustrated graphically in Fig. 2.4. The inventories, NFI1 to NFI3, were carried out during three distinct cycles, each of which ran for 3 years and included field-data collection on the full set of sampling units. A new, continuous mode of field-data collection was introduced in NFI4 in which the original panel is split into nine annual panels in the form of interpenetrating grids, each covering the entire country, and field data is collected from all sampling units of an annual panel within one calendar year.

2.4.4 NFI Panels for Auxiliary Data Collection

In the NFI, manual stereo-image interpretation has always been used to exclude plots that are clearly 'non-forest' from the field-data collection. This minimises the very high cost of accessing plot centres in some of the mountainous regions in the country.

In NFI2, a densified square grid with eight sampling units per 2 km^2 was installed for the manual stereo-image interpretation and the assessment of auxiliary data. These units have been used in a two-phase estimation procedure to partly mitigate

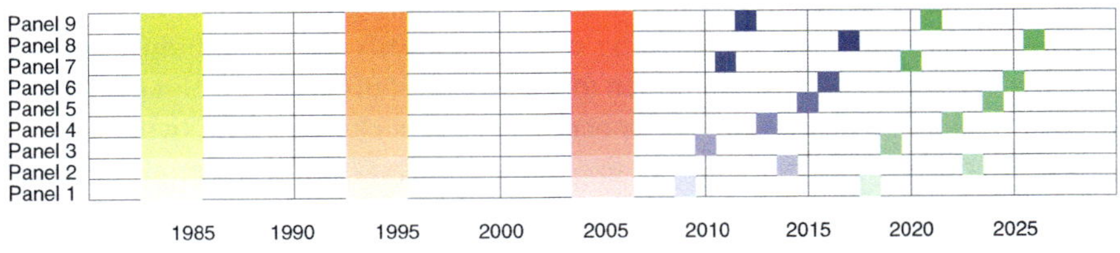

Fig. 2.4 Graphical representation of the inventory cycles for field-data collection in the NFI with the calendar years of the field-data collection on the x-axis. The colours yellow, orange, red, blue and green refer to NFI1 to NFI5, respectively, and the different shades of each colour to the nine annual panels

increased sampling error due to the 50% reduction in the original sampling grid. Manual stereo-image interpretation was repeated in NFI3 on the same densified grid.

Currently, the auxiliary data used in the two-phase estimation procedures is collected periodically and derived semi-automatically over the entire country (Chap. 7). The relevant aspects of integrating the auxiliary data into the estimation procedures are: (a) GPS measurements during field-data collection to identify the exact position of sampling unit centres; (b) identification and production of suitable response data from the vast repositories of auxiliary raw data; (c) calibration of optimised regression models for the standard output production; and (d) analyses and case studies for the custom-tailored estimation results for small areas.

These aspects of auxiliary data integration are outlined in Sect. 2.5.2.

Annual Panels The system of field data collection in NFI4 was made with the following objectives and considerations in mind:

- Change the NFI survey into a Project with a continuous, regular budget
- Promote the availability of timely information, if needed
- Remain within the same overall budget as in the periodic system
- Give more priority to estimating change
- Increase flexibility so as to be able to adapt field protocols at any time
- Maintain the existing panels of permanent plots

At the statistical sampling design level, two main aspects were investigated: (1) the selection of sampling units for annual panels from the existing square grid of terrestrial plots in order to minimise that the travel time and distance between sampling units. A rotating system allows the concentration of the sampling in a different region of the country each year and thus should reduce the cost of travel to a minimum. Using annual interpenetrating square grids requires far more travelling since each grid covers the entire country. With nine annual panels, this corresponds to a plot density of one plot per 18 km^2. The travel costs involved in maintaining the basic idea of annual interpenetrating panels, each covering the entire country, can be reduced by selecting clusters of two to nine neighbouring sampling units from the existing grid to form panels consisting of systematically distributed clusters. The within-cluster density corresponds then to the original density of one plot per 2 km^2, while the cluster density is one cluster per 32 km^2 for a cluster size of two plots and one cluster per 162 km^2 for a cluster size of nine plots. With the larger cluster size, however, sizeable regions may not be covered at all in specific years.

An interpenetrating, non-clustered square grid configuration was finally chosen for the NFI (Fig. 2.5). The relative position of the annual square sampling grids was selected such that the spatial spread of sampling units is maximised when subsequent annual sampling grids are pooled together at the estimation stage of the inventory. For example, if the positions of the annual panels 3 and 6 were to be interchanged, the pooled sampling units of annual panels 1–3 would be arranged in lines with short distances between sampling units in the North-South direction and large distances between sampling units in the East-West direction. The pooled sampling units of annual panels 4–6 would then be clearly arranged in clusters of three sampling units.

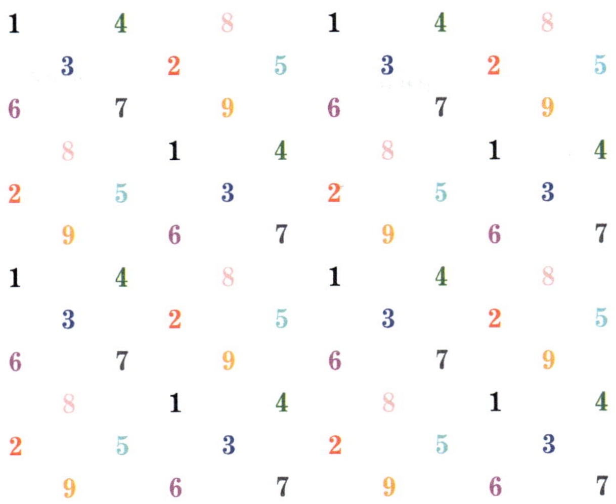

Fig. 2.5 Layout of the nine annual panels of the NFI

Under the periodic system, field data from about 7000 sampling units was collected every 10 years. Under the new annual system, assuming an approximately equal overall budget, field data from approximately 700 sampling units can be collected every year. The final decision was to maintain the entire sample of about 7000 permanent plots and to create 9 annual panels with about 7500 plots re-measured every year. The resulting interval between plot re-measurements is 9 years.

A major point of discussion has been the length of time between the re-measurements of the permanent plot. With long intervals in-between, more changes in the population remain unobserved, and the time when a specific change happens cannot be detected precisely. Many European NFIs have switched to annual inventory systems with permanent plots that are re-measured every 5 years.

Remeasuring the Swiss permanent plots every 5 years would mean ignoring about 50% of the original sample as field data, which could only be collected from about 700 plots per year under the budget constraints. Hence, the trade-off is essentially between reducing the existing sample of permanent plots by 50% or more, and detecting changes less precisely because the plots are measured less frequently. In the NFI, the interval between plot measurements has always been ten (or more) years. Moreover, various (external) data users expressed a strong interest in maintaining as many permanent plots as possible. Therefore, it was decided to create nine annual panels.

Pooled Samples With annual panels that each cover the entire study area, one option is to pool consecutive years together into a larger sample for the estimations. The general effects of pooling panels and some important considerations are:

- The precision of estimates increases with the number of pooled annual panels, but at the price of a less precise assignment of the estimates to a specific point in time (calendar year)

- The trade-off between the two effects depends heavily on effective and/or supposed changes in the population, which may differ markedly between target variables and area domains of interest
- Pooled samples tend to smooth peaks of cyclic change in the population and detect trends later than annual panels
- Consecutive (unpooled) annual panels may suggest changes in the population, whereas in fact only random sampling effects are observed

Since the introduction of the annual system of field-data collection, the NFI produced intermediate results (NFI4a) using the pooled sample of the three annual panels from 2009 to 2011, intermediate results (NFI4b) using the pooled sample of the five annual panels from 2009 to 2013, and finally, results (NFI4) from the pooled sample of the nine annual panels from 2009 to 2017.

2.5 Estimation Procedures

2.5.1 *Basic Estimators*

Under the systematic sampling design of the NFI,

$$\hat{T}_G^{(X)} = \lambda_C \sum_{j \in G} y_X(\omega_j) \tag{2.27}$$

is an exact design-unbiased estimator of the total $T_G^{(X)}$ of the local density function $y_X(\omega)$ over G (Mandallaz 2008), and therefore the total of target variable X in P_F, where λ_C denotes the size (surface area) of the basic cell of the systematic grid.

The number of sampling units m_G in the sampling frame G is a random number under systematic sampling, with two consequences. First, the estimator

$$\hat{Y}_G^{(X)} = \frac{1}{m_G} \sum_{j \in G} y_X(\omega_j) \tag{2.28}$$

for the true mean spatial density $Y_G^{(X)} = T_G^{(X)}/\lambda_G$ of $y(\omega)$ over G is not strictly design-unbiased because $\hat{Y}_G^{(X)}$ is a ratio of two random variables. Second, conditioning on the realised sample size m_G, $\hat{T}_G^{(X)}$ is biased whenever $\frac{\lambda_G}{\lambda_C} \neq m_G$.

For this reason, $\hat{T}_G^{(X)} = \lambda_G \hat{Y}_G^{(X)}$ has been used to estimate totals since NFI2.

Because the NFI sampling frame extends over the entire country, the mean spatial density $Y_G^{(X)}$ of a local density $y_X(\omega)$ over the entire frame G is almost never of interest. Many population parameters of interest can be expressed, however, as the ratio of the totals of two local densities $y_X(\omega)$ and $y_Z(\omega)$, both defined over the entire frame G. In this case

$$\hat{R}_G^{(X/Z)} = \frac{\hat{T}_G^{(X)}}{\hat{T}_G^{(Z)}} \tag{2.29}$$

is used to make an asymptotic design-unbiased estimate of the unknown population parameter $R_G^{(X/Z)}$.

Post-sampling Strata The NFI uses stratified estimators for country estimates because the standard output tables are formatted so that estimates are always presented for the entire country and for a set of the main regions of interest, which are usually the five production regions of Switzerland. Other sets of primary interest used in the output tables could be cantons, biogeographic regions and forest districts.

The estimates within these post-sampling strata of known surface area are produced with the estimators given above. To maintain *additivity of regional estimates with the estimate for the country* within output tables, the post-stratified estimator

$$\hat{T}_G^{(X)} = \sum_{h=1}^{H} \hat{T}_h^{(X)} = \sum_{h=1}^{H} \lambda_h \hat{Y}_h^{(X)} = \sum_{h=1}^{H} \frac{\lambda_h}{m_h} \sum_{j \in h} y_X(\omega_j) \tag{2.30}$$

is used for estimating totals, and the combined ratio estimator

$$\hat{R}_G^{(X/Z)} = \frac{\sum_{h=1}^{H} \hat{T}_h^{(X)}}{\sum_{h=1}^{H} \hat{T}_h^{(Z)}} \tag{2.31}$$

for estimating ratios $R_G^{(X/Z)}$.

The disadvantage of this approach is that the estimates for the entire country vary (slightly) numerically in the output tables according to the set of regions used for the repartitioning of the results. The numerical differences between the estimates are, however, very small because the strata and sample sizes are reasonably large, and all estimates remain design-unbiased. Technically, the minimum differences in the estimates reflect the different sources of auxiliary data used for estimation. For total growing stock, for instance, for which the relative standard error of the estimates with different sets of post-sampling strata is around 1%, the difference between estimates is less than 0.1%.

Variance Estimators The NFI variance estimators are those derived under the assumption of an independent distribution of the sampling units (Sect. 2.4.1). No alternatives have yet been implemented. A few, mostly undocumented, attempts have confirmed that the spatial auto-correlation between sampling units is very low for most target variables. The NFI standard error estimates can be assumed to have a tendency to overestimate the true error under the systematic sampling. This makes it less likely that stakeholders will interpret the results over-optimistically. The estimators are documented in Sect. 20.8.2 and follow the recommendations in the literature.

2.5.2 Use of Auxiliary Data

The NFI uses auxiliary data made available on a dense grid in an estimation procedure known as *two-phase (double) sampling for stratification*. The current sampling grid from which auxiliary data is derived has a density of one sampling unit per hectare and includes sampling units located both within and outside the forest.

The primary auxiliary data is a recently produced forest cover map (Chap. 7). Additional data includes: (a) vegetation height characteristics retrieved from digital stereo images (Chap. 7), where the main variables are means, medians and upper quantiles derived on circular plots 500 m^2 in area; (b) the proportion of coniferous trees in the total canopy cover, recently derived from four-band aerial images (Chap. 7) and (c) other spatial data, such as elevation, aspect and slope, obtained from digital terrain models. All raw data is provided by swisstopo, the Swiss Federal Office of Topography, and is prepared for use in the standard estimation procedures of the NFI.

Regression-tree modelling was used to calibrate the post-strata categories. The growing stock estimation was optimized (Pulkkinen et al. 2018). The estimation procedure can be understood as model-assisted estimation with an ANOVA-type regression model in a two-phase sampling framework, where the first-phase sample of sampling units is selected uniformly over the entire territory of Switzerland, and the second-phase subsample of sampling units with field data is a systematic subsample of the first-phase sample. In the derivation of the estimators, the assumption is a simple random sampling (without replacement) of second-phase plots from the first-phase sample of plots with auxiliary data. The inferential framework and estimators were introduced in NFI2 (Köhl 1994, 2001). Slight adaptations to the methods used for model building and for variance estimation have been implemented recently for NFI4. The estimators are given in Sect. 20.8.2.

A few aspects are worth noting:

- The two-phase estimation procedure is a so-called *model-assisted estimation technique* (Särndal et al. 2003; Mandallaz 2008), which leads to approximate design-unbiased estimates regardless of how well the model fits the data.
- Positional matching of sampling units for field-data collection and for auxiliary data collection is assumed and a homogeneous source of auxiliary data must be used. Modern automatic classification systems usually meet this assumption better than earlier manual image interpretation, and the assumption is becoming easier to fulfil as technology progresses.

- The primary gain in the context of the NFI is explicit and implicit total forest area estimation, where implicit forest area estimation is involved whenever the total of a target variable over the entire forest is estimated. This procedure divides the territory into a post-stratum of *presumed forest* and a post-stratum of *presumed non-forest*. This markedly increases the precision of the total estimates, even for target variables, such as lying deadwood, that are otherwise not strongly correlated with the auxiliary data.
- Auxiliary data related to vegetation height further increases the precision of the total and mean spatial density estimates for target variables, such as growing stock and biomass, which are clearly related to vegetation height. While gains in precision are achieved when estimating growing stock over the entire forest area, the gain may be marginal if the growing stock is estimated over certain area subdomains, such as private forest, or for certain subpopulations, such as beech trees, if auxiliary data indicating these subdomains and subpopulations is not available. For example, the relative standard errors of the single-phase estimates of total growing stock for the entire country in forest and private forest is 1.5% and 2.7%, respectively. Under two-phase sampling, the sampling error is reduced by 40% for all forest land, but by only 15% for the private forest land.
- The double sampling post-sampling strata have been optimised for growing stock estimation, but they are used for virtually all target variables and standard NFI outputs for the following reasons: (a) estimates for area subdomains, such as private and public forests, or for subpopulations, such as broadleaved and coniferous trees, would otherwise not remain numerically additive; (b) the effort to calibrate and adopt a model for each target variable is considerable; and (c) auxiliary data for area subdomains and subpopulations are not readily available.

2.5.3 Domain Estimation

Estimates for parts of the population distinguished according to certain criteria are regularly provided in the standard result tables of the NFI. From a statistical point of view, the related terminology and methods are not always unambiguous. In this section, we therefore provide a short overview of specific aspects of domain estimation in forest inventories.

Subpopulations We use the term *subpopulation* at the population element level, i.e. for population elements (trees) with a certain characteristic. Hence, when estimating growing stock, separate estimates are often required for the subpopulations of living broadleaved and living coniferous trees. In the infinite population approach to forest inventories, the actual population and population elements of interest become obscured at the estimation stage of the survey.

Technically, the handling of subpopulations is straightforward with subpopulation indicator variables applied at the element level in the derivation of plot-level local densities. There is no immediate difficulty in the estimation of totals and spatial means of rare subpopulations unless the local density is zero for most sampling units. The sample size itself remains constant.

A special case is the estimation of the mean of target variables at the population element level, for example the mean stem volume (of a given species). Under the infinite population approach, the estimator is a ratio, with an estimate for the total volume (of a given species) in the numerator and an estimate for the total number of stems (of a given species) in the denominator.

Area Subdomains In survey planning in general, planned and unplanned domains are commonly distinguished. Planned domains are domains of high interest and relevance for optimised allocation of sampling efforts, whereas unplanned domains correspond to domains of lower interest and relevance at the estimation stage of the survey.

The main area domains of interest in the NFI are the five production regions in Switzerland. The surface area of each of these regions is known, and independent estimates are produced for each of them. Stratified estimators are then used to combine these estimates to get an overall estimate for the entire country.

Most area domains of interest are of unknown size, but the membership of plot centres with area domains is known. The local density of target variables with respect to area domain is simply set to $y_D^{(X)}(\omega) = \iota_D(\omega)\, y_F^{(X)}(\omega)$, where $y_F^{(X)}(\omega)$ denotes the local density of target variable X with respect to the forest land F and $\iota_D(w)$ indicates plots with centre in area domain D.

The approach applied in the NFI involves associating the entire plot, with all its population elements (trees), with area domains through a point decision at the plot centre. This approach is robust, cost efficient in the field, and facilitates computations and algorithms at the estimation stage of the survey. In many national forest inventories, a more complex field protocol is used: sample plots are partitioned into subplots, and associations between area domains and population elements (trees) are observed and recorded for each subplot. The advantage of this approach is higher accuracy in domain estimation, at least in theory. In practice it comes at the price of increased complexity at the sampling and estimation stages of the inventory. Therefore, the system applied in the NFI has so far been left unchanged.

2.5.4 Sampling Error and Confidence Intervals

All NFI estimates in the standard output tables are accompanied by the corresponding *standard error* estimate, which is defined as the square root of the estimated variance of the estimate.

Non-sampling Errors Non-sampling errors, such as measurement and registration errors or frame imperfection effects, are not included in the reported standard error of the estimates. However, the magnitude of some of these errors is known from the approximately 5% of field plots independently re-measured by a second field team (Chap. 21). Moreover, the accuracy of some of the models used in NFI, such as single-stem volume functions, has been investigated (Chap. 12).

In this context, the two-stage sampling and estimation procedure of the NFI for stem volume (growing stock and biomass) is worth noting. In contrast to many other (national) forest inventories, the NFI subsample of tariff trees is used not only to fit tariff functions, but also to correct bias in tariff functions. The two-stage estimates are only at first glance less precise than the single-stage estimates, in which inference is based on the large sample of tally trees and associated stem volume predictions. Although the two-stage sampling inference is based on a much smaller subsample of tariff trees, the stem volumes are accurate. In this sense, the two-stage estimation for growing stock can be understood as a method that integrates the remaining uncertainties associated with single-stem volume models into the reported sampling error.

Confidence Intervals The standard output tables of the NFI contain the estimate $\hat{\theta}$ of the unknown population parameter θ together with a second value, the *absolute standard error* $\sqrt{\hat{\mathbb{V}}\langle\hat{\theta}\rangle}$ or the *relative standard error* $\frac{\sqrt{\hat{\mathbb{V}}\langle\hat{\theta}\rangle}}{\hat{\theta}} \times 100$ of the estimate.

The absolute standard error $\sqrt{\hat{\mathbb{V}}\langle\hat{\theta}\rangle}$ is an estimate of the unknown standard deviation, according to which the estimates $\hat{\theta}$ would be distributed under reiterations of the survey in the same (unchanged) population and with the same sampling design. It is not possible from a single survey to determine whether the estimate $\hat{\theta}$ is greater or less than the true value θ for the population parameter of interest, and whether $\hat{\theta}$ is close to θ. However, we can be confident at a $1-\alpha$ level – again under repeated surveys – that the proportion of surveys for which the true value $\hat{\theta}$ is within the lower and upper bounds of the respective *1 – α confidence interval* $[\hat{\theta} - z_{1-\alpha/2}\sqrt{\widehat{V\langle\theta\rangle}}, \hat{\theta} + z_{1-\alpha/2}\sqrt{\widehat{V\langle\theta\rangle}}]$ is $1-\alpha$, where $z_{1-\alpha/2}$ is the $1-\alpha/2$ quantile of the standard normal distribution, so that $z_{1-\alpha/2} = 2$ is frequently used when calculating the 95% confidence interval.

2.6 Estimation of Change

The permanent plot (annual panel) sampling design of the NFI facilitates the estimation of *net change* and *components of change* between inventories. However, there are some specific issues in the derivation of variables and in the interpretation of results that need special attention and are discussed in this section.

Net change is the general technical term for the difference between two population states. More precisely, it is the difference between the value of a population parameter of interest, for example the total growing stock in a forest, in two consecutive inventories.

Under the infinite population approach, net change in the total of a target variable over a sampling frame G is understood as

$$\Delta_T = T_2 - T_1 = \int_G y_2(\omega)d\omega - \int_G y_1(\omega)d\omega \tag{2.32}$$

$$= \int_G (y_2(\omega) - y_1(\omega))d\omega = \int_G \Delta_y(\omega)d\omega \tag{2.33}$$

Where $\Delta_y(\omega) = y_2(\omega) - y_1(\omega)$ denotes the difference at point ω between the local density of the target variable in the second inventory at t_2 and the local density of the target variable in the first inventory at t_1.

$\Delta_y(\omega)$ is only available if data collection is repeated at exactly the same sample points (*inventory with permanent plots*). However, Δ_T may alternatively be estimated with two independent samples, with data collected at t_1 and t_2, resulting in the estimates T_1 and T_2 (*inventories with temporary plots*).

2.6.1 Time Schedule Effects

In the NFI, as in many other national and large-area forest inventories, the temporal sequence of plot measurements within and between inventory cycles is complex. NFI field-data collection starts in March and ends in October to take into account the large proportion of mountain forest in Switzerland. In NFI1, NFI2 and NFI3 field data was collected over three successive years in a periodic sampling system. Data collection in NFI4 was done in a continuous (annual) sampling system spread over nine consecutive years (Fig. 2.4). The annual panel field data is again collected between March and October, and it is planned to continue using this approach in NFI5 and beyond.

Each annual panel with data collected in two successive inventory cycles can be used to estimate change between plot re-measurements (Fig. 2.6). The transition from periodic to annual sampling led to the number of years between inventories NFI3 and NFI4 being different for different annual panels. For instance, annual panel 5 could be used to estimate the average NFI1 state for the combined years 1983–1985, the NFI5 state for 2024, or intermediate NFI2, NFI3 and NFI4 states (Fig. 2.6). It could also be used to estimate the average change between NFI1 and NFI2 for the combined years 1983–1985 to 1993–1995, the average change between NFI3 and NFI4 for the combined years 2004–2006 to 2015, or the intermediate changes between NFI2 and NFI3, as well as between NFI4 and NFI5.

For the NFI4 output tables, between three and five annual panels were pooled for the intermediate reporting and nine for the final reporting. This is an intuitive

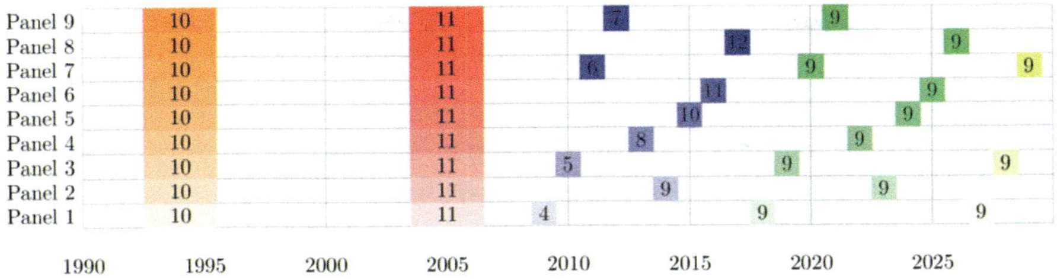

Fig. 2.6 Graphic representation of the inventory cycles for field-data collection with measurement of change in the NFI with the calendar years on the x-axis. The colours yellow, orange, red, blue and green refer to NFI1 to NFI5, respectively, and the different shades of each colour to the nine annual panels. The approximate number of years since the last inventory cycle is given for each cell

approach to produce estimates that include the full extent of the field data sets collected during the annual inventory panel.

The result tables provide estimates of *states in individual inventory cycles* and of *change between inventory cycles* without any clear-cut associations with points in time. This can be problematic for users, e.g. some international agencies, requiring estimates for particular calendar years. Estimates of change are, furthermore, typically expressed as average *annual changes* when analysing and comparing forest growth between regions and over time.

2.6.2 Dates and Calendar Years

In the NFI, the exact dates of field-data collection are known for each sampling unit. This information can be used to calculate the mean date of a given annual panel or of pooled annual panels. This mean date is, however, of little practical use and is therefore not computed or published as part of the NFI standard output routines.

Changes in target variables for a population rarely occur continuously and linearly over time, which is why results for a specific date are always biased, irrespective of the averaging method used. Moreover, the complete data set from a representative sample is needed to produce estimates.

Even estimates produced with data from a single annual panel refer to an average *state in that calendar year* and never to a specific date, and have, so far, not been published in the NFI. This situation extends to estimates produced with data collected over multiple years. The NFI solution is to label result tables with: (a) the inventory cycle for which the plot measurements were used, and (b) the calendar years during which the field data was collected.

2.6.3 Vegetation Periods

For any forest inventory, the *average annual growth, or increment, of trees* is important information. In Switzerland, the length and dates of the growing season for trees vary considerably from one calendar year to the next, but also depend on the vegetation zones and various other site and stand characteristics.

For a given sampling unit, field-data collection may take place before, during or after the in-situ growing season of tally trees in a given calendar year. The exact date of the plot measurements is known, but the time interval between re-measurements, expressed in calendar years (or months or days), does not necessarily represent the number of tree-growing seasons between field visits. This effect is amplified in the NFI, where field-data collection between inventory cycles is not synchronised to a constant time interval between plot re-measurements for all sampling units. A sampling unit measured in May in one inventory may be re-measured on a different date of the year in the next inventory.

To estimate the average annual growth of trees, the *number of vegetation periods* $\Delta_v(\omega)$ *between plot re-measurements* is needed. In the NFI, this value is approximated by assuming that: field data collected before June corresponds to the period in a given calendar year before the start of the tree-growing season; a quarter of the growing season will have happened already if field data is collected in June, half in July and three-quarters in August. The growing season of trees will have almost finished if field data is collected later than August.

Using an approximation for a target variable like growing stock seems reasonable when deriving the number of vegetation periods and transforming the observed total change between two inventories into an average annual change, or more precisely, into an average change per growing season. The situation is much less clear for other target variables, such as estimates of area domain sizes, for example the proportion of private forest, and quantities of deadwood. Values for these variables can change at any time during the calendar year. The average annual change in these variables may therefore be better defined as the number of calendar years between the field visits than as the number of vegetation periods. To maintain consistency among target variables, however, NFI annual changes are always annualised with the number of vegetation periods between plot re-measurements.

2.7 Net Change with Independent Samples

Here we briefly describe one method for estimating the net change between arbitrary calendar years, which is not currently in routine use in the NFI. The basis for the estimation is the continuous mode of field-data collection in which a representative sample is obtained each calendar year.

Estimates of states for any calendar year since 2009 and estimates of net changes between any two calendar years are, in principle, readily available. Using this method for estimating change with independent samples has two basic effects:

- The resulting state estimates are not very precise because the annual samples are small.
- The two samples used for net change estimation are independent of each other, with field data collected on different annual panels. As a result, (a) estimates of change components, such as gains and losses, are not available, and (b) the precision of the net change estimates is relatively low.

The second point follows from the general statistical principle that the variance of a difference between two independent random variables, here the state estimates $\hat{T}_1$ and $\hat{T}_2$ in the first and second inventory, is given by

$$\mathbb{V}\langle\hat{\Delta}_T\rangle = \mathbb{V}\langle\hat{T}_2 - \hat{T}_1\rangle = \mathbb{V}\langle\hat{T}_2\rangle + \mathbb{V}\langle\hat{T}_1\rangle - 2\mathbb{C}\langle\hat{T}_1, \hat{T}_2\rangle \tag{2.34}$$

where $\mathbb{C}\langle\hat{T}_1, \hat{T}_2\rangle$ is the co-*variance between the two state estimates.* $\mathbb{C}\langle\hat{T}_1, \hat{T}_2\rangle$ is zero in the case of independent samples (with temporary plots) and positive in the case of an annual panel survey (with permanent plots).

2.7.1 Pooled Annual Panels

Field data collected over two or more annual panels is frequently combined into a single pooled sample for estimation in national forest inventories. This has the advantage of increasing the precision of the estimates because of the size of the pooled sample is larger. The disadvantage is the increased fuzziness in the assignment of the results to a specific point in time, for example a calendar year. From a mathematical point of view, the optimal trade-off depends on the target variable, as well as other considerations such as:

- Estimates based on pooled annual panels cannot be expected to be reliable for specific calendar years if the population abruptly changes, for example following a massive windthrow of trees or extensive forest fires
- A pooled sample may provide reasonable estimates for specific calendar years if there is a linear or no change in the population
- In a long-term monitoring program such as a national forest inventory, the average (smoothed) states are likely to be considered sufficient for requirements such as deciding on national policy
- With pooled annual panels, emerging trends cannot be detected early.

We conclude that the NFI sampling system with nine annual panels offers various ways for detecting, monitoring and understanding states and trends in the population, although this may be occasionally at the price of extensive and complex data analysis. As already mentioned, the reports from the NFI are produced with the entire sample of the inventory cycle. Thus, nine annual panels are pooled together for the NFI4 results.

2.7.2 Average Annual Change

The number of years between states, and therefore the transformation of net change between two inventories into an average annual change, is obvious with independent samples. Instead of using calendar years, it may be reasonable to approximate the number of calendar years between the inventories by the *number of vegetation periods*, as explained in Sect. 2.6.3. With the calculated mean dates $\bar{a}_1$ and $\bar{a}_2$ of the respective state estimates and the mean number of vegetation periods $\Delta_v = \bar{a}_2 - \bar{a}_1$ between the two inventories, the *average annual change* is $\hat{\Delta}_T/\Delta_v$ with variance $\mathbb{V}\langle\hat{T}_2 - \hat{T}_1\rangle/\Delta_v^2$.

2.8 Estimation of Change Using Annual Panels

Estimating net change from a sample of permanent, re-measured sampling units is very efficient because the co-variance between the two state estimates in Eq. 2.34 is maximised. In practice, the co-variance and variance terms are not estimated separately because a representative sample of the difference $\Delta_y(\omega) = y_2(\omega) - y_1(\omega)$ between the local density of the target variable in the second inventory at t_2 and the local density of the target variable in the first inventory at t_1 is directly measured for sampling units ω in the annual panel.

Net change can also be estimated over three or more inventories. To estimate net change over three inventories, for instance, it is treated simply as the plot-level difference, which is the sum of two already derived changes $y_2(\omega) - y_1(\omega)$ between the inventories at t_2 and t_1 and of $y_3(\omega) - y_2(\omega)$ for the inventories at t_3 and t_2.

2.8.1 The Role of Area Domains

NFI estimates are always produced according to *reference domains* and *area domains*. A typical reference domain is the *accessible forest without shrub forest*. A set of area domains partitions the reference domain into subcategories of interest, such as forest ownership categories.

Reference and area domains play an intuitive and prominent role in the description of forest states and net changes. In the estimation of change, there are different options for the treatment of reference and area subdomains, depending on the user's information needs.

We introduce

$$\iota_t^{(e)}(\omega) = \begin{cases} 1 & \text{if } \omega \in e \\ 0 & \text{else} \end{cases} \tag{2.35}$$

to indicate whether, in inventory t, the sampling unit at point ω is inside or outside some area domain e.

In the NFI, the method usually used to derive the local density of the net change in a target variable between two inventories t_2 and t_1 is

$$\Delta_y(\omega) = \iota_1^{(r)}(\omega)\,\iota_2^{(r)}(\omega)\,\iota_2^{(e)}(\omega)\,(y_2(\omega) - y_1(\omega)) \tag{2.36}$$

where r denotes the reference domain, such as the accessible forest without shrub forest, and e denotes some area domain category, such as private forest. The resulting estimates refer to the net change in the target variable in areas that are considered accessible forest without shrub forest in both inventories, according to the ownership categories observed in the second inventory.

Another method for deriving net change in sampling units is

$$\Delta_y(w) = \iota_2^{(r)}(\omega)\iota_2^{(e)}(\omega)y_2(\omega) - \iota_1^{(r)}(\omega)\iota_1^{(e)}(\omega)y_1(\omega) \tag{2.37}$$

This method is used to estimate the *overall net change according to reference and area domains*. $\Delta_y(w)$ is zero for sampling units that lie outside the reference domain in both inventories. The change estimates include all target variable gains and losses due to increases and decreases in the geographic extent of the reference and area domains. In the NFI, this method is optionally applied in net change estimation.

Both of the approaches described above, and their resulting estimates, make intuitive sense and can be correctly understood by the user if clearly specified in output tables.

Components of Change in Area Domain There are no technical difficulties in estimating change in target variables due to gains and losses in the reference or area domain between inventories. For instance

$$\Delta_y(w) = \iota_2^{(r)}(\omega)\iota_1^{(r)}(\omega)\iota_2^{(e)}(\omega)\left(1 - \iota_1^{(e)}(\omega)\right)y_2(\omega) \tag{2.38}$$

generates local densities used for estimating the overall net change in the target variable with respect to area domain e that is due to an increase in the area extent of e between inventories. Because result tables tend to become quite complex and large, these components of change at the area domain level are normally not calculated and reported in the NFI standard output routines.

Notes on Estimating Area Domain Changes First, it is conceptually important to distinguish between target variable change caused by *area domain (plot-level) change* between inventories, described in this section, and target variable change caused by *population element (tree-level) change* between inventories, covered in Sect. 2.9.

Reference and area domains of the NFI are always defined for the point ω, the plot centre, which is why target variable gains and losses according to reference and area domains always include all population elements associated with the sampling unit at ω. In contrast, the field protocol of a forest inventory may allow the partitioning of sampling units – usually sectors or segments of circular plots – into different area domains, such as land-use classes or forest ownership categories.

Both approaches are common. The advantages of the NFI approach are that the field protocol is relatively simple and field-data collection is restricted to sampling units with their centres in forests. The subplot or plot segments approach, on the other hand, requires field measurements on sampling units at the forest boundary with their centre outside forests. An elaborate protocol is therefore required to register area-domain proportions and the association of population elements (trees) with area domains. The downside of the NFI method is that the association of population elements (trees) with area domains is relatively coarse. However, this is only the case for sampling units partitioned into different area domains. In a national forest inventory, differences in the effects of using the two approaches for obtaining overall estimates can be assumed to be negligible.

2.8.2 Average Change Per Unit Area

For comparison reasons, the net change between inventories is often expressed as an average change per unit area, i.e. an average density of the target variable per hectare of forest.

While the mean spatial density parameter is easily computed and clearly defined when estimating states, this is not necessarily the case when estimating change. Because the reference domain and the area domains of interest may change between inventories, *there is no unambiguous definition and understanding of the average change per unit area between inventories.*

Such estimates are produced in the NFI, but only with local densities defined according to Eq. 2.36. In this case, net change is estimated for the land that is part of the reference domain in both inventories according to the area-domain categories used in the second inventory. The disadvantage of this approach is that changes in the target variable due to area gains and losses of the reference domain is not included in the change estimates.

2.8.3 Average Annual Change

In this section, we describe two approaches for calculating the *average annual change*. The first method was applied in the NFI under the periodic inventory system (NFI1 to NFI3) and is known from the forest inventory literature. The second method possibly copes better with the range of time intervals between plot re-measurements in NFI4 became wider than in NFI3 with the transition from the periodic to the continuous (annual) inventory system.

Traditional Estimator The traditional estimator for the average annual change requires a simple modification of the local density function

$$\Delta_y^{(MOR)}(\omega) = \frac{\Delta_y(\omega)}{\Delta_v(\omega)} \tag{2.39}$$

where $\Delta_y(\omega)$ is the local density of the change in the target variable between the two inventories at point ω, and $\Delta_v(\omega)$ is the number of vegetation periods between the two inventories at point ω.

A conspicuous property of this local density function $\Delta_y^{(MOR)}(\omega)$ is that the implicit mean number of vegetation periods between the two inventories $\Delta_v^* = \hat{\Delta}_T / \hat{\Delta}_{T/a}*$ varies depending on the target variable for which the change components are estimated. $\hat{\Delta}_T$ denotes the estimate for the *total change between inventories*, for example the mean of plot values under simple random point sampling, and $\hat{\Delta}^*_{T/a}$ denotes the estimate of the *average annual change between inventories* with the local density function $\Delta_y^{(MOR)}(\omega)$ and under the same sampling design.

Transition Estimator The aims of the estimator of average annual change are: (a) to use the same mean number of vegetation periods between inventories for all target variables, and (b) to use an average number of vegetation periods between inventories, which has an intuitive interpretation.

With a set of H post-sampling strata, the *mean number of vegetation periods* for plots located in an area domain e is calculated as

$$\Delta_v = \frac{\sum_{h=1}^{H} \lambda_h n_h^{-1} \sum_{j=1}^{n_h} \iota_e(\omega) \Delta_v(\omega)}{\sum_{h=1}^{H} \lambda_h n_h^{-1} \sum_{j=1}^{n_h} \iota_e(\omega)} \tag{2.40}$$

where λ_h is the surface area and n_h the number of sampling units in post-sampling stratum h, and $\iota_e(\omega)$ is the area domain indicator variable. The resulting estimator of the average annual change is $\hat{\Delta}_{T/a} = \hat{\Delta}_T / \Delta_v$. Because Δ_v is a sampling and measurement design constant, the variance of $\hat{\Delta}_{T/a}$ is $V\langle\hat{\Delta}_{T/a}\rangle = V\langle\hat{\Delta}_T\rangle / \Delta_v^2$.

Discussion The traditional and the new estimator are identical if the number of vegetation periods is the same for all plots ($\Delta_\nu(\omega) = \Delta_\nu$), as it is (approximately) between NFI1 and NFI2, NFI2 and NFI3, and NFI4 and NFI5. Between NFI3 and NFI4, the new estimator has the advantage that the number of vegetation periods is always the same for a given area domain of interest and intuitively interpreted as *the total change in the target variable measured over all plots in the respective area domain of interest, divided by the total number of vegetation periods over which this change has been measured on these plots.*

2.9 Change in the Population of Trees

In Sect. 2.8, we discussed the estimation of change and change components at the plot level. However, forest inventories, especially those with permanent plots, also require estimations of changes in the population of trees. In this section we describe methods for estimating these changes.

2.9.1 *Definitions and Notation*

The definition, analysis and estimation of specific change components between the state of a population of trees at t_1 and t_2, with $t_1 < t_2$, requires a tree-level analysis of change effects.

Population The population is defined as consisting of living and dead trees, and the notation is a bit tricky. A population survivor tree is a tree that is eligible in both inventories. It is not necessarily a living tree; it may be dead at t_2, or at t_1 and t_2.

We use $\mathcal{P} = \mathcal{P}_1 \cup \mathcal{P}_2$ to denote the *population of eligible trees* at t_1 and t_2, and define:

- The subpopulations of living trees at time points t_1 and t_2 *as* $\mathcal{P}_{l.1}, \mathcal{P}_{l.2}$, and of dead trees as $\mathcal{P}_{d.1}, \mathcal{P}_{d.2}$, with $\mathcal{P}_1 = \mathcal{P}_{l.1} \cup \mathcal{P}_{d.1}$ and $\mathcal{P}_2 = \mathcal{P}_{l.2} \cup \mathcal{P}_{d.2}$
- $\mathcal{S}_{12} = \mathcal{P}_1 \cap \mathcal{P}_2$, the *subpopulation of survivor trees* between t_1 and t_2 with
  - $\mathcal{S}_{l.12} = \mathcal{P}_{l.1} \cap \mathcal{P}_{l.2}$, the subpopulation of trees living at both time points,
  - $\mathcal{S}_{d.12} = \mathcal{P}_{d.1} \cap \mathcal{P}_{d.2}$, the subpopulation of dead trees at both time points,
  - $\mathcal{M}_{12} = \mathcal{P}_{l.1} \cap \mathcal{P}_{d.2}$, the subpopulation of *mortality trees*, i.e. living at t_1 and dead at t_2, with $\mathcal{S}_{12} = \mathcal{S}_{l.12} \cup \mathcal{S}_{d.12} \cup \mathcal{M}_{12}$
- $\mathcal{D}_{12} = \mathcal{P}_1 \backslash \mathcal{P}_2$, the subpopulation of cut and mortality trees between t_1 and t_2, with
  - $\mathcal{D}_{12}^{(A)}$, the subpopulation of trees lost between t_1 and t_2 because they are located inside the domain of interest at t_1 and outside at t_2, e.g. because of a shift in the domain boundary,

- $\mathcal{D}_{12}^{(L)}$, the subpopulation of trees lost due to (natural) in-situ destruction (decomposition) between t_1 and t_2,
- $\mathcal{D}_{12}^{(U)}$, the subpopulation of trees lost because of removal (and usage) from the population between t_1 and t_2,
- $\mathcal{D}_{l.12} = \mathcal{P}_{l.1} \backslash \mathcal{P}_{l.2} = \mathcal{D}_{l.12}^{(A)} \cup \mathcal{D}_{l.12}^{(L)} \cup \mathcal{D}_{l.12}^{(U)} \cup \mathcal{M}_{12}$, the subpopulation of trees lost from the population $\mathcal{P}_{l.1}$ of living trees,
- $\mathcal{D}_{d.12} = \mathcal{P}_{d.1} \backslash \mathcal{P}_{d.2} = \mathcal{D}_{d.12}^{(A)} \cup \mathcal{D}_{d.12}^{(L)} \cup \mathcal{D}_{d.12}^{(U)}$, the subpopulation of trees lost from the population $\mathcal{P}_{d.1}$ of dead trees, with $\mathcal{D}_{12} = (\mathcal{D}_{1.12} \cup \mathcal{D}_{d.12}) \backslash \mathcal{M}_{12} = \mathcal{D}_{12}^{(A)} \cup \mathcal{D}_{12}^{(L)} \cup \mathcal{D}_{12}^{(U)}$

- $\mathcal{E}_{12} = \mathcal{P}_2 \backslash \mathcal{P}_1$, the *subpopulation of ingrowth trees* between t_1 and t_2, with
  - $\mathcal{E}_{12}^{(A)}$, the subpopulation of trees located outside the domain of interest at t_1 and inside at t_2,
  - $\mathcal{E}_{12}^{(I)}$, the subpopulation of trees reaching population eligibility, such as the minimum dbh threshold, between t_1 and t_2,
  - $\mathcal{E}_{l.12} = \mathcal{P}_{l.2} \backslash \mathcal{P}_{l.1} = \mathcal{E}_{l.12}^{(A)} \cup \mathcal{E}_{l.12}^{(I)}$, the subpopulation of newly eligible trees in population $\mathcal{P}_{l.2}$ of living trees
  - $\mathcal{E}_{d.12} = \mathcal{P}_{d.2} \backslash \mathcal{P}_{d.1} = \mathcal{E}_{d.12}^{(A)} \cup \mathcal{E}_{d.12}^{(I)} \cup \mathcal{M}_{ld.12}$, the subpopulation of newly eligible trees in population $\mathcal{P}_{d.2}$ of dead trees, with $\mathcal{E}_{12} = \mathcal{E}_{l.12} \cup \mathcal{E}_{d.12} = \mathcal{E}_{12}^{(A)} \cup \mathcal{E}_{12}^{(i)}$.

We did not specify the length of time between t_1 and t_2 and related effects of trees becoming intermediate members of the population for some periods, such as from t_e to t_d, with $t_1 < t_e < t_d < t_2$ between t_1 and t_2. The population is periodically observed with permanently installed sample plots, which means such intermediate changes in population are, by definition, not detected and not included in the resulting change estimates.

Sample of Permanent Plots In a forest inventory with permanent plots, the tally trees have to be classified during field-data collection according to the definitions of population survivor, population cut and mortality, and population ingrowth trees. The first-inventory tally trees must be re-identified in the second inventory.

The classification of the tally trees into subsamples is similar (in terminology) to the classification of population trees into subpopulations, with some artefact trees (Eriksson 1995a). We use a different notation for trees in the sample of permanent plots. For a given sampling unit at ω or the entire sample of permanent plots, the samples $P_1(\omega)$ and $P_2(\omega)$ of tally trees at t_1 and t_2 can be partitioned into the following subsamples:

- $S_{12}(\omega)$ sample survivor trees registered at t_1 and t_2,
- $C_{12}(\omega)$ sample cut trees removed from the population at t_d, with $t_1 < t_d < t_2$,

- $M_{12}(\omega)$ sample mortality trees removed from the population of living trees, and transferred to and remaining in the population of dead trees at t_d, with $t_1 < t_d < t_2$.
- $I_{12}(\omega)$ sample ingrowth trees, not fulfilling population eligibility at t_1 (population ingrowth tree) and immediate association with the sampling unit when population eligibility is reached at $t_1 < t_e < t_2$,
- $O_{12}(\omega)$ sample ongrowth trees, not fulfilling population eligibility at t_1 (population ingrowth tree), and either (a) reaching first population eligibility and later sampling unit association at $t_1 < t_e < t_2$ ($O_{12.\ a}(\omega)$), or (b) immediate association with the sampling unit at $t_1 < t_e < t_2$, together with population eligibility ($O_{12.\ b}(\omega)$),
- $N_{12}(\omega)$ sample nongrowth trees, fulfilling population eligibility at time t_1 (population survivor tree), but reaching sampling unit association later at time $t_1 < t_e < t_2$.

Note that sample ongrowth trees exist under angle-count sampling but not under the NFI plot configuration with two nested, concentric circles of fixed size.

2.9.2 *Change Components*

The conceptual difference between subpopulations of trees relevant for describing and defining change and change components in the population is subtle, and the related subsamples of trees are observed and measured on permanent plots.

Definitions The forest inventory in Switzerland has a long tradition and unique understanding of change assessment and relevant change components. The pioneering work, both in theory and in practice, is the *méthode de contrôle* (Biolley 1901).

Biolley introduced periodic, full-census tree measurements in forests in the Jura valleys of Canton Neuchâtel about 130 years ago. He divided the forest into compartments (*divisions*), each with a surface area of approximately 10–20 ha. Measurements were taken with callipers and have continued to be taken every 5–15 years ever since. The diameters of any removed trees are assessed at the time of felling and a local tariff function applied for estimating stem volumes. The inventory is restricted to trees with $d_{1.3} \geq 17.5$ cm.

The analysis of change, especially growth, starts from the simple fact that the state in the second inventory is the result of the gains and losses that have occurred since the first inventory, where gains refer to either the growth of trees already measured in the first inventory and to the ingrowth of new trees, while losses arise from tree felling, removal and mortality.

In our symbolic notation, Y represents the total of some target variable or the per hectare mean density, for example the stem volume of trees, in the compartment. The subscript denotes the respective change component with a notation referring to the subpopulations of trees introduced in Sect. 2.8.

Biolley's change components of interest are

$$\begin{array}{ll} \text{net change } (\textit{augmentation/diminuition}) & \hat{=} Y_{\mathcal{P}_{l.2}} - Y_{\mathcal{P}_{l.1}} \\ \text{depletion } (\textit{matériel exploité}) & \hat{=} Y_{\mathcal{D}_{l.12}} \\ \text{eligibility } (\textit{passage à la futaie}) & \hat{=} Y_{\mathcal{E}_{12}} \\ \text{total growth } (\textit{accroissement total}) & \hat{=} Y_{\mathcal{P}_{l.2}} - Y_{\mathcal{P}_{l.1}} + Y_{\mathcal{D}_{l.12}} \end{array}$$

as well as gross increment without eligibility (*accroissement du matériel initial*), defined as $Y_{\mathcal{P}_{l.2}} - Y_{\mathcal{P}_{l.1}} + Y_{\mathcal{D}_{l.12}} - Y_{\mathcal{E}_{l.12}}$.

Depletion is defined at the moment of loss and *eligibility* at the moment of gain. In other words, *depletion* is the amount of wood at the moment t_d of loss, and *eligibility* is the amount of wood at the moment t_e of trees reaching *eligibility*, which is the $d_{1.3}$ threshold of 17.5 cm in the case of Biolley's inventory system. The English terms have been taken from Eriksson (1995a) and Mandallaz (2008). These authors use the same definitions for the components of change as Biolley.

Beers and Schmid-Haas Components of Change In an inventory with permanent plots, the change components are derived for each sampling unit, based on the classification of tally trees into 'survivor', 'cut', 'mortality', 'ingrowth', 'ongrowth' and 'nongrowth' trees. The original definition and estimation of change components with permanent plot inventories seems to come from Beers (1962). Schmid-Hass adopted and developed the system of conducting forest inventories with permanent plots (*Kontrollstichprobe*) for Switzerland more than 50 years ago (Schmid-Haas 1983). The system is in wide use for inventories at the forest enterprise (regional) level and uses a single fixed-area circle for tree association. The relevant components of change are, derived at plot level,

$$\begin{array}{ll} \text{net change} & \hat{=} y_{P_{l.2}}(\omega) - y_{P_{l.1}}(\omega) = y_{l.2} - y_{l.1} \\ \text{cut and mortality} & \hat{=} y_{C_{l.12}}(\omega) + y_{M_{12}}(\omega) = c_{l.1} + m_1 \\ \text{ingrowth} & \hat{=} y_{I_{l.12}}(\omega) = i_{l.2} \\ \text{gross increment} & \hat{=} y_{S_{l.2}}(\omega) - y_{S_{l.1}}(\omega) + y_{I_{l.12}}(\omega) = s_{l.2} - s_{l.1} + i_{l.2} \end{array}$$

The respective subset of tally trees is given in the subscript and follows the notation given in Sect. 2.9.1.

There are two specific problems with periodic re-measurement of tally trees in permanent plot sampling, for which various solutions have been proposed (Grosenbaugh 1958; Martin 1982; Van Deusen et al. 1986; Roesch et al. 1989, 1991; Eriksson 1995a). First, the definitions for *cut and mortality*, *ingrowth* and *gross increment* are not exactly identical to the change components introduced by Biolley, i.e. *depletion*, *eligibility* and *total growth*. The reason is that neither the exact time points t_d and t_e of depletion and eligibility, nor the value of the target variables at these time points, are known under periodic sampling. Therefore, *cut and mortality* underestimates *depletion* and *gross increment* underestimates *total growth* by the (unknown) amount of growth in the subpopulation of cut and mortality trees between t_1 and t_d.

A second problem arises under sampling designs in which the extrapolation factors of sample trees change over time due to an increment in $d_{1.3}$. This is the case for angle-count sampling, as well as the nested (concentric) fixed-area plot design of the NFI.

Eriksson's Components of Change Eriksson's components of change are using the same abbreviated notation as before:

net increase (net change)	$\hat{=} y_{l.2} - y_{l.1}$
depletion (cut and mortality)	$\hat{=} c_{l.d} + m_d$
growth of cut and mortality	$\hat{=} c_{l.d} + m_d - c_{l.1} - m_1$
eligibility (ingrowth)	$\hat{=} i_{l.e}$
growth of ingrowth	$\hat{=} i_{l.2} + o_{b.l.2} - i_{l.e} - o_{b.l.e}$
survivor growth	$\hat{=} s_{l.2} + n_{l.2} + o_{a.l.2} - s_{l.1}$

where *total growth* is equal to the sum of *survivor growth*, *growth of cut and mortality* and *growth of ingrowth* (Eriksson 1995a). The target variable values $c_{l.d}$ and m_d at the time of cut and mortality are not observed and are assumed to be predicted with a growth model.

Eriksson's method provides, as in the methods of Biolley, Beers and Schmid-Hass, additive components in the sense that, at sample plot or domain level, *net increase* is numerically equal to the sum of *total growth* and *eligibility*, minus *depletion*.

This method is appealing because no back-prediction of target variables is needed for *nongrowth trees* (see below) and that additivity is guaranteed over more than two successive measurements.

NFI Components of Change A disadvantage of Eriksson's approach is that there is a relatively large jump in the local density whenever a nongrowth tree appears in the sample. This has a negative effect on the precision of the overall estimates of change components, i.e. of total growth. In the NFI approach, this effect is mitigated by backdating the growth of nongrowth trees as follows

net change	$\hat{=} y_{l.2} - y_{l.1}$
cut and mortality	$\hat{=} c_{l.d} + m_d$
growth of cut and mortality	$\hat{=} c_{l.d} + m_d - c'_{l.1} - m'_1$
ingrowth	$\hat{=} i_{l.2}$
survivor growth	$\hat{=} s_{l.2} + n_{l.2} - s'_{l.1} - n'_{l.1}$

In addition, *gross increment* is defined as the sum of survivor growth, growth of cut and mortality, and ingrowth.

The target variable values $c_{l.d}$ and m_d at time t_d of loss for cut and mortality trees, and the target variable value $n'_{l.1}$ for nongrowth trees at time t_1 are not known. Appropriate models for tree-level predictions of basal area, and the related volume and biomass, have been developed in the NFI (Chap. 12).

The prime symbol in $c'_{l.1}$ and m'_1, as well as $s'_{l.1}$ and $n'_{l.1}$ indicates that these values are calculated with values taken from t_1 and extrapolation factors taken from t_d and t_2, respectively.

The change components *cut and mortality*, *ingrowth*, *survivor growth* and *gross increment* can be considered unbiased, but they are not numerically exactly additive with *net change*. In other words,

$$y_{l.2} - y_{l.1} \neq s_{l.2} + n_{l.2} - s'_{l.1} - n'_{l.1} + i_{l.2} - c_{l.d} - m_d. \tag{2.41}$$

In practice, the non-additivity of these components of change is not relevant.

A disadvantage and challenge of this approach is that models for the back-prediction of nongrowth trees are needed.

Subpopulations of Living and Dead Trees Change components are traditionally defined only for the subpopulation of living trees. In Fig. 2.7, we graphically show an immediate extension to the subpopulation of dead trees, at least to the subpopulation of standing dead trees.

Lying dead trees are defined and measured in the NFI. At the moment of status change, however, trees may change position, leaving or newly appearing in the sample. The assessment of change components in the subpopulation of dead trees is, therefore, less precise than in the subpopulation of living trees. An extension of the NFI4 field protocol introduces all relevant parameters for, in principle, unbiased estimation of change components in the subpopulation of living and dead trees.

2.9.3 Derivation of Change Components for Tariff Trees

Under two-stage sampling, the general form of local densities for the estimation of change and change components is (compare with Eqs. 2.11 and 2.12)

$$y_2 - y_1 = \left(\hat{y}_2 - \hat{y}_1\right) + \left(r_2 - r_1\right) \tag{2.42}$$

where the notation has been simplified slightly. $\hat{y}$ denotes a local density based on tariff function stem volumes, and r a local density based on the residual between volume and tariff function stem volumes, available for the subset of second-stage tariff trees only. The subscripts indicate the two successive inventories.

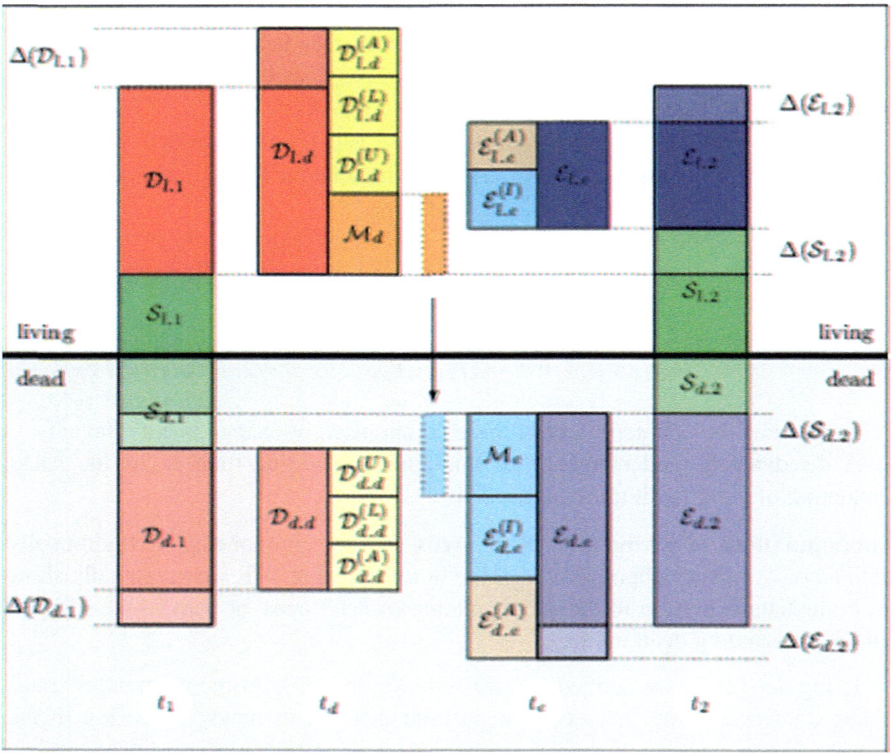

Fig. 2.7 Graphic representation of the subpopulations of trees relevant for estimating change between two inventories at t_1 and t_2. The subpopulations are defined in Sect. 2.9.1. Δ denotes the increment (or decrement) within subpopulations for certain variables, such as basal area, stem volume and biomass. The block in the middle, which changes the status from living (orange) to dead (cyan), refers to mortality trees, which remain in the overall population and are losses in the subpopulation of living trees and gains in the subpopulation of dead trees

Change at Tally Tree and Tariff Tree Level The full expression of change between two inventories under two-stage sampling is

$$y_2 - y_1 = \left(\sum_{i \in \dot{S}_{1\cap 2}} \left(\frac{\hat{X}_{2.i}}{\pi_{2.i}} - \frac{\hat{X}_{1.i}}{\pi_{1.i}} \right) \right) + \left(\sum_{i \in \dot{S}_{2\setminus 1}} \frac{\hat{X}_{2.i}}{\pi_{2.i}} \right) - \left(\sum_{i \in \dot{S}_{1\setminus 2}} \frac{\hat{X}_{1.i}}{\pi_{1.i}} \right)$$
$$+ \left(\sum_{i \in \ddot{S}_{1\cap 2}} \left(\frac{R_{2.i}}{\pi_{2.i} p_{2.i}} - \frac{R_{1.i}}{\pi_{1.i} p_{1.i}} \right) \right) + \left(\sum_{i \in \ddot{S}_{2\setminus 1}} \frac{R_{2.i}}{\pi_{2.i} p_{2.i}} \right) - \left(\sum_{i \in \ddot{S}_{1\setminus 2}} \frac{R_{1.i}}{\pi_{1.i} p_{1.i}} \right). \tag{2.43}$$

The first three sums represent changes in the first-stage sample of tally trees, with $\dot{S}_{1\cap 2} = \dot{S}_1 \cap \dot{S}_2$ denoting the set of first-phase tally trees remaining in the sample (and in the population) between the two inventories, and $\dot{S}_{2\backslash 1}$ and $\dot{S}_{1\backslash 2}$ denoting new and lost members of the first-stage sample of tally trees. The same basic subsets of trees remaining in the sample, becoming new members of the sample, or disappearing from the sample also occur in the second-stage sample of tariff trees. These three subsets are denoted by $\ddot{S}_{1\cap 2}, \ddot{S}_{2\backslash 1}$, and $\ddot{S}_{1\backslash 2}$, respectively, and it should be kept in mind that $\ddot{S}_1 \subseteq \dot{S}_1$ and $\ddot{S}_2 \subseteq \dot{S}_2$ are always fulfilled by definition.

The estimation of change components is a bit more complex than immediately revealed by Eq. (2.43). *First-stage tally trees* include:

- Tally trees in the first sum of Eq. (2.43), which are trees that have survived in the sample (and in the population);
- Tally trees in the second sum of Eq. (2.43), which fall into the categories ingrowth, ongrowth and nongrowth trees. Various options exist regarding how these trees should be treated in change estimation;
- Tally trees in the third sum of Eq. (2.43), which are losses from the sample. Basically, these are trees that have been removed from the population through cut and mortality.

Likewise, second-stage tariff trees include:

- Tariff trees in the fourth sum of Eq. (2.43), which are trees that have survived in the sample (and in the population);
- Tariff trees in the fifth sum of Eq. (2.43), which are trees in the second inventory that may be population survivor or population ingrowth trees;
- Tariff trees in the sixth sum of Eq. (2.43), which have become lost since the first inventory and may be population survivor or population cut or mortality trees.

Survivor Growth The growth of population survivor trees is of major interest because most of the overall population growth is in these trees.

By specifying the different subsets of tally and tariff trees in the sample, the *difference in the residual part for population survivor trees* turns out to be

$$(r_2 - r_1)_{|\text{survivors in population}} = \sum_{\substack{i \in \dot{S}_{1\cap 2} \\ i \in \ddot{S}_{1\cap 2}}} \left(\frac{R_{2.i}}{\pi_{2.i} p_{2.i}} - \frac{R_{1.i}}{\pi_{1.i} p_{1.i}} \right)$$
$$+ \sum_{\substack{i \in \dot{S}_{1\cap 2} \\ i \in \ddot{S}_{2\backslash 1}}} \frac{R_{2.i}}{\pi_{2.i} p_{2.i}} - \sum_{\substack{i \in \dot{S}_{1\cap 2} \\ i \in \ddot{S}_{1\backslash 2}}} \frac{R_{1.i}}{\pi_{1.i} p_{1.i}} + \sum_{\substack{i \in \dot{S}_{2\backslash 1} \\ i \in \ddot{S}_{2\backslash 1}}} I_{P_{1\cap 2}.i} \frac{R_{2.i}}{\pi_{2.i} p_{2.i}} \tag{2.44}$$

where the two subscripts under the sigma sign indicate the respective subsamples of tally and tariff trees, and indicates nongrowth tally trees (population survivor trees).

The fourth sum of Eq. (2.44) refers to residuals after selecting nongrowth tally trees as tariff trees.

The components in the second and third sums in Eq. (2.44) only occur when the subselection of tariff trees is not permanent, i.e. when a tariff tree chosen (by chance) in the first inventory is not automatically re-measured as a tariff tree in the second inventory. In the NFI, this is the case for tally trees with $d_{1.3}$ <60 cm. The selection of second-stage tariff trees in inventories NFI2, NFI3 and NFI4 (years 2009–2014) was carried out independently from the selection in the previous inventory (Sect. 2.3).

These components lead to an increased variance in the change estimates because even a considerable volume residual for an unusually formed stem can remain quite stable between inventories. It is thus of negligible consequence for inference, but it can carry considerable weight if observed in only one of the two inventories.

If tariff trees are selected independently on occasions one and two, another estimator may be used. We know that $P\langle i \in \ddot{S}_{1\cap 2}\rangle = p_{1.i}p_{2.i}$ and the residual component for population survivor trees can be estimated separately for the subset of sample survivor and nongrowth trees as

$$(r_2 - r_1)_{|\text{survivors in population}} = \sum_{\substack{i \in \dot{S}_{1\cap 2} \\ i \in \ddot{S}_{1\cap 2}}} \frac{R_{2.i} - R_{1.i}}{\pi_{2.i}p_{1.i}p_{2.i}} + \sum_{\substack{i \in \dot{S}_{2\backslash 1} \\ i \in \ddot{S}_{2\backslash 1}}} I_{P_{1\cap 2}.i} \frac{R_{2.i} - \tilde{R}_{1.i}}{\pi_{2.i}p_{2.i}} \tag{2.45}$$

Only re-measured tariff trees are used in the first sum, whereas back-predicted, tree-level residuals $\tilde{R}_{1.i}$ are assumed for nongrowth tariff trees in the second sum. In practice, this estimator has a larger variance than the estimator given in Eq. (2.44) because the subset of tariff tress selected in both inventories is small.

The NFI has a model for $\tilde{R}_{1.i}$ back-prediction, and the residual component for population survivor trees is then calculated according to

$$\begin{aligned}(r_2 - r_1)_{|\text{survivors in population}} = & \sum_{\substack{i \in \dot{S}_{1\cap 2} \\ i \in \ddot{S}_{1\cap 2}}} \frac{R_{2.i} - R_{1.i}}{\pi_{2.i}p_{2.i}} \\ & + \sum_{\substack{i \in \dot{S}_{1\cap 2} \\ i \in \ddot{S}_{2\backslash 1}}} \frac{R_{2.i} - \tilde{R}_{1.i}}{\pi_{2.i}p_{2.i}} + \sum_{\substack{i \in \dot{S}_{2\backslash 1} \\ i \in \ddot{S}_{2\backslash 1}}} I_{P_{1\cap 2}.i} \frac{R_{2.i} - \tilde{R}_{1.i}}{\pi_{2.i}p_{2.i}}.\end{aligned} \tag{2.46}$$

The first two sums provide together a prediction for the residual component in sample survivor trees, and the third sum provides the same prediction for nongrowth trees. Added together, these three sums provide a prediction for the change in the residual component in population survivor trees. This approximate estimator has the tendency to slightly underestimate the true change in the residual component because of the modelled (smoothed) back-predictions $\tilde{R}_{1.i}$. The estimator is, therefore, robust with a lower variance than the estimators in Eqs. 2.44 and 2.45.

Overall, for two-stage change estimation and tariff-tree selection, we conclude that:

- The residual correction in the estimation of change (growth) is, in principle, needed to ensure the state estimates with the estimates of change are unbiased (and numerically additive);
- The precision of the estimation change in the residual component depends on the method of second-stage tariff tree selection;
- The efficiency of the design-based estimators increases with a larger proportion of re-measured tariff trees;
- For small domains and small sample sizes, back-predicting residuals as, well as omitting the estimation of change in the residuals and restricting change estimation to first-stage trees, may be an option.

Appendix

Table 2.1 Tariff tree inclusion probabilities

d	p_old	p_new	Sector small	Sector small rounded	Sector large	Sector large rounded
12	0.04050	0.04050	16.200	16		
13	0.04753	0.04753	19.013	19		
14	0.05513	0.05513	22.050	22		
15	0.06328	0.06328	25.313	25		
16	0.07200	0.07200	28.800	29		
17	0.08128	0.08128	32.513	33		
18	0.09113	0.09113	36.450	36		
19	0.10153	0.10153	40.613	41		
20	0.11250	0.11250	45.000	45		
21	0.12403	0.12403	49.613	50		
22	0.13613	0.13613	54.450	54		
23	0.14878	0.14878	59.513	60		
24	0.16200	0.16200	64.800	65		
25	0.17578	0.17578	70.313	70		
26	0.19013	0.19013	76.050	76		
27	0.20503	0.20503	82.013	82		
28	0.22050	0.22050	88.200	88		
29	0.23653	0.23653	94.613	95		
30	0.25313	0.25313	101.250	101		
31	0.27028	0.27028	108.113	108		

(continued)

Table 2.1 (continued)

d	p_old	p_new	Sector small	Sector small rounded	Sector large	Sector large rounded
32	0.28800	0.28800	115.200	115		
33	0.30628	0.30628	122.513	123		
34	0.32513	0.32513	130.050	130		
35	0.34453	0.34453	137.813	138		
36	0.14580	0.15428	137.813	138	10.980	11
37	0.15401	0.17098	137.813	138	22.110	22
38	0.16245	0.18896	137.813	138	34.099	34
39	0.17111	0.20829	137.813	138	46.987	47
40	0.18000	0.22904	137.813	138	60.817	61
41	0.18911	0.25126	137.813	138	75.631	76
42	0.19845	0.27502	137.813	138	91.473	91
43	0.20801	0.30039	137.813	138	108.387	108
44	0.21780	0.32744	137.813	138	126.417	126
45	0.22781	0.35623			142.492	142
46	0.23805	0.38683			154.734	155
47	0.24851	0.41932			167.730	168
48	0.25920	0.45377			181.509	182
49	0.27011	0.49025			196.100	196
50	0.28125	0.52883			211.534	212
51	0.29261	0.56960			227.840	228
52	0.30420	0.61263			245.050	245
53	0.31601	0.65799			263.195	263
54	0.32805	0.70576			282.305	282
55	0.34031	0.75604			302.414	302
56	0.35280	0.80889			323.555	324
57	0.36551	0.86440			345.759	346
58	0.37500	0.92265			369.060	369
59	0.37500	0.98373			393.493	393
60	1.00000	1.00000			400.000	400

New (p new) and old (p old) tariff tree inclusion probabilities according to diameter (d) classes with respective exact sector openings on the small (sector small) and on the large (sector large) plot, both also with a rounded to integer version

References

Beers TW (1962) Components of forest growth. J For 60(4):245–248

Biolley B (1901) Le jardinage cultural. Schweiz Z Forstwes 52(6):97–104

Biolley H (1920) L'Aménagement des forêts par la méthode expérimentale et spécialement la méthode du contrôle. Attinger Frères, Neuchâtel

Cordy CB (1993) An extension of the Horvitz—Thompson theorem to point sampling from a continuous universe. Stat Probab Lett 18(5):353–362

Düggelin C, Keller M (eds) (2017) Schweizerisches Landesforstinventar – Feldaufnahme Anleitung 2017. Eidgenössische Forschungsanstalt für Wald Schnee und Landschaft, Birmensdorf

Eriksson M (1995a) Compatible and time-additive change component estimators for horizontal-point-sampled data. For Sci 41(4):796–822

Eriksson M (1995b) Design-based approaches to horizontal-point-sampling. For Sci 41(4):890–907

Gregoire T (1982) The unbiasedness of the mirage correction procedure for boundary overlap. For Sci 28:504–508

Gregoire T, Valentine H (2007) Sampling strategies for natural resources and the environment, Applied Environmental Statistics. Chapman & Hall, Boca Raton

Grosenbaugh L (1958) Point-sampling and line-sampling: probability theory, geometric implications, synthesis. Occ. Paper 160, USDA For. Serv. South. For. Exp. Stn

Gurnaud A (1886) La Sylviculture française et la méthode du Contrôle. Imprimerie Paul Jacquin, Besançon

Horvitz DG, Thompson DJ (1952) A generalization of sampling without replacement from a finite universe. J Am Stat Assoc 47:663. https://doi.org/10.2307/2280784

Kaiser L (1983) Unbiased estimation in line-intersect sampling. Biometrics 39:965–976

Köhl M (1994) Statistisches Design für das zweite Schweizerische Landesforstinventar: ein Folgeninventurkonzept unter Verwendung von Luftbildern und terrestrischen Aufnahmen, Mitteilungen der Eidgenössischen Forschungsanstalt für Wald, Schnee und Landschaft, vol 1. Eidgenössische Forschungsanstalt für Wald Schnee und Landschaft

Köhl M (2001) Inventory concept NFI2. In: Brassel P, Lischke H (eds) Swiss National Forest Inventory: methods and models of the second assessment. Swiss Federal Research Institute WSL, Birmensdorf. http://www.lfi.ch/publikationen/publ/methoden-en.php

Lanz A (2000) Optimal sample design for extensive forest inventories. Doctoral Thesis, ETH Zürich, Switzerland

Madigan D, Bartlett P, Bühlmann P et al (2014) Statistics and science – a report of the London workshop on the future of the statistical sciences

Mahrer F (1976) Abgrenzungsnormen zur Erfassung der Waldfläche im Schweizerischen landesforstinventar (LFI). In: Beiträge zur Bestimmung des Begriffes „Wald". Bericht Eidg. Anstalt für das Forstliche Versuchswesen, 167, Birmensdorf, Switzerland

Mandallaz D (2008) Sampling techniques for forest inventories. Chapman & Hall/CRC, New York

Mandelbrot B (1967) How long is the coast of Britain? Statistical self-similarity and fractional dimension. Science 156(5):636–638

Martin GL (1982) A method for estimating ingrowth on permanent horizontal sample points. For Sci 28(1):110–114

Paschedag I, Zinggeler J (2001) Forest transportation system survey. In: Brassel P, Lischke H (eds) Swiss National Forest Inventory: methods and models of the second assessment. Swiss Federal Research Institute WSL, Birmensdorf, pp 162–196

Pulkkinen M, Ginzler C, Traub B, Lanz A (2018) Stereo-imagery-based post-stratification by recursive partitioning in Swiss National Forest Inventory. Remote Sens Environ 213:182–194

Roesch FA, Green EJ, Scott CT (1989) New compatible estimators for survivor growth and ingrowth from remeasured horizontal point samples. For Sci 35(2):281–293

Roesch FA, Green EJ, Scott CT (1991) Compatible basal area and number of trees estimators from remeasured horizontal point samples. For Sci 37(1):136–145

Särndal CE, Swensson B, Wretman J (2003) Model assisted survey sampling. Springer Science & Business Media, New York

Schmid-Haas P (1983) Swiss Continuous Forest Inventory: twenty years' experience. In: Bell JF, Atterbury T (eds) Renewable resource inventories for monitoring changes and trend. Proc., 15–19 August 1983, Corvallis, OR, SAF 83–14. pp 133–140

Schwyzer A, Lanz A (2010) Verjüngungserhebung im schweizerischen Landesforstinventar. In: Jahrestagung der Sektion Forstliche Biometrie und Informatik im Deutschen Verband der Forstlichen Forschungsanstalten (22.) und der AG Ökologie und Umwelt in der Internationalen Biometrischen Gesellschaft – Deutsche Region, Klaus Römisch, TU Desden Fachrichtung Forstwissenschaften Tharandt. Arne Nothdurft, Forstliche Versuchs- und Forschungsanstalt Baden-Württemberg (FVA) Freiburg. Uwe Wunn, Forschungsanstalt für Waldökologie und Forstwissenschaft Reinland-Pfalz in Lippstadt, Göttingen, Die Grüne Reihe. ISSN: 1860-4064

Van Deusen PC, Dell TR, Thomas CE (1986) Volume growth estimation from permanent horizontal points. For Sci 32(2):415–422

Part III
Remote Sensing

Chapter 3
Remote Sensing Data Sources

Christian Ginzler

Abstract Remote sensing techniques and data have been used in the Swiss National Forest Inventory (NFI) since its beginnings in the 1970s. Over the decades, the image data used has changed from first analogue black and white aerial images to digitised RGB true-colour images, and then to the digital colour infrared (CIR) sensor data in use today. In addition to applying aerial stereo-image-based interpretation on different sampling grids, area-wide data sets were developed and produced within the framework of the NFI4. In addition to aerial images, satellite images are also increasingly being used.

3.1 Introduction

Remote sensing techniques and data have been used in the Swiss NFI (NFI) since its beginnings in the 1970s. Over the decades, the image data used has changed from first analogue black and white aerial images to digitised RGB true-colour images, and then to the digital colour infrared (CIR) sensor data in use today. In addition to applying aerial stereo-image-based interpretation on different sampling grids, area-wide data sets were developed and produced within the framework of the NFI4.

In NFI1, aerial images served mainly to determine the forest area and therefore the forest sample-plot locations. It was also used to measure reference-point data for the subsequent assessments of samples in the field survey (Mahrer and Vollenweider 1983; Brassel and Lischke 2001).

In NFI2, the number of sample plots interpreted using analogue aerial images was increased eightfold, and the number of expensive field surveys could therefore be reduced (Köhl 1994). The catalogue of attributes measured in the aerial images was extended. The measured landscape attributes were selected to treat the forest as a discrete landscape element. If a sample plot in NFI1 and NFI2 was located in an area

C. Ginzler (✉)
Remote Sensing, Swiss Federal Research Institute WSL, Birmensdorf, Switzerland
e-mail: christian.ginzler@wsl.ch

C. Fischer, B. Traub (eds.), *Swiss National Forest Inventory – Methods and Models of the Fourth Assessment*, Managing Forest Ecosystems 35,
https://doi.org/10.1007/978-3-030-19293-8_3

with no or only a few woody species, it was classified as *non-forest* and no further attributes were measured or recorded. If a sample plot evolved from *non-forest* to *forest*, no information about the previous land cover was recorded even though such information could have been relevant for subsequent NFI surveys.

NFI3 took into account dynamic landscape changes in space and time (Wildi and Brassel 1999). In addition to data on forest area, land-cover attributes were also determined in detail. The attributes were recorded irrespective of whether the sample plot was classified as forest or non-forest. The *forest/non-forest* decision was one of many products of aerial image interpretation (Mathys et al. 2005). Regarding the methodology and technology used, considerable improvements were made between the aerial image interpretation in NFI1 and that in NFI2. All data became available in digital form, from the oriented stereo models to topographic maps. The results of the interpretation and the decisions of the interpreters have been collected and fully documented in a geo-database.

The methods of aerial image interpretation used in NFI4 correspond to those used in NFI3 to a large degree (Hastedt et al. 2009). However, to be more flexible to apply different forest definitions, forest boundary lines are now mapped on a 100 × 100 m plot area and a *potential forest use* attribute is interpreted for *non-forest* plots. In the case of land uses that clearly do not permit forest use, the attribute *non-forest* is assigned. This makes it possible to apply forest definitions that differ from the Swiss definition (e.g. with a smaller crown cover), and to therefore obtain data comparable and harmonised with international standards. All other attributes are measured on the traditional 50 × 50 m area, as in previous Swiss inventories. The interpretation is carried out on the terrestrial grid, where the distance between the sample plots is 1.4 km. Since 2008, Leica ADS40/80 false-colour infrared images have been available, which are well suited for the interpretation of vegetation data. Moreover, aerial image interpretation is now performed completely within the framework of a Geographic Information System (ArcGIS with the extension Stereo Analyst).

The objectives of the aerial image interpretation are to:

- Classify plots as *forest* or as *non-forest* (*forest/non-forest* decision according to NFI criteria to optimise fieldwork).
- Assess the composition of land-cover classes regardless of the *forest/non-forest* decision.
- Identify reference points (coordinates, description) close to the centre of each sample plot. These reference points are used by the field survey teams to locate the sample-plot centre (SPC) of each plot.
- Generate training and validation data for area-wide products.

The results enable statistical analyses of certain variables for specific domains. Tables and maps showing means over domains can be produced. The availability of digital images and advances in methods of photogrammetry and remote sensing have enabled the development of workflows to derive not only sample data but also area-wide products with forest-related information. This area-wide data offers the potential for the spatially explicit analysis of forests. It can be used to quantify forest functions in cases where spatially explicit information is crucial, for example in assessments of the protective function of forests or biodiversity. The main aim of developing area-wide data sets, however, is to integrate the data as auxiliary information in the two-phase estimation process to replace dense sampling during aerial image interpretation.

3.2 Data Sources

The ideal data for aerial image interpretation involving questions related to forestry and vegetation ecology is in the form of colour infrared (CIR) images. Vegetation has a maximum absorption in the blue and red bands of the electromagnetic spectrum. In the near-infrared band, more than 80% of the incident radiation is reflected. When the spectral bands of green, red and near-infrared are combined, these images enable a well-differentiated interpretation of trees and vegetation (Hildebrandt 1996).

The Swiss Federal Office of Topography (swisstopo) acquires digital aerial images covering the whole country every 3 years: every 6 years in summer (leaf-on conditions) and every 6 years in spring or autumn (leaf-off conditions). Since 2005, swisstopo has used a Leica Geosystems airborne digital sensor (ADS) camera to acquire the digital images. In contrast to frame cameras, the ADS camera is based on push-broom technology in which the sensor scans the earth's surface and produces long image strips backwards, nadir and forwards, simultaneously (Fig. 3.1).

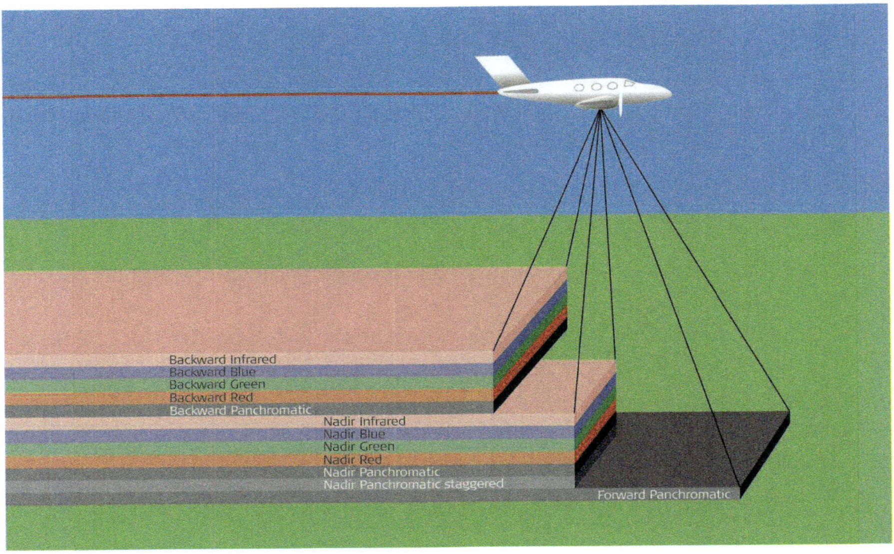

Fig. 3.1 Configuration of multiple linear charge-coupled device (CCD) arrays for the Leica airborne digital sensor (ADS). (Source: Leica Geosystems)

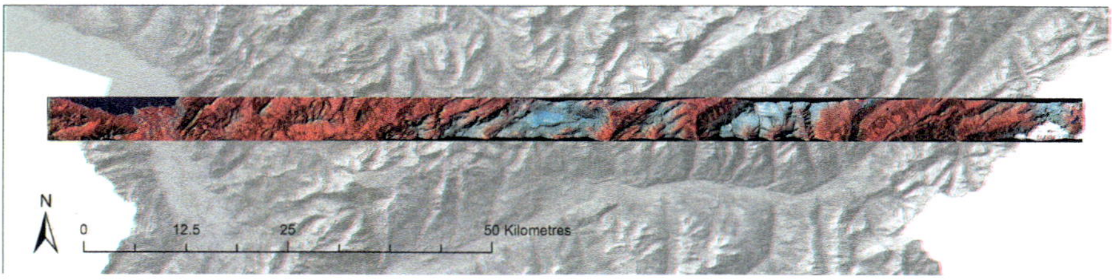

Fig. 3.2 Example of a colour infrared (CIR) airborne digital sensor (ADS) image strip (one image) with a ground coverage of 125 × 5 km

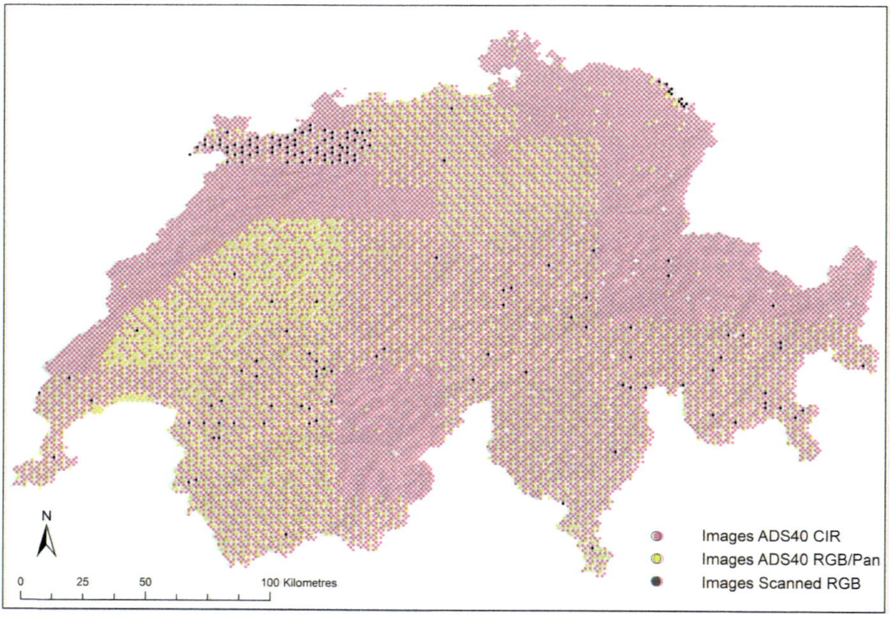

Fig. 3.3 Image material used in NFI4 aerial-image interpretation

Thus, the image strips can be visualised in three dimensions and digital surface models can be calculated.

Images from before 2005 are available as scanned true-colour (RGB) RC30 aerial frame images. Since 2005, the images have been available as digital ADS40 true-colour/panchromatic (RGB/Pan) stereo-image strips.

Colour infrared (CIR) stereo-image strips created with an updated sensor head for the ADS80 sensor have been available since 2008 (Fig. 3.2). The ADS40/80 images have a mean ground resolution of 25 cm (scale of ~1:15'000) in the Swiss Plateau, Jura and Pre-Alps and 50 cm (scale of ~1:30'000) in the Alps.

In NFI4, colour infrared images were used for aerial-image interpretation whenever possible (Fig. 3.3). More than 89% of the samples were interpreted with CIR ADS80 images, 20% with RGB/Pan images and less than 1% with scanned analogue

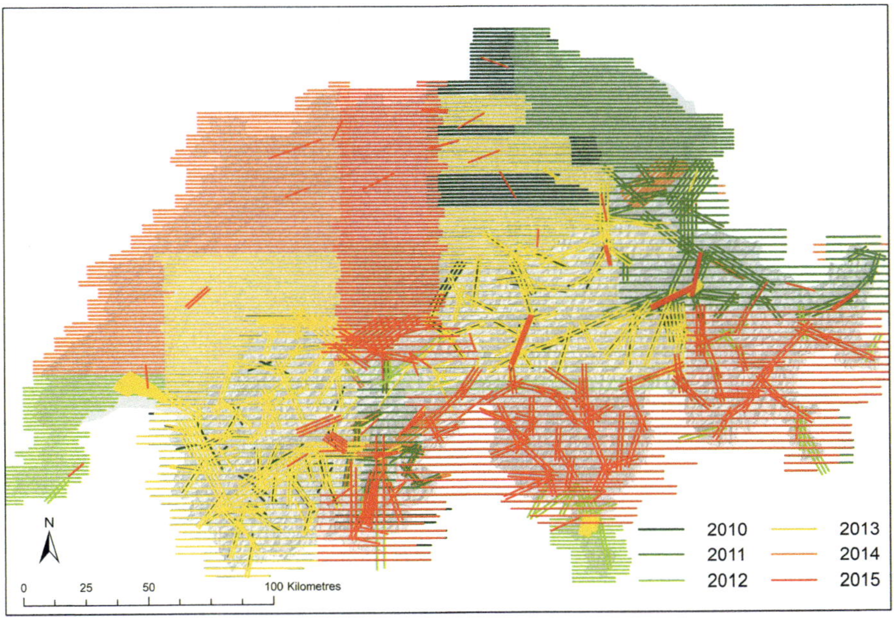

Fig. 3.4 Trajectories of ADS80 image strips used in NFI4 for area-wide products. Areas with denser strips have a resolution of 25 cm, whereas areas with sparser strips, mainly in mountainous regions, have a resolution of 50 cm

RGB images. Tests have shown that the type of image has no influence on the *forest/non-forest* decision. However, small gaps in the forest can be identified better with the newer images.

ADS80 CIR image data from the years 2010 to 2015, taken in the leaf-on months between May and September, were used to derive the area-wide products (Fig. 3.4)

References

Brassel P, Lischke H (eds) (2001) Swiss national forest inventory: methods and models of the second assessment. Swiss Federal Research Institute WSL, Birmensdorf

Hastedt H, Ginzler C, Meile R (2009) GIS-gestützte 3D-Luftbildinterpretation im Landesforstinventar der Schweiz. Geomatik Schweiz, Géomatique Suisse, Geomatica Svizzera 107(12/2009):596–599

Hildebrandt G (1996) Fernerkundung und Luftbildmessung für Forstwirtschaft, Vegetationskartierungen und Landschaftsökologie. Germany, 676 S

Köhl M (1994) Statistisches Design für das zweite Schweizerische Landesforstinventar: Ein Folgeinventurkonzept unter Verwendung von Luftbildern und terrestrischen Aufnahmen. Mitt Eidgenöss Forsch anst Wald Schnee Landsch 69(1):1–141

Mahrer F, Vollenweider C (1983) Das Landesforstinventar der Schweiz, Bericht 247. Eidgenössische Anstalt für das forstliche Versuchswesen EAFV, Birmensdorf

Mathys L, Wildi O, Brassel P (2005) LFI3-Das Landesforstinventar ist auch ein Landschaftsinventar. Inf bl Forschbereich Landschaft Landsch 64:1–4

Swisstopo (2017) The topographic landscape model TML. Available online: https://www.swisstopo.admin.ch/en/knowledge-facts/topographic-landscape-model.html. Accessed on 14 Sept 2017

Wildi O, Brassel P (1999) Die Landschaftsinventur – Ein Informationssystem für die "Landschaft Schweiz". Infb Forschbereich Landschaft 44:1–4

Chapter 4
Remote Sensing Data Management

Rolf Meile

Abstract The stereo GIS application is driven by aerial-image metadata as well as administration data like task lists and priority settings. All metadata including the interpretation results are stored in a spatial database. For analytic purposes the data is accessible by means of Structured Query Language (SQL), GIS tools or any statistical software.

4.1 Introduction

In addition to the digital aerial-image strips, administration data for steering the GIS application and aerial-image metadata is supplied. These two data sources are maintained in a database along with the final results of the image interpretation. An overview of the dataflow processes is depicted in Fig. 4.1.

The applied approach separates the large aerial-image data sets (>250 Terabytes stored on file servers) from their metadata and sample-plot task lists. As a consequence, the GIS application can dynamically load only the required stereo-image data for the current sample plot to the local desktop repository. This reduces latency and improves application loading speed, as well as providing maximum flexibility because the complete aerial-image data is available from all clients all the time. Additional pre-setting data is loaded from administration tables in order to prepare the interpretation process. The actual data collection then starts with an interpreter choosing a sample plot for interpretation. After all necessary information is measured and interpreted, the results are crosschecked and finally stored in the database. This raw data originating from stereo-image interpretation is used in further steps of the Swiss NFI (NFI), such as preselecting field-sample plots or deriving values for the data warehouse (Sect. 20.3).

R. Meile (✉)
GIS, Swiss Federal Research Institute WSL, Birmensdorf, Switzerland
e-mail: rolf.meile@wsl.ch

C. Fischer, B. Traub (eds.), *Swiss National Forest Inventory – Methods and Models of the Fourth Assessment*, Managing Forest Ecosystems 35,
https://doi.org/10.1007/978-3-030-19293-8_4

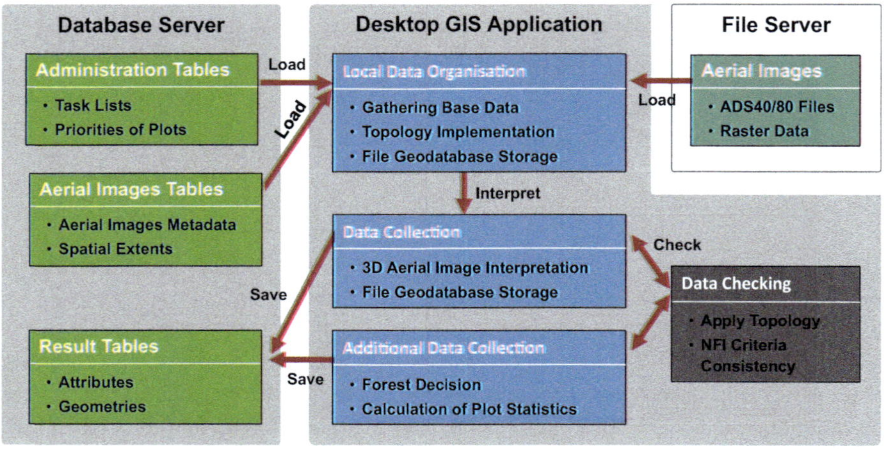

Fig. 4.1 Dataflow processes initialised by the GIS application

4.2 Database Model

A relational database model in third normal form (Codd 1971) serves as a solid base for pre-setting data, aerial-image metadata and result data from the sample-plot interpretation. Figure 4.2 shows an overview of these three data classes, along with the associated entities/tables of the database. Categorised attributes in the tables are constrained with additional lookup tables (not shown in Fig. 4.2) using foreign keys.

Over the past 20 years, the model has been adapted and extended several times (Hastedt et al. 2009). Some adjustments have recently been made for the upcoming NFI5 interpretations. Roughly 4.5 million sample plots have been interpreted since NFI1, and source data from NFI1-NFI4 is thus available instantly. One of the main goals of database development in recent years has been to keep all interpretation data in the same structure over time. Fridman et al. (2014) describes an identical approach for the Swedish National Forest Inventory. This continuation greatly simplifies access to the raw data time-series. At the same time, it means that derivations can reliably access information for further aggregation and calculated products.

4.3 Controlling Processes and Resulting Data

To initiate the process of aerial-image interpretation, the GIS application uses pre-setting data from the database. The specific steps of this process are:

- The sample plots are randomly selected for each interpreter from the annual subpanel to avoid spatial clustering of the plots assigned to an individual interpreter. The selection can be controlled by setting priorities of certain plots using

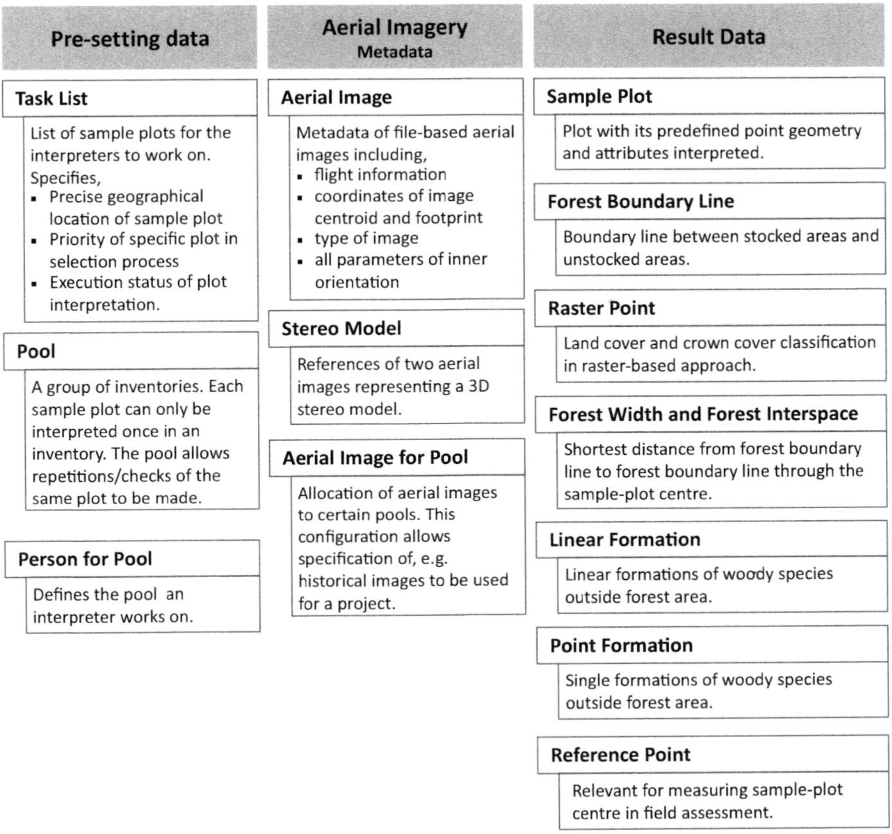

Fig. 4.2 Classes of data used for aerial-image interpretation of sample plots. All data is stored in the NFI database

attributes pre-stored in the task-list table. Individual interpreters can, for example, be assigned to a special training inventory.
- Some attributes in the task-list table influence which attributes are used in the subsequent interpretation process.
- Every plot in the sample of each inventory can only be interpreted once (Fig. 4.2). Pools of inventories have been formed to allow repeated interpretation of the same plot by different interpreters for, e.g. quality control or training. These pools usually include the main inventory plus inventories for double or triple interpretation of individual sample plots. Repeated interpretation for quality control purposes is conducted on approx. 4% of the sample.
- Each pool is configured with its own aerial-image references. Current project pools are usually connected with recent images.
- Interpreters are allocated to pools. They work on the pre-configured sample plots available in their respective pools.

- The task-list table is used to keep track of sample plots already interpreted or temporarily aborted by a user.

The final handling of result data is a major concern for data management. Therefore, an approach has been developed that involves modelling and storing spatial data sets with foreign key relationships. A parent–child relationship exists that links the sample-plot table to all other spatial tables. This ensures the referential integrity of all results collected. Additionally, spatial and non-spatial data of this type can be accessed by means of Structured Query Language (SQL), GIS software or any other client capable of connecting to a database.

References

Codd EF (1971) Further normalization of the data base relational model. IBM Research Report, San Jose, California RJ909

Fridman J et al (2014) Adapting national forest inventories to changing requirements – the case of the Swedish National Forest Inventory at the turn of the 20th century. Silva Fenn 48(3)

Hastedt H, Meile R, Ginzler C (2009) ArcGIS Applikation zur 3D Datenerfassung im Landesforstinventar der Schweiz. In: Strobl J, Blaschke T, Griesebner G (eds) 2009: Angewandte Geoinformatik 2009. Beiträge zum 21. AGIT-Symposium Salzburg. 21. AGIT-symposium, Salzburg, 2009, pp 8–13. https://doi.org/10.5169/seals-236647

Chapter 5
Stereo GIS Application for Aerial-Image Interpretation

Barbara Schneider

Abstract The aerial-image application is used to digitise the forest boundary line (where present), ground cover on 25 points, single and linear formations of woody species as well as three reference points on each sample plot. Based on the forest boundary line and the ground cover points, it is determined whether the sample plot is a forest plot (forest/non-forest decision). The aerial-image application has been developed at WSL as an ArcGIS Desktop extension.

5.1 Introduction

The aerial-image application has been developed at WSL as an ArcGIS Desktop extension for ArcMap 10.x (ESRI 2018) (Fig. 5.1). This extension includes the workflow for the interpreters and the digitisation and visualisation tools. Moreover, an ArcMap Editor extension is used to react to editing events. To visualise stereo-image pairs and to collect data on these images, Hexagon Stereo Analyst (HEXAGON 2018) is used, which is also an ArcMap extension. Communication to the task lists and image metadata stored in the Oracle database (Chap. 20) is achieved using the Devart OracleClient in conjunction with SQL. Data can be stored unversioned because multiple users never access the same features. The development environment is Microsoft Visual Studio, and the programming language is C#.

5.2 Toolbar for Stereo GIS Application

The following commands are available on the toolbar:

B. Schneider (✉)
Remote Sensing, Swiss Federal Research Institute WSL, Birmensdorf, Switzerland
e-mail: barbara.schneider@wsl.ch

C. Fischer, B. Traub (eds.), *Swiss National Forest Inventory – Methods and Models of the Fourth Assessment*, Managing Forest Ecosystems 35,
https://doi.org/10.1007/978-3-030-19293-8_5

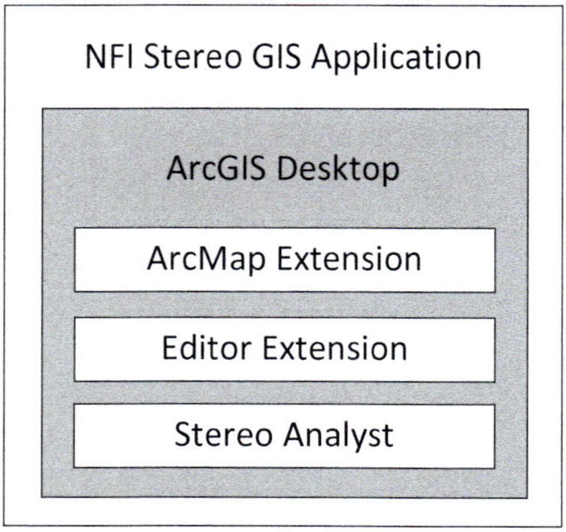

Fig. 5.1 System architecture of the aerial-image application

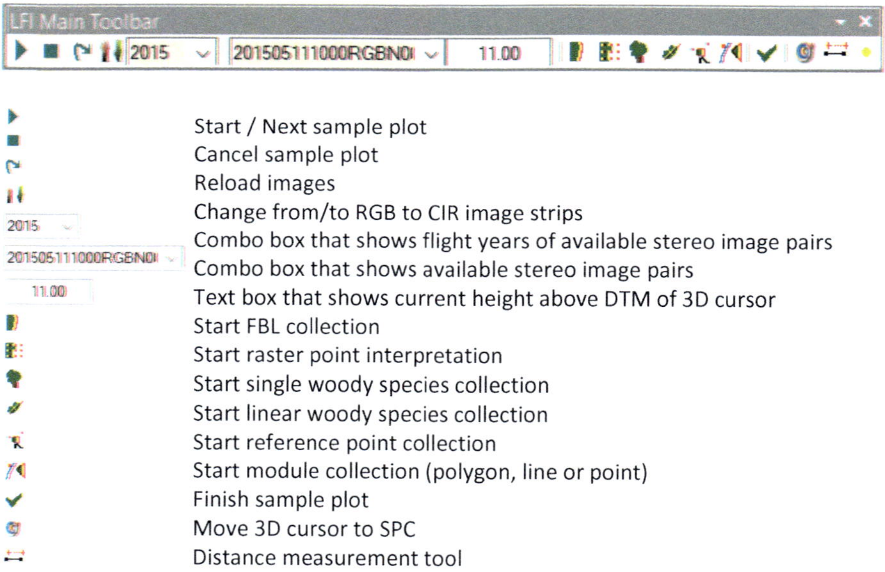

Fig. 5.2 Toolbar for the stereo GIS application used for NFI aerial-image interpretation

To start data collection, the interpreter clicks on the button *Start/Next sample plot* to select a random sample plot from the sample-plot pool (Fig. 5.2). It is possible to stop working on a sample plot and continue collecting data later. The image-data management system and the process of collecting data for the different variables are described in Sects. 5.3, 5.4, and 5.5. When data collection is complete, the interpreter clicks on the button *Finish sample plot* and the data is written to the database.

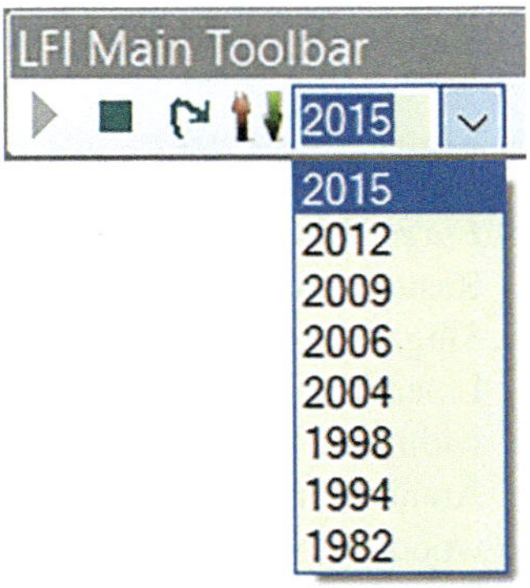

Fig. 5.3 Pull-down menu in the NFI stereo GIS application, showing the flight years of the available stereo images covering the sample plot of interest

5.3 Image-Data Management

The image data is located on a file server, and metadata for the images is stored in the database. Hence, it is possible to select suitable images for the current sample plot using a spatial query. As a result, the application creates the most nadir colour infrared (CIR) stereo-image pair from the most recent year and automatically loads it into the stereo window. The image stereo pair can be changed before any data is collected. For image-strip data (Leica ADS), changing between RGB and CIR images is allowed during interpretation (button). In order to guarantee traceability of the interpretation, the selection of different image pairs or different flight years is only allowed before interpretation starts (Fig. 5.3).

5.4 Terrain Data, Topographic Maps and Feature Data Management

All object heights that are measured or calculated during interpretation are based on the most recently available digital terrain model (DTM). For Switzerland, a high resolution DTM with a ground resolution of 2 × 2 m, based on LiDAR acquisition and photogrammetry, is available (Artuso et al. 2003). A 1:25’000 topographic map is additionally loaded into the ArcMap window.

Feature collection is completed using special NFI editor tools that use the ArcGIS Editor and a particularly configured File Geodatabase (FileGDB). The FileGDB considers a feature class for each single feature type and its corresponding symbology, for example *forest boundary lines* and raster points. It additionally considers two relationship classes for woody species and a topology class. When a sample plot is loaded into the application, this FileGDB is automatically generated and loaded into the ArcGIS interface.

5.5 Data Collection

The following variables are collected during stereo-image interpretation:

- *Forest boundary line* (used for 'forest/non-forest' decision)
- Raster points (used for 'forest/non-forest' decision)
- Single formations of woody species outside forest area: type, height and radius
- Linear formations of woody species outside forest area: type, height, length and width
- Areal formations of woody species outside forest area: measured like linear woody species formations in conformity with the largest width
- Reference points: for each sample plot classified as 'forest', three reference points are identified and measured. These reference points are ideally within a radius of 200 m from the sample-plot centre.

To collect the features listed above, the stereo interpreters use the Intergraph Topomouse (Fig. 5.4). The most frequent actions the interpreters perform are mapped to the mouse buttons; for example, features such as the *forest edge* are digitised using the command AddFeatureVertex, and raster points are identified using the command MoveToXY (Raster). The z-coordinate of the 3D cursor is controlled by the Topomouse wheel.

5.6 Data Checking

The interpreted data can be checked at any time during interpretation. The option to check the data is offered after the raster point measurements have been completed. Data checking includes:

- Features in the FileGDB are checked for correct topology; for example, *forest boundary lines* must not cross themselves and the centre of a single woody element must be inside the sample-plot area.
- An error in the sample-plot centre (SPC) assignment appears when the SPC is not set to be inside or outside the forest boundary.
- An error appears for a raster point if its measurement and interpretation are missing.
- Suggestions are provided for discrepancies on raster-point attributes and neighbouring feature elements (e.g. *forest boundary line* or linear woody species formations).

Errors and suggestions are displayed in red in the NFI Data Errors and Suggestions dialogue. If errors are included in the current sample-plot data set, the plot cannot be finished and saved in the database until they have been fixed.

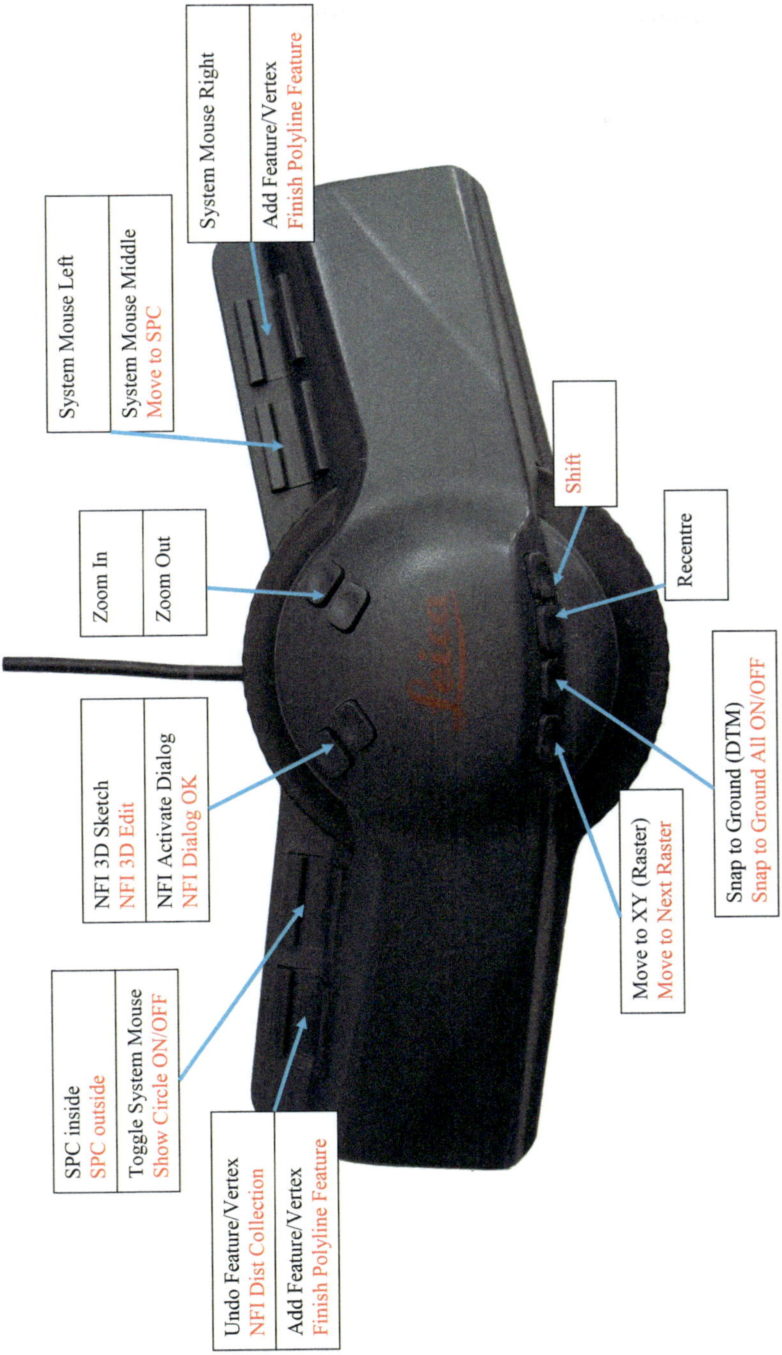

Fig. 5.4 The Intergraph Topomouse is the main interface used for data collection. The main interpretation tasks are mapped to buttons on the Topomouse (in red: option together with the shift button)

References

Artuso R, Bovet S, Streilein A (2003) Practical methods for the verification of countrywide terrain and surface models. In: Proceedings of the ISPRS Working Group III/3 workshop XXXIV–3/W13. 3-D reconstruction from airborne laserscanner and InSAR data, Dresden, Germany

ESRI (2018) ArcGIS desktop: release 10. Environmental Systems Research Institute, Redlands

HEXAGON (2018) Stereo analyst for ArcGIS: release 10. Hexagon Geospatial, Madison

Chapter 6
Variables on the Sample Plot Captured by the Stereo GIS Application

Christian Ginzler

Abstract Guided by the Stereo GIS Application, different measurements are carried out in the aerial stereo-images in order to determine a 'forest/non-forest' decision. The measurements try to make the forest decision as reproducible as possible. In addition to these measurements, expert interpretations are also carried out.

6.1 Forest Boundary Line

The forest boundary line (FBL) determines the border between stocked and unstocked areas. It makes it possible to measure or calculate the forest width and interspace. These values are needed to make a 'forest/non-forest' decision for each sample plot (Ginzler et al. 2005).

The FBL is the longest possible line connecting all stocking elements which form the forest edge and/or which stand outside the forest edge but are still close to the stocked area (horizontal distance between tree tops $\leq$25 m). On the aerial stereo-images, the FBL is measured by connecting crown top centres.

Stocking elements classified as belonging to the forest are woody species that are interpreted as being used for forestry purposes. The surroundings are taken into account in deciding whether a woody species is a stocking element. The minimum height of a stocking element is 3 m, with the exception of shrubs. Shrubs in mountainous areas have no minimum height.

The FBL may cross the following landscape elements:

- Bridges, if there is forest underneath them
- Buildings and installations belowground or at ground level (e.g. water reservoirs, low walls, avalanche-control structures and bunkers)
- Roads $\leq$6 m wide
- River beds $\leq$6 m wide

C. Ginzler (✉)
Remote Sensing, Swiss Federal Research Institute WSL, Birmensdorf, Switzerland
e-mail: christian.ginzler@wsl.ch

C. Fischer, B. Traub (eds.), *Swiss National Forest Inventory – Methods and Models of the Fourth Assessment*, Managing Forest Ecosystems 35,
https://doi.org/10.1007/978-3-030-19293-8_6

- Roadside ditches
- Passing places, turning places and wide bends on roads
- Places used for timber storage
- Recreational facilities (e.g. forest cabins, picnic areas, parking places, camping sites)
- Seedling nurseries (i.e. forest plantations)
- Landslides, avalanches and timber corridors where logs have slid down the mountainside
- Meadows, pastures and fields
- Other clearings (e.g. damp sites, screes and cliffs)
- Cut areas, such as areas affected by erosion, fire, windfall or avalanches, as well as afforestation and regeneration sites
- Park forests with forest use

The FBL **may not** cross the following landscape elements:

- Roads >6 m wide
- River beds >6 m wide
- Water bodies such as ponds or lakes >6 m wide
- Railways (even when covered), funicular railways, industrial railways and similar elements, as well as ski-lift tracks
- Buildings and installations standing above ground level that are not related to forest use
- Combinations of streams and roads if the stream bed (erosion width) runs parallel to the road and the combined width is >6 m

The FBL is digitised as a line in the aerial stereo-image (Fig. 6.1). Convex forest boundary lines are digitised exactly by connecting neighbouring stocking elements, meaning that no stocking elements may be excluded. Concave forest boundary lines connect stocking elements that are as far apart as possible, but not more than a 25 m horizontal distance apart. The non-forest area may be reduced by excluding stocking elements.

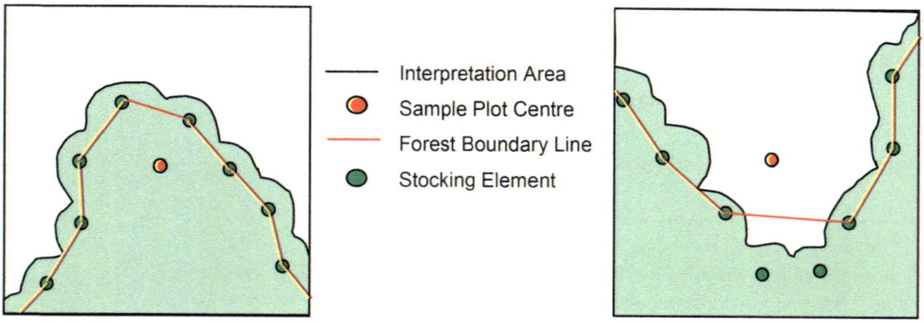

Fig. 6.1 Digitised representation of a forest boundary line (FBL) on a convex (left) and a concave (right) forest edge

Fig. 6.2 Examples of an incorrect and a correct forest boundary line (FBL) for a forest with adjacent stocking elements (fruit trees) located outside the forest area. The fruit trees have been incorrectly included in the FBL in the image on the left. In the image on the right, the fruit trees have been correctly excluded from the FBL, whereas the clearing situated South of the sample-plot centre (SPC) has been correctly included in the forest area, in conformity with the rules for concave FBL delineation

6.1.1 Adjacent Stocking Elements Close to a Forest

Adjacent stocking elements located outside the forest area are only enclosed in the FBL if the elements are used for forest purposes and located within a 25 m radius of the forest area. Other kinds of stocking elements, such as fruit trees, are not included within the FBL (Fig. 6.2).

Adjacent stocking elements outside the closed forest area are also included in the FBL if the rest of the forest edge is outside the sample plot as long as the stocking element itself is located within a 25 m radius of the forest area.

6.1.2 Special Cases

The following areas are treated differently from other elements: cutting areas, afforestation areas, regeneration areas, burned areas (also for damaged areas in general, e.g. due to rockfall, avalanche, debris flow, flood or slump), and areas devastated by storms (Fig. 6.3).

These areas are considered to be temporarily unstocked, meaning that they belong to the forest area even though they do not contain stocking elements. If an FBL is delineated, such areas are located within the FBL. Instead of crown centres, stumps are used to digitise the FBL (Fig. 6.4). Standing deadwood is considered a stocking element even if the tree species can no longer be identified. Fallen trees and trees

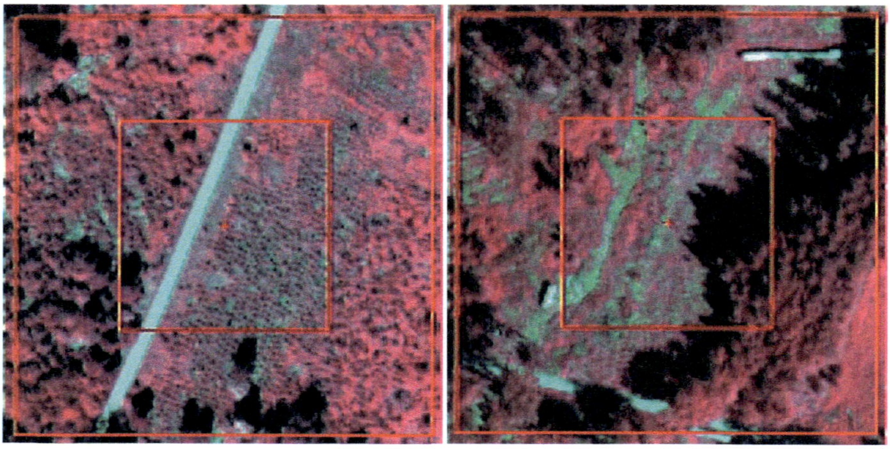

Fig. 6.3 Special cases for the delineation of the forest boundary line. Left: afforestation. Right: cut area

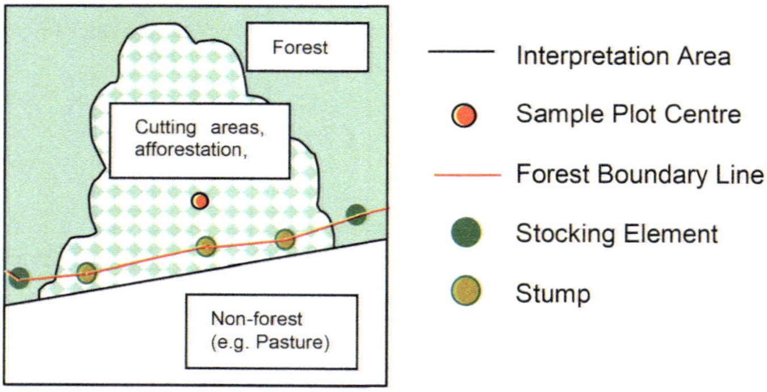

Fig. 6.4 Special cases: FBL delineation on cutting, afforestation, regeneration, burned and storm areas located next to 'non-forest'

with no forest purpose, such as fruit trees, park trees and Christmas trees growing in agricultural areas, are not considered stocking elements.

If no stumps or standing deadwood are present, for example in burned, reforestation and regeneration areas, the FBL is delineated according to the interpreter's expert knowledge.

Delineation of the FBL in these special cases follows the rules for concave and convex forms.

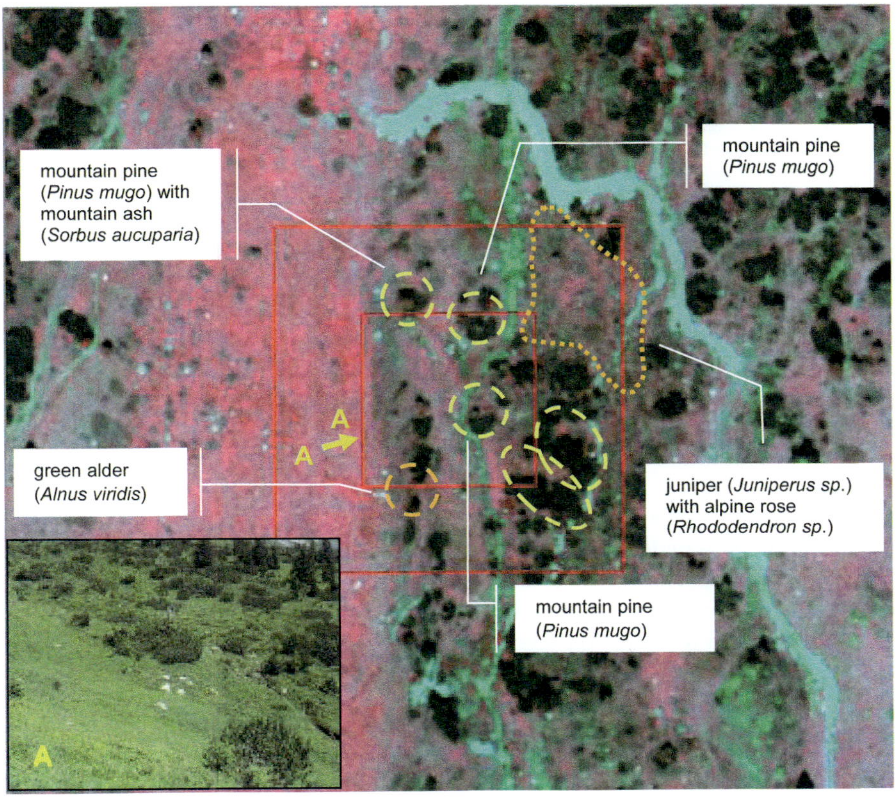

Fig. 6.5 Sample plot with a shrub-forest appearance, featuring mountain pine as the dominant species

6.1.3 Forest Boundaries with Shrub Vegetation

Shrubs (bushes) are woody plants that reach a height of 0.5–5.0 m when mature. They are branched at the base and have a shrubby habit or appearance.

A forest is classified as 'shrub forest' if at least two-thirds of the plot area is covered by wooded elements with shrub morphology. Green alder (*Alnus viridis*) and mountain pine (*Pinus mugo*) are the most common shrub species in Switzerland, while coppice forests of hazel or similar stockings are found less frequently (Brändli 2010). Shrub forests are mostly found in the Alps. In sample plots that look shrubby, the FBL can be digitised using stocking elements <3 m in height (Fig. 6.5).

Fig. 6.6 Examples of how forest width and forest interspace are calculated

6.1.4 Location of Sample-Plot Centre Relative to the Forest Boundary Line

After delineation of the FBL, the interpreter decides whether the sample-plot centre (SPC) is positioned on the forest side (inside) or on the non-forest side (outside) of the FBL. If the SPC is located inside the FBL, the software calculates the forest width, defined as the shortest distance from FBL to FBL through the SPC (Fig. 6.6 left). If the sample-plot centre is located outside the FBL, the software calculates the forest interspace, which is also defined as the shortest distance from FBL to FBL through the SPC but traverses the non-forest rather than the forest area (Fig. 6.6 right). The forest width and the forest interspace are needed for the 'forest/non-forest' decision.

If the SPC is located outside the FBL and the interspace is $\leq$25 m, then the forest width must be determined as well. In this case, the width is also the shortest distance from FBL to FBL, with its extension going through the SPC (Fig. 6.7). This can be calculated in the application, but must be done manually.

6.2 Raster Points

Crown coverage (CC) is an attribute needed for the 'forest/non-forest' decision. To estimate the crown coverage, 25 raster points are laid over the interpretation area (50 × 50 m). The type of land cover and the elevation are interpreted and measured on a regular grid with 5 × 5 points each 10 m apart (Fig. 6.8). The raster points are classified into 11 different land-cover classes.

6.2.1 Height of Objects at Raster Points

The GIS application automatically moves the 3D cursor to each of the raster points, which are located 25 m above the surface of the digital terrain model (DTM). The interpreter places the 3D cursor on the surface of the object below this position. The

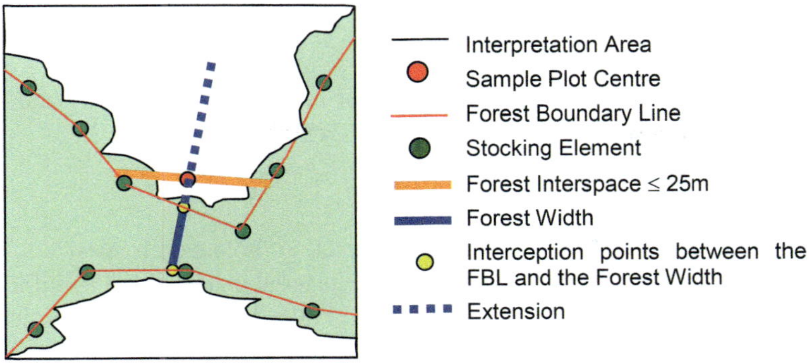

Fig. 6.7 Example of how a forest interspace ≤25 m and the corresponding forest width are determined. The sample-plot centre (SPC) is located outside the FBL, so the forest interspace is calculated by the stereo GIS application. Because the forest interspace is ≤25 m, the interpreter must manually measure the shortest forest width from FBL to FBL with its extension going through the SPC

Fig. 6.8 Raster points laid over the interpretation area to assess land cover

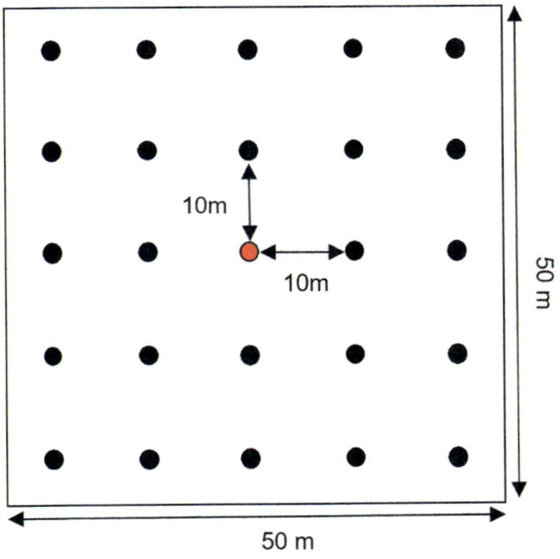

cursor must be placed on the surface of the first object touched from above, for example on a tree branch over a road.

The object height (m) at each raster point is calculated by the application as the difference between the elevation of the 3D cursor (m a.s.l.) and the elevation of the DTM.

6.2.2 Land Cover

Eleven land cover classes are distinguished for each raster point. The land cover classes are as follows:

- **Broadleaved tree** (tree with broad leaves)
- **Coniferous tree** (all conifers except larch)
- **Larch** (deciduous conifer (*Larix* sp.), important in protection forests)
- **Shrub** (woody plants with a shrubby habit). In aerial images it can be difficult to distinguish between shrubs and small trees. For individual woody species, the decision is based on the habit. In cases of uncertainty, the object is interpreted as a tree at low elevations and as a shrub at higher elevations
- **Herbaceous vegetation** (forbs and grasses). Open arable farmland is classified as herbaceous vegetation, which means that the season when the aerial photo was taken is not important
- **Rock** (rock outcrop with a minimum surface area of 625 m^2, equivalent to one quarter of the 50 × 50 m interpretation area)
- **Sand, stone, bare soil** (loose stones covering a maximum area of 625 m^2, bare soil and bare sand)
- **Paved surface** (artificial objects with zero height). These completely or partially sealed built surfaces include asphalt, concrete, cobblestones and surfaces covered with stones or slabs
- **Constructed object/building** (object with height >0 m that is permanently connected to the ground)
- **Water body** (open bodies of still or running water)
- **Glacier, firn** (surfaces with a minimum area of 2500 m^2 (equivalent to the interpretation area) that are predominantly covered with ice and permanent snow)

6.2.3 Level of Certainty

Every raster point must be attributed to a land-cover class. If the type of land cover is not clear because, for example, shadows are present or the elevation cannot be measured, then the attribute code 'unsure' is assigned in addition to the land-cover class and the best-guess elevation measurement. In these cases, the raster point is attributed to a land-cover class based on the most probable class according to the surrounding area.

6.2.4 Position Relative to Forest Boundary Line

After the position of the sample-plot centre is determined in relation to a forest boundary line, the position (inside or outside the forest) of every raster point is calculated.

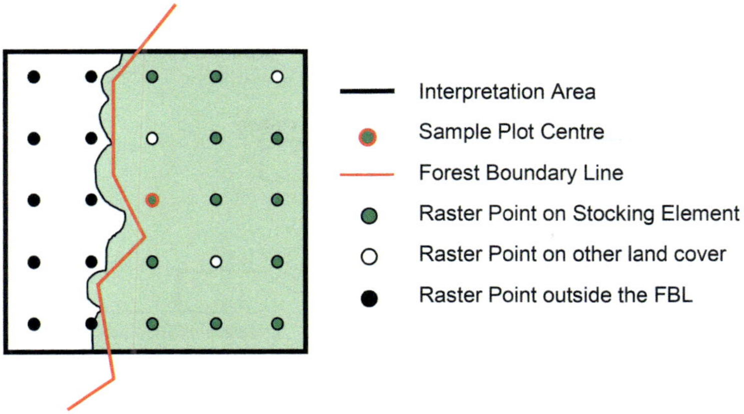

Fig. 6.9 Example of the calculation of crown coverage: 10 raster points are located outside the FBL and 15 raster points are located inside. Of these 15 raster points, 12 are positioned on stocking elements with an object height ≥3 m (shown in green). Therefore, the crown coverage calculated for the sample plot is 80%

6.2.5 Crown Coverage

The crown coverage (CC) of a sample plot is defined as the number of raster points located on the forested side of the FBL that are classified as broadleaved tree, coniferous tree, larch or shrub and have a height ≥3 m (<3 m is acceptable for shrubs) in relation to the total number of raster points on the forest side of the FBL. Raster points on the non-forest side of the FBL are not considered in the calculation of crown coverage (Fig. 6.9).

6.3 'Forest/Non–forest' Decision

In order to decide whether a sample plot, in particular its sample-plot centre (SPC), is 'forest' or 'non-forest', the following decision diagram (Fig. 6.10) is applied, according to NFI criteria:

The following conditions must be fulfilled for a sample plot to be classified as 'forest':

- Crown coverage must be at least 20% in the part of the interpretation area defined by the FBL. Exceptions are afforestation, regeneration, cutting, burned and storm-affected areas.
- The shortest distance from FBL to FBL through the SPC must be ≥25 m. This minimum forest width depends on crown coverage (Fig. 6.11).

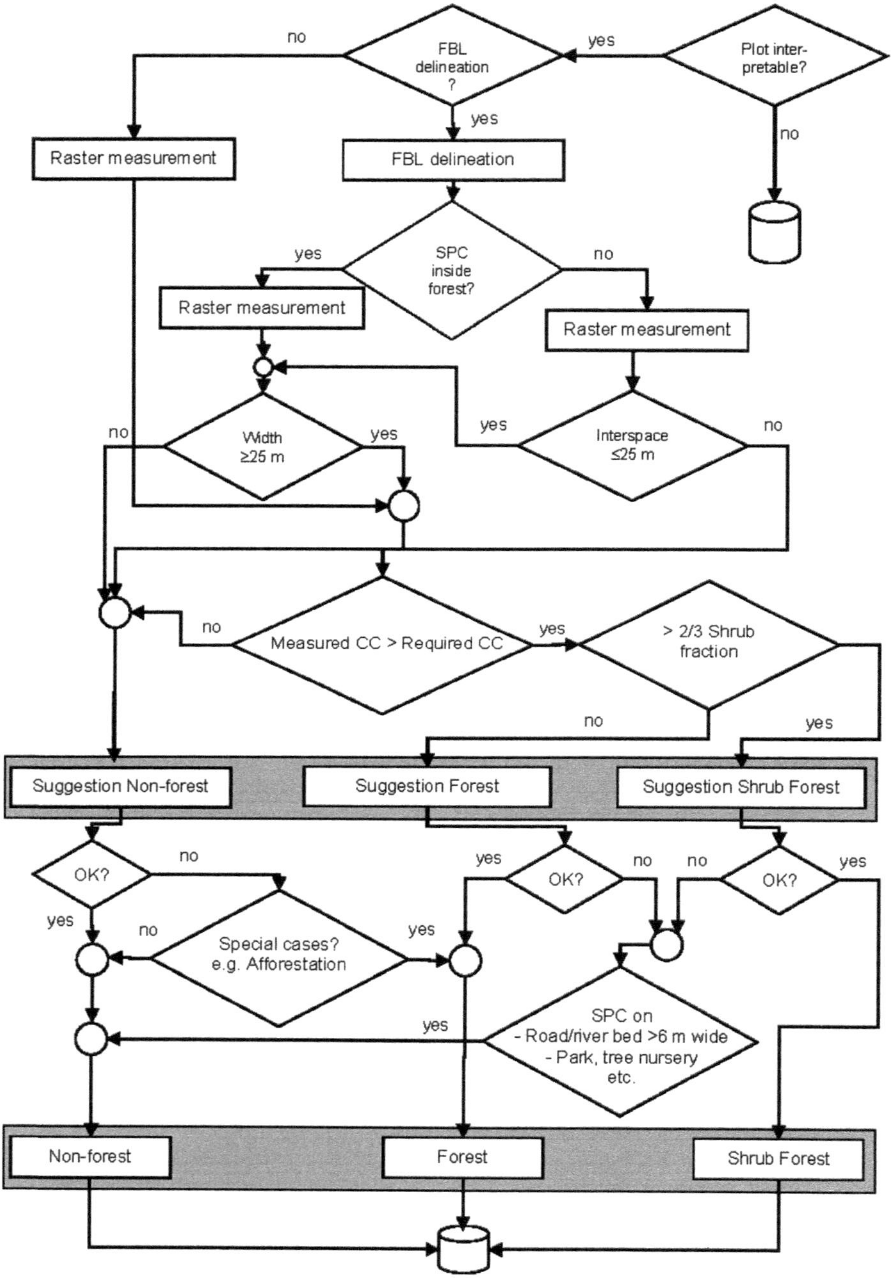

Fig. 6.10 'Forest/non-forest', decision diagram

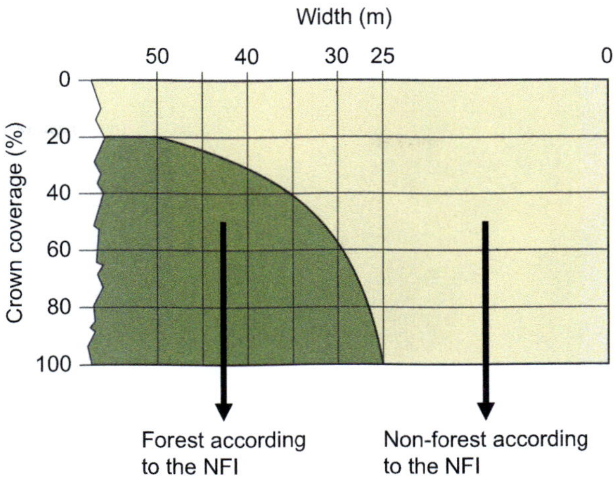

Fig. 6.11 Degree of coverage/width criteria for the 'forest/non-forest' decision. The smaller the degree of coverage, the wider the forest has to be for the sample plot to be classified as forest

- The shortest distance from FBL to FBL through the sample-plot centre in the non-stocked part of the interpretation area (interspace) must be <25 m.
- The land use of the potential forest plot is forest use. Sample plots in areas such as urban parks, orchards and cemeteries are not considered forest plots.
- The SPC must not be located on the following land-use classes even if the land-use class is surrounded by forest:
  - Road >6 m wide
  - River bed >6 m wide
  - Railway track
  - Ski-lift track
  - Garden, including in settlement areas
  - Tree nursery
  - Park (Fig. 6.12)

Examples of forest/non-forest decisions are provided in Fig. 6.13.

In the example on the left in Fig. 6.13

1. SPC is not located on an excluded land-use class
2. SPC is located on the stocked side of the FBL
3. Degree of coverage criterion: 9 out of 12 raster points are positioned on a stocking element inside the FBL; the degree of coverage is 75%, and the forest width must therefore be $\geq$27 m. If the width is smaller, => non-forest decision. If the width is larger, => forest decision.

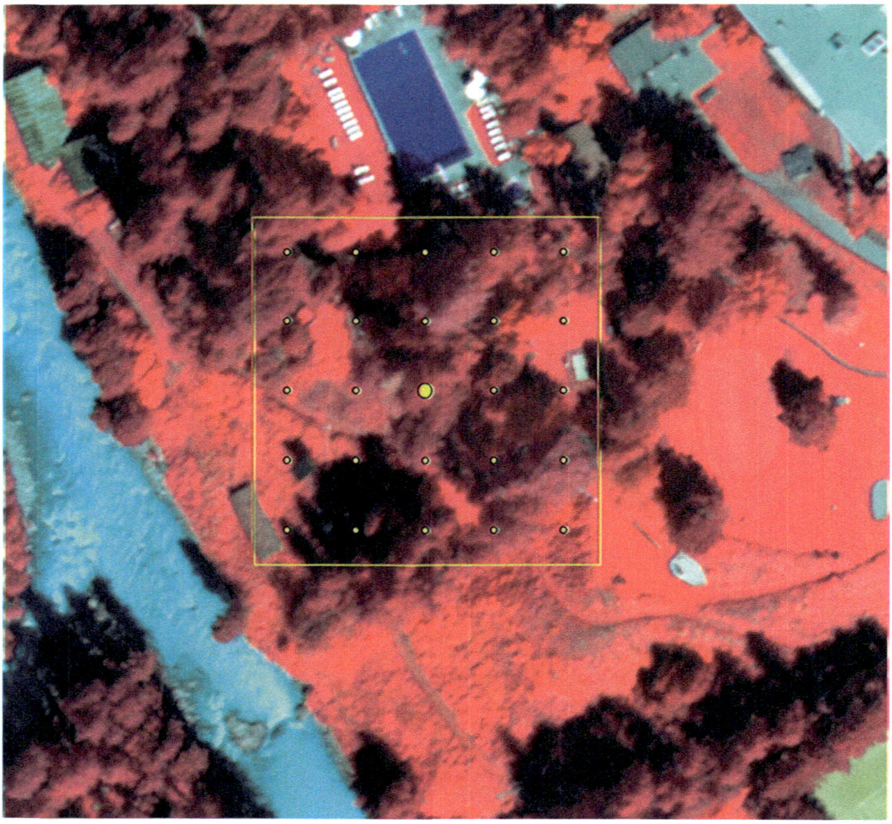

Fig. 6.12 Example of a sample-plot centre in a park (large yellow dot). This sample plot would not be classified as forest because the land use is not forest

In the example on the right in Fig. 6.13

1. SPC is not located on an excluded land-use class
2. SPC is located outside the forest
3. Degree of coverage criterion: 13 out of 16 raster points are on a stocking element inside the FBL, the degree of coverage is 81%, and the width must therefore be ≥26.3 m. If the width is smaller, => forest decision. If the width is larger, => non-forest decision.

- Land use

  As there is no global agreement on the definition and classification of land use, classification systems vary among and within countries and are not always compatible (Tomppo et al. 2010).

  The attribute 'land use' is independent of the NFI definition of forest in that forest width, forest interspace and crown coverage are irrelevant, and it is partly independent of the 'forest/non-forest' decision. This attribute is assessed only

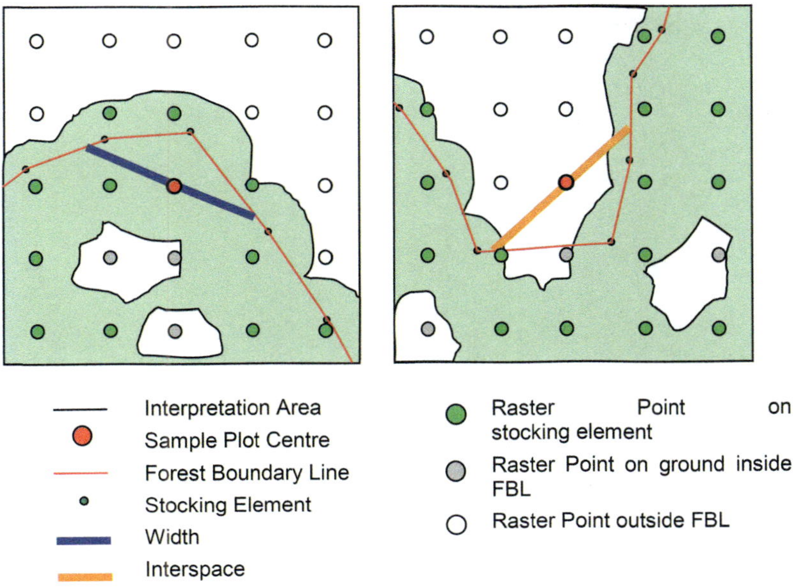

Fig. 6.13 Examples of forest/non-forest decisions. Forest width is calculated as the shortest distance from FBL to FBL through the SPC when the SPC is located on the stocked side of the FBL (left). Interspace is calculated as the shortest distance from FBL to FBL through the SPC when the SPC is located outside the forest (right)

through aerial-photo interpretation. Land use is classified as either 'potential forest use' or 'other use':

– **Potential forest use**: sample plots that have at least one stocking element and can be used for forestry. These areas are considered to be currently unstocked. They are mostly plots located in agricultural areas or pastures that potentially experience ingrowth when abandoned.
– **Other use**: sample plots that cannot be used for forestry. These plots are located on open water or in high mountain regions or glaciers. This comprises all sample plots with a sample-plot centre located:

- On buildings
- On a road >6 m wide
- On a river bed >6 m wide
- On a railway track
- On a ski-lift track
- In a garden, including in settlement areas
- In a tree nursery
- In a park

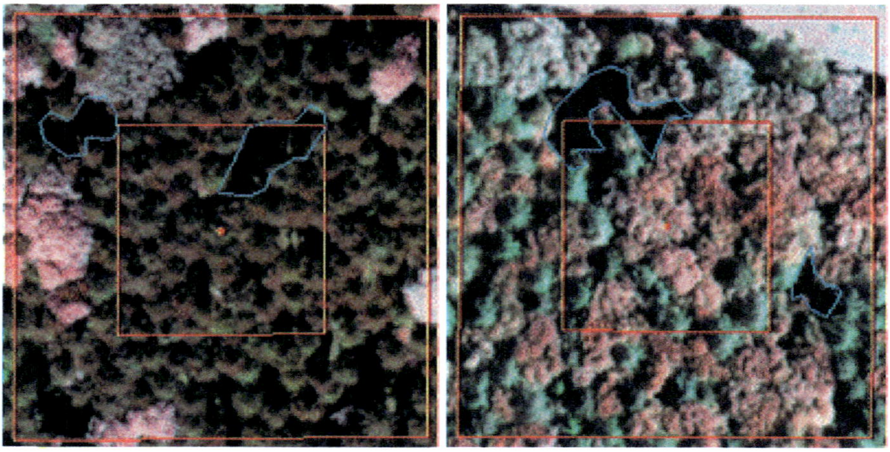

Fig. 6.14 Examples of sample plots with forest gaps (areas outlined in blue)

6.4 Forest Gaps

Gaps inside the forest that have an area of at least 10 × 10 m are mapped (Fig. 6.14). The land cover of the gap can be herbaceous vegetation, rock, sand, stone, bare soil, paved surface, building or water. Forest gaps are digitised along the throughfall of the trees. The crown coverage within a gap must not exceed 20%.

References

Brändli U-B (ed) (2010) Schweizerisches Landesforstinventar. Ergebnisse der dritten Erhebung 2004–2006. Eidgenössische Forschungsanstalt für Wald, Schnee und Landschaft WSL, Birmensdorf. Bern, Bundesamt für Umwelt, BAFU. Switzerland

Ginzler C, Bärtschi H, Bedolla A, Brassel P, Hägeli M, Hauser M, Kamphues M, Laranjeiro L, Mathys L, Uebersax D, Weber E, Wicki P, Zulliger D (2005) Luftbildinterpretation LFI3. Interpretationsanleitung zum dritten Landesforstinventar. Eidg. Forschungsanstalt für Wald, Schnee und Landschaft WSL, Birmensdorf

Tomppo E, Gschwantner T, Lawrence M, McRoberts RE (eds) (2010) National Forest Inventories. Springer Netherlands, Dordrecht

Chapter 7
Area-Wide Products

Christian Ginzler, Bronwyn Price, Ruedi Bösch, Christoph Fischer, Martina L. Hobi, Achilleas Psomas, Natalia Rehush, Zuyuan Wang, and Lars T. Waser

Abstract Methods and workflows for creating area-wide products using remote sensing methods have been developed for various reasons. For the sample-based estimates in the Swiss National Forest Inventory (NFI), area-wide data sets are used for the two-phase estimations, which have substantially lower estimation errors than one-phase inventories. Using area-wide data sets, it is possible to skip dense manual interpretation of the stereo-images and thus save resources. Further, many functions of the forest, such as biodiversity and protection against natural hazards, can be described and quantified better with area-wide spatial data than with field plot data.

Introduction Methods and workflows for creating area-wide products using remote sensingmethods have been developed for various reasons. For the sample-based estimates in the Swiss NFI (NFI), area-wide data sets are used for the two-phase estimations, which have substantially lower estimation errors than one-phase inventories. Using area-wide data sets, it is possible to skip dense manual interpretation of the stereo-images and thus save resources. Further, many functions of the forest, such as biodiversityand protection against natural hazards, can be described and quantified better with area-wide spatial data than with field plot data. The most basic area-wide data set is the vegetation height model, which is calculated from the aerial stereo-image data. Based on the vegetation height model, the NFI forest cover map and the growing stock data set are produced and the first methodological developments for forest-cover changes are completed. The forest-cover type is based on the

C. Ginzler (✉) · B. Price · R. Bösch · A. Psomas · Z. Wang · L. T. Waser
Remote Sensing, GIS, Swiss Federal Research Institute WSL, Birmensdorf, Switzerland
e-mail: christian.ginzler@wsl.ch

C. Fischer · N. Rehush
Scientific Service NFI, Swiss Federal Research Institute WSL, Birmensdorf, Switzerland

M. L. Hobi
Stand Dynamics and Silviculture, Swiss Federal Research Institute WSL, Birmensdorf, Switzerland

C. Fischer, B. Traub (eds.), *Swiss National Forest Inventory – Methods and Models of the Fourth Assessment*, Managing Forest Ecosystems 35,
https://doi.org/10.1007/978-3-030-19293-8_7

same aerial images used for stereo-interpretation and for the vegetation height model. The individual area-wide data sets are presented below.

Objectives of the area-wide products:

- Provide spatial data sets on forest-related variables
- Generate area-wide data sets on vegetation height, independent of the land use
- Generate auxiliary data for two-phase sampling to stratify and reduce estimation errors in the terrestrial variables

7.1 Vegetation Height Model

Vegetation height models (VHMs) provide a three-dimensional view of the forest canopy. They can be used to analyse and assess various forest characteristics. These include: biomass or timber volume (Price et al. 2017) in forest management, stand structure (Zellweger et al. 2013, 2014) in a forest biodiversity context, and crown cover or canopy-gap density (Zielewska-Büttner et al. 2016) in protection forests. In the Swiss NFI (NFI), the VHM is used to derive a binary forest cover map for stratification. The VHM is then used to further stratify the forest in the forest cover map.

Remotely sensed information about canopy height can be gathered from laser scanning, aerial stereo-images, interferometric synthetic aperture radar (SAR) data or stereo satellite data (Gao 2007). Stereo data provides a three-dimensional picture of the forest and is used to model the forest canopy surface by calculating a digital surface model (DSM). In combination with a digital terrain model (DTM) derived, for example, from laser scanning, a VHM depicting the vegetation height above-ground is generated by subtracting the DTM from the DSM. In Switzerland, aerial stereo-image data is acquired during routine mapping cycles every 6 years under leaf-on conditions. As the ground model can be assumed to be more-or-less stable over time, an updated VHM of the whole of Switzerland is produced every 6 years.

The automated workflow that was used to generate the nationwide DSM with a spatial resolution of 1×1 m, based on photogrammetric image matching, is described below. Leaf-on aerial images, acquired in summer (May–September) by swisstopo with a Leica ADS80 digital camera, were used as input data for stereo matching. These images have a ground sample distance (GSD) of 0.25 m in the lowlands of Switzerland and 0.5 m in high mountain areas. Image matching for DSM generation was carried out with the software *SocetSet 5.6 Next-Generation Automatic Terrain Extraction* (NGATE) by BAE Systems. With NGATE, matching is performed for every pixel using area-matching and edge-matching algorithms (Zhang and Gruen 2006; DeVenecia et al. 2007). For the Swiss DSM, two different matching strategies were parameterised: (a) the *ADS steep* strategy, which is optimised for single features such as trees and for abrupt changes in elevation such as steep rocks, and (b) the *ADS flat* strategy, which is optimised for homogeneous areas, such as meadows and glaciers (Ginzler and Hobi 2015, 2016).

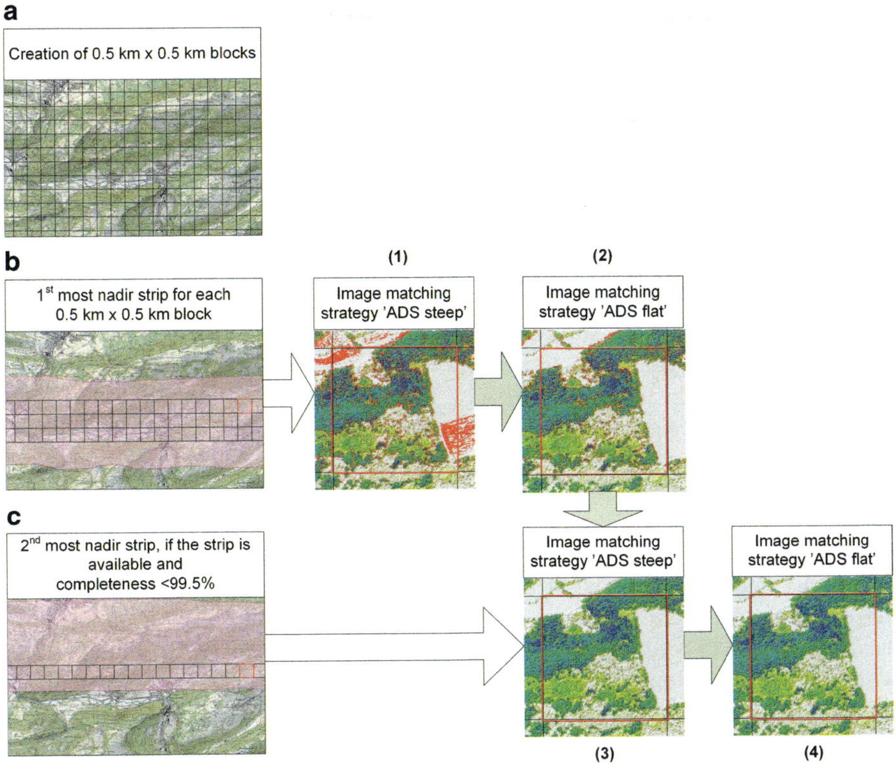

Fig. 7.1 Image-matching workflow used to create a digital surface model (DSM): (**a**) rasterise into 0.5 × 0.5 km blocks; (**b**) select the strip with the smallest distance from nadir to the centre of the block; and (**c**) select the strip with the second smallest distance from nadir to the centre of the block. (1) The result after the matching process for the block using the first most nadir stereo strip and the *ADS steep* image-matching strategy. Red areas in the image could not be matched. The completeness for this block is 84%. (2) The result after additionally applying the *ADS flat* matching strategy. At 91%, the completeness threshold is still not fulfilled. (3) The result after matching using the second-most nadir strip and the ADS steep strategy (completeness 96%), and (4) the final DSM after applying the ADS flat strategy, with a matching completeness of 96%. (Figure taken from Ginzler and Hobi 2015)

To distribute the calculations over a range of different processing machines, the country was divided into 165,500 blocks, each 0.5 × 0.5 km in area (Fig. 7.1). This block size was chosen to fit the 500 × 500 m sampling grid of the photo interpretation of NFI3. It proved to be a good size in terms of computing power. To avoid edge effects, a 100-m-wide buffer was added to all sides of the blocks, resulting in an area of 0.7 × 0.7 km for image matching. Up to two image strips were available for each block, owing to an overlap of about 50%. For the image matching, the image strip with the smallest distance from nadir to the centre of the block (first most nadir strip) was used first, followed by the strip with the second smallest distance from nadir to the centre of the block (second most nadir strip). The second most nadir strip was not always available and was only used in blocks where the completeness threshold of 99.5% was not fulfilled using the first most nadir strip. Image matching was done with only one image pair at any given time, which made it easier to control the number of matched pixels. The two colour-infrared (CIR) stereo-images (backward 16° and nadir 0°) were used in combination for matching over the whole of Switzerland because they were most successful, especially for forested and high-alpine glacier regions.

The percentage of successfully matched points (completeness) was 97.9%; 54.0% of the country was matched with a GSD of 0.25 m and 45.5% with a GSD of 0.5 m. Even when only one image strip and one strategy was used, a completeness of 94.9% was achieved. The additional processing of the first image strip with a second strategy led to a gain in completeness of 1.7%. If the second strip was used and both matching strategies applied, the gain was 1.3%. After the automatic matching, manual correction was performed for 0.1% of the matched blocks, in particular for those with cloud coverage because they were difficult to match.

To test the accuracy of the nationwide DSM of Switzerland, two reference data sets were used: (a) topographic survey points of the national triangulation network (N = 198); and (b) stereo measurements (N = 195,784) within the NFI framework, used to distinguish various land-cover types (Mathys et al. 2006). An overall median accuracy of 0.04 m with a normalised median absolute deviation (NMAD) of 0.32 m was found using the topographic survey points (Ginzler and Hobi 2015). The second reference data set, based on stereo measurements, enabled an agreement assessment for different land-cover types. Sealed surfaces had the lowest median (−0.04 m for a 0.25 m GSD and 0.05 m for a 0.5 m GSD) and NMAD values (0.52 m for a 0.25 m GSD and 0.77 m for a 0.5 m GSD) (Ginzler and Hobi 2015). Most of the other land-cover classes, such as shrub, herb, rock, visible earth, buildings and glaciers had NMAD values <1.5 m. The three forest categories coniferous tree, broadleaved tree and larch were the most difficult to model, as indicated by NMAD values between 1.76 and 3.94 m.

To test the performance of the VHM for forest applications and assess its potential use in NFI, the VHM values were compared with field measurements of trees (Ginzler and Hobi 2015, 2016). The tree heights of single trees were measured during NFI4 with a Vertex ultrasonic hypsometer. A total of 3109 trees were measured between 2009 and 2013 and compared to the maximum height values of the VHM within a 2.5 m radius around each stem. The resulting median value for broadleaved trees at a GSD of 0.25 m was −0.56 m with a NMAD of 3.34 m, and for coniferous trees −1.86 m with a NMAD of 2.65 (Ginzler and Hobi 2015). In the mountains, where the GSD was 0.5 m, broadleaved trees still had a lower median value than conifers, but the NMAD was lower.

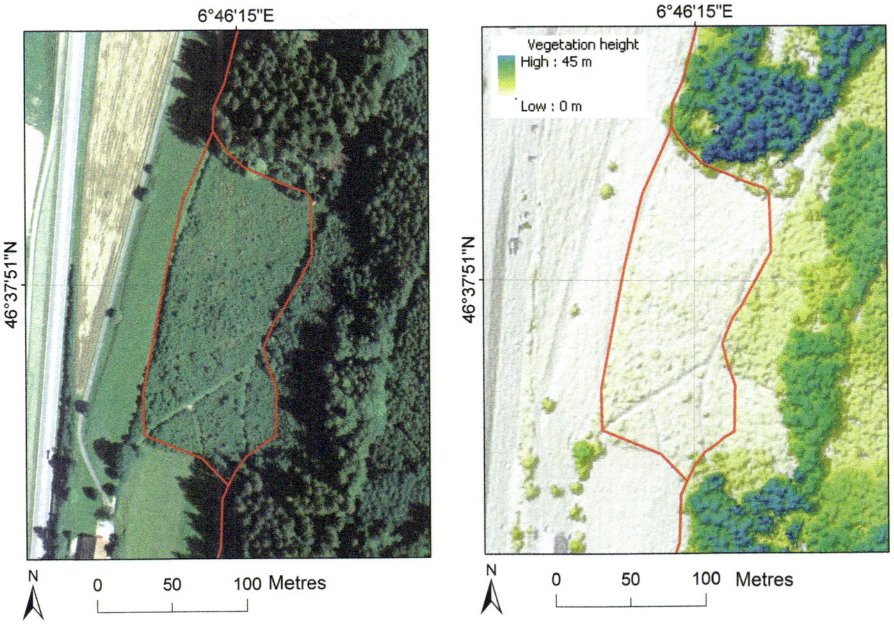

Fig. 7.2 Example of a temporarily unstocked forest area, which remains identified as forest according to the NFI forest definition. Left: true-colour ortho-image with the temporarily unstocked forest in the centre (red polygon). Right: corresponding VHM model with no trees in this area. The forest information for this area was taken from swisstopo's Topographic Landscape Model (TLM) and the figure from Waser et al. (2015)

7.2 Forest-Cover Map

According to Tomppo et al. (2010) and Barrett et al. (2016), forest-cover mapping is an important source of information for assessing woodland resources and a key issue for any national forest inventory. One particular challenge in mapping forest cover is the fact that the term *forest* is frequently defined similarly to *woodland* by the remote sensing community, and thus a simplified and generalised definition of forest is applied. In contrast, a detailed wall-to-wall forest-cover map was generated for the entire area of Switzerland, applying the more specific definition of forest required for the NFI. Thus, the focus was on implementing the four criteria of minimum height, minimum crown coverage, minimum width and land use in a comprehensible, geometrically clearly defined way, which also facilitates the production of a nation-wide comparable forest cover map. Land-use information is required to identify temporarily unstocked areas, which remain identified as forest according to the NFI forest definition (Fig. 7.2). As the land-use criterion could not be obtained from the VHM or from the aerial images, it was implemented using cover information from the Topographic Landscape Model (TLM) of Swisstopo (2017).

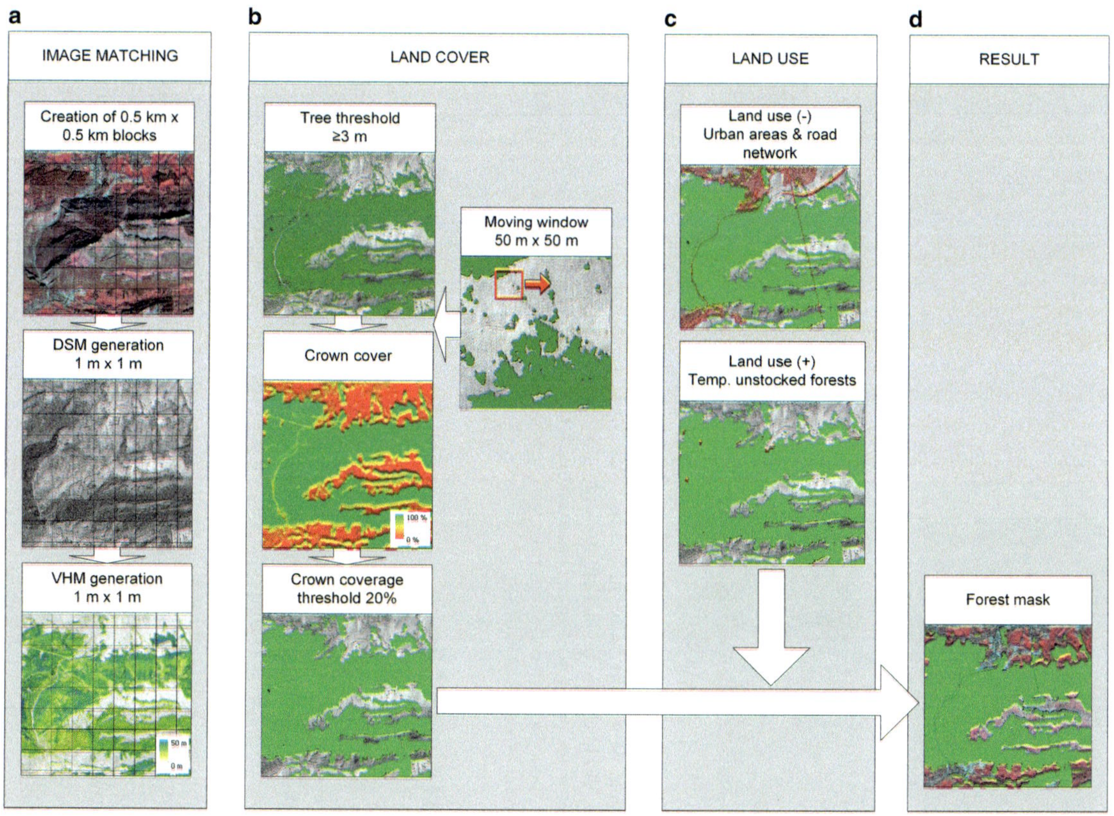

Fig. 7.3 Workflow of forest-cover mapping showing the main steps: (**a**) computation of the VHM by image matching; (**b**) creation of a preliminary forest-cover map based on land cover and the thresholds and the criteria for the NFI forest definition (tree height ≥3m, crown coverage ≥20% and width ≥25 m); (**c**) application of the land-use criterion to remove forest on other land, such as orchards and urban parks, and to add temporarily unstocked forests, such as windthrow and harvesting, using the TLM; and (**d**) calculation of a final forest-cover map that implements all criteria of the forest definition, as used in the stereo-image-based interpretation of the NFI. (Figure taken from Waser et al. 2015)

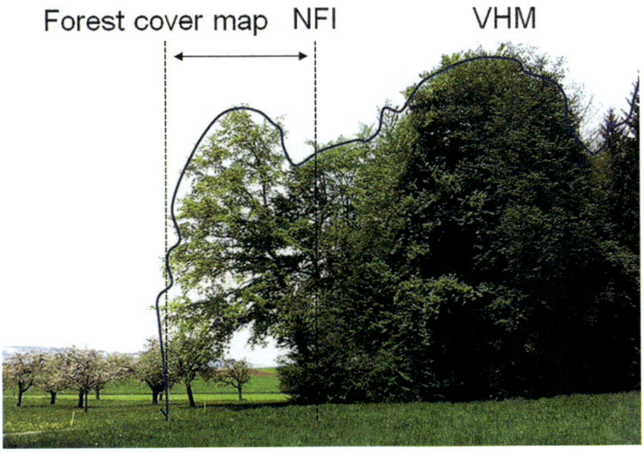

Fig. 7.4 Overall accuracy at the forest edge is affected by the different forest edge definitions applied by the forest-cover map, which is based on the VHM (tree crown determines edge), and the NFI (tree bottom determines edge). (Figure taken from Waser et al. 2015)

The workflow is highly automated and based on a moving-window approach using the VHM. A detailed description of this approach has been published by Waser et al. (2015), and only the main steps are given here. Figure 7.3 gives an overview of the input data sets and the main steps in the workflow.

The forest-cover map was validated with 9984 NFI plots ('forest/non-forest' decision) and had an overall accuracy (OA) of 0.97, with a forest overestimation of 10.1% and an underestimation of 3.8%. Sources of error in the forest-cover map were mainly related to the images used for the VHM, the simple processing of the forest mapping, in particular regarding the complexity of forest borders, and inconsistencies in the forest definitions applied. The coarse image resolution of 50 cm used to generate the VHM in areas above 1500 m a.s.l. reduces the quality of the forest-cover map in these areas and results in generally lower accuracies for mountainous regions. This is in contrast to the Plateau and Jura regions, where the forest-cover map is solely based on images with a 25 cm resolution. Moreover, although the quality of the aerial images is high and allows for robust image matching, the resulting VHM is more error prone at forest borders. Additionally, the presence of oblique off-nadir objects, resulting from the central perspective of the images, means that hidden areas cannot be matched and are thus interpolated using an algorithm (Ginzler and Hobi 2015). Another source of error is related to the forest edge: although the forest definition implemented in this approach is considered a good approximation, it is still not entirely congruent with the NFI forest definition obtained from terrestrial surveys. For example, the forest edge is defined by the position of the bottom of trees in NFI but by the position of crown (canopy cover) in the VHM. Figure 7.4 illustrates that the largest differences between the forest-cover map and the terrestrial NFI plots occur at forest borders. These differences are larger for broadleaved trees.

The forest-cover map is valuable because it guarantees replicable and objective results. For example, the nationwide wall-to-wall forest- cover map will enable densification of the existing 1.4 km NFI terrestrial sample-plot grid for smaller areas of interest. The forest-cover map may also enable further optimisation of the planning and implementation of cost-effective terrestrial surveys. The 10% commission error rate of the forest-cover map means that solely relying on it would result in unnecessary non-forest plot visits. Nonetheless, distinctly non-forest plots on the forest-cover map can be excluded during the fieldwork planning process. The accuracy of the estimates of the terrestrial variables is improved by using information from the forest-cover map when applying model-assisted estimators.

7.3 Tree Type Map

Precise and regularly updated information on the state, change and distribution of tree types is essential for many applications, not only those related to forestry. Tree type maps are also an important source of information for assessing woodland resources and are a valuable output of the NFI.

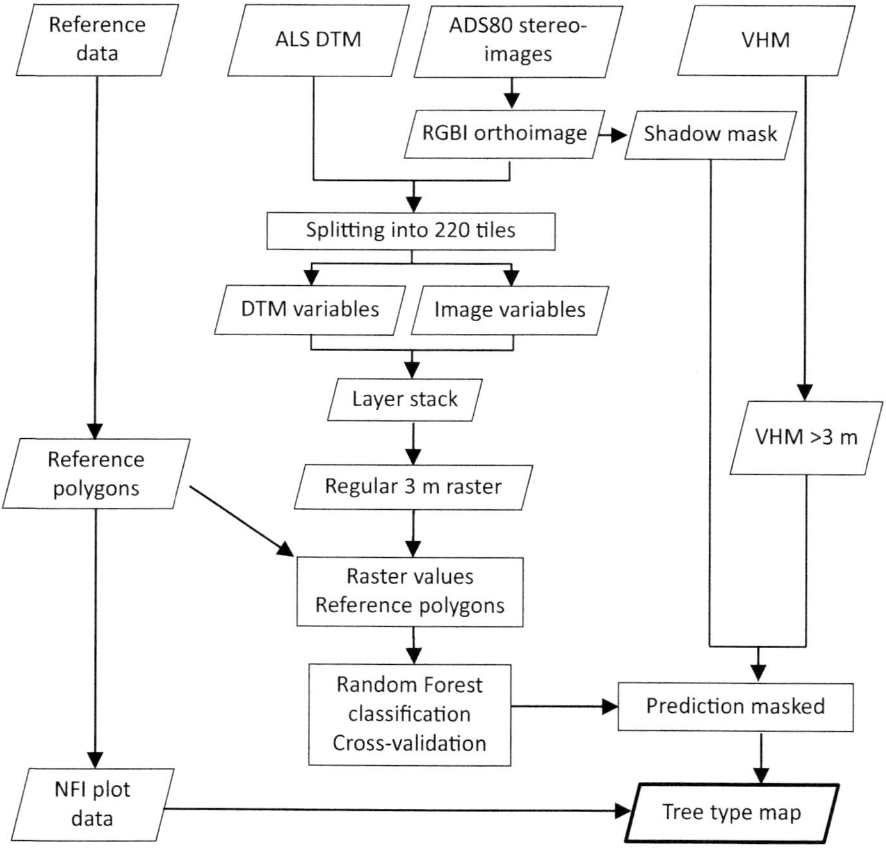

Fig. 7.5 Methodological workflow of the tree-type mapping approach

Waser et al. (2017) developed a novel and highly automated approach for generating an area-wide tree type map for the whole of Switzerland that consists of the two tree classes, broadleaved and coniferous, with a resolution of 3 × 3 m. The workflow of this approach is given in Fig. 7.5 and includes data pre-processing and preparation of the classification, computation of variables, classification and prediction of the broadleaved and coniferous trees, and finally statistical analysis and accuracy assessment. The entire classification approach is implemented in the statistics software R 3.3.2 (R Core team 2017).

The input data sets used in this approach were: airborne digital sensor (ADS) images, the digital terrain model (DTM) from airborne laser scanning (ALS) surveys made between 2001 and 2015, and the VHM (Ginzler and Hobi 2015). The ADS images were acquired from 2010 to 2015 during the leaf-on period, from the beginning of June to the end of August. The collection of reference data for the two tree types, broadleaved and coniferous, was based on delineations of the ADS ortho-image mosaic, independent of the regular stereo-image interpretation of the sample plots. In total, 185,240 polygons were digitised by experienced interpreters using standard GIS software. The stereo-image-based interpretation of the sample plots, with land-cover classes broadleaved, coniferous and larch, was used as an independent data set to validate the tree type map. Preparation of the remote sensing data involved orthorectification of the ADS image to obtain a layer stack of the four spectral bands and derived variables at a 1 m resolution, and generation of a shadow mask using the intensity values from the image bands.

Broadleaved and coniferous trees were mapped on the country level, using supervised random forest classification (RF), a widely used ensemble classifier that produces multiple decision trees using a randomly selected subset of training samples and variables (Breiman 2001). To guarantee high computing performance, the whole of Switzerland was divided into 220 equally sized tiles, each of which corresponds to one sheet of the national topographic map of Switzerland (scale 1:25'000). Explanatory variables were derived from the ADS ortho-images and from the DTM data, and consisted mainly of original image bands, vegetation indices, elevation and aspect. These explanatory variables were stacked and, using this 1-m-resolution layered stack, new variables (e.g. mean and standard deviation) were calculated at a 3-m resolution. The resulting 3-m-resolution raster cells were labelled using the reference data and exploited for training and validation. Means and standard deviations extracted from the raster cells were used for the random forest (RF) classification. The classifications of the 220 tiles were each tenfold cross-validated five times. The final tree type map was the product of the single predictions of the 220 tiles merged into one data set. All pixels of the VHM lower than 3 m were then masked out (Fig. 7.6).

The accuracy assessment of the tree type map based on the delineated training polygons revealed high overall model accuracies in the range of 0.95–0.99. Comparison of the tree type map with the NFI stereo-image-based interpretation of broadleaved and coniferous trees showed that the predicted values were generally positively biased towards coniferous and negatively biased towards broadleaved trees (Table 7.1).

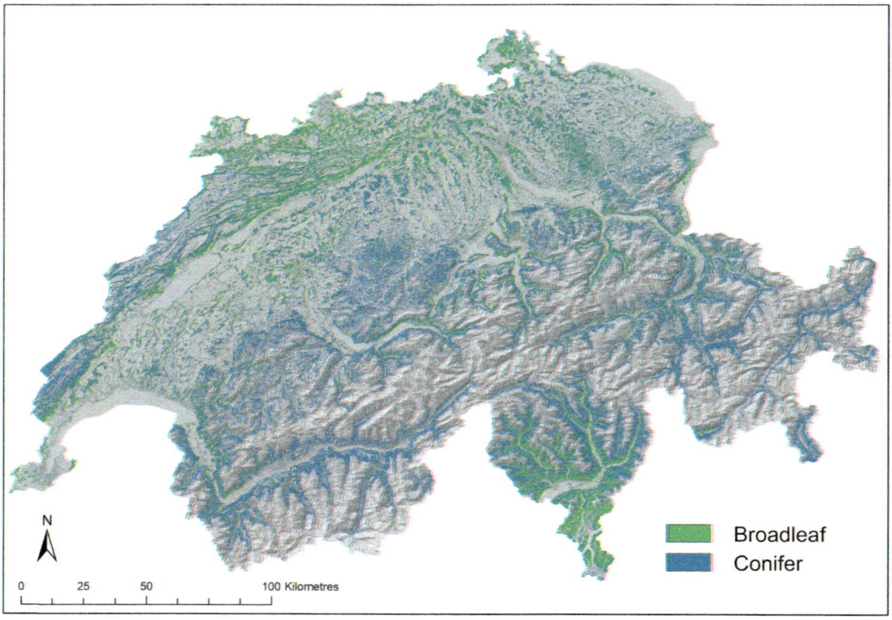

Fig. 7.6 Tree type map of Switzerland

Table 7.1 Confusion matrix of reference image interpretation and the forest type map

		NFI stereo-image-based interpretation		
		Broadleaved	Coniferous	
Tree type map	Broadleaved	110,204	1016	111,220
	Coniferous	2792	71,228	74,020
		112,996	72,244	

Waser et al. (2017) found that underestimation of the broadleaved fraction mainly occurs in mixed forests across all parts of the country, with a higher concentration in the Plateau, the Jura and the Pre-Alps. Underestimation increases with decreasing elevation and is therefore more distinctive for broadleaved stands, especially in areas with a West- or North-exposure.

In Fig. 7.7 the smallest deviations ($\pm 10\%$) in the fraction of broadleaved trees between the two data sets are depicted in green. These are mainly in the Alps, with a few outliers in the valleys, where red represents an overestimation of broadleaved trees and blue an overestimation of coniferous trees. The largest (negative) deviations in the broadleaved fraction indicate an overestimation of coniferous trees (blue) and are concentrated in the eastern and central parts of Switzerland and the Southern Alps. These deviations occur for many reasons related to factors such as topography, i.e. aspect and elevation, and expert interpretations. The interpretation experts mostly interpreted broadleaved classes correctly using the stereo-images, even in

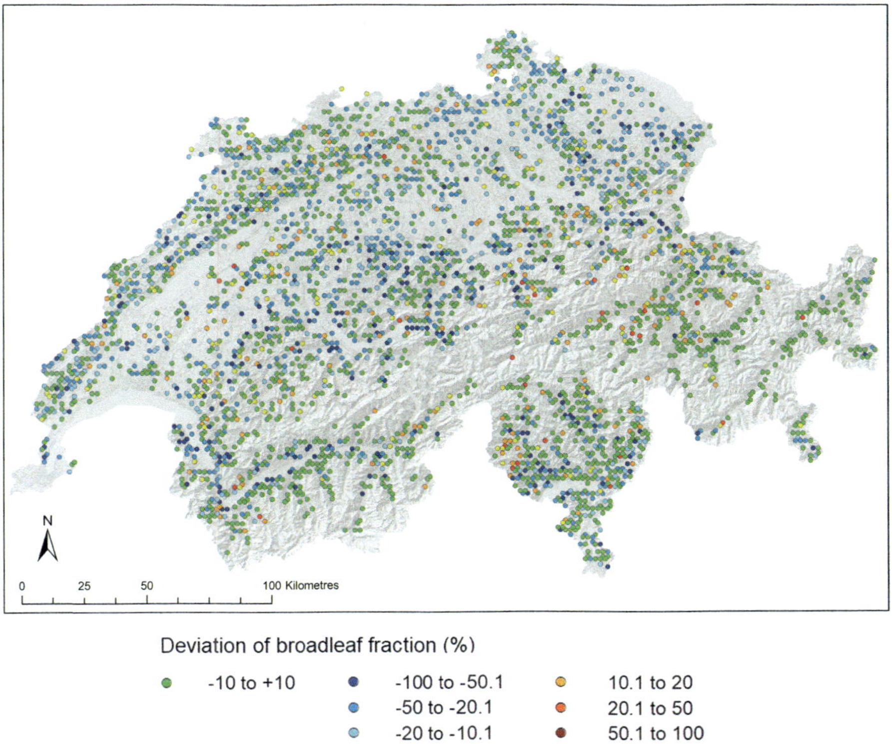

Fig. 7.7 Deviation of predicted (tree type map) from observed (NFI image interpretation) broadleaved fraction in the interpretation areas

shadowed conditions, but in the classification approach they sometimes classified them as coniferous trees.

The tree-type mapping approach could be considerably improved by reducing overestimations of coniferous trees. As expected, the date and time of data acquisition influenced the phenology and illumination of the images, which led to high spectral variability. This greatly affected the resulting tree type map, in particular regarding broadleaved trees. Thus, the few image strips acquired in early June under mostly leaf-off conditions at higher elevations were not optimal for mountainous regions. Figure 7.8 illustrates that topography, and consequently shadows, in the ADS images (top left) affected the tree-type mapping approach and resulted in an overestimation of coniferous trees (bottom left). This overestimation was reduced substantially (bottom right) by using additional winter images (leaf-off conditions in broadleaved stands) from Sentinel-2 satellite images (top right). The level of detail of the mapped tree types is lower, as a consequence of the coarser spatial resolution (10 m) of the Sentinel-2 data, but is still considered sufficient for most purposes.

Overall, the tree-type mapping approach presented here is appropriate for wall-to-wall products for entire countries. It is flexible and can be processed with a

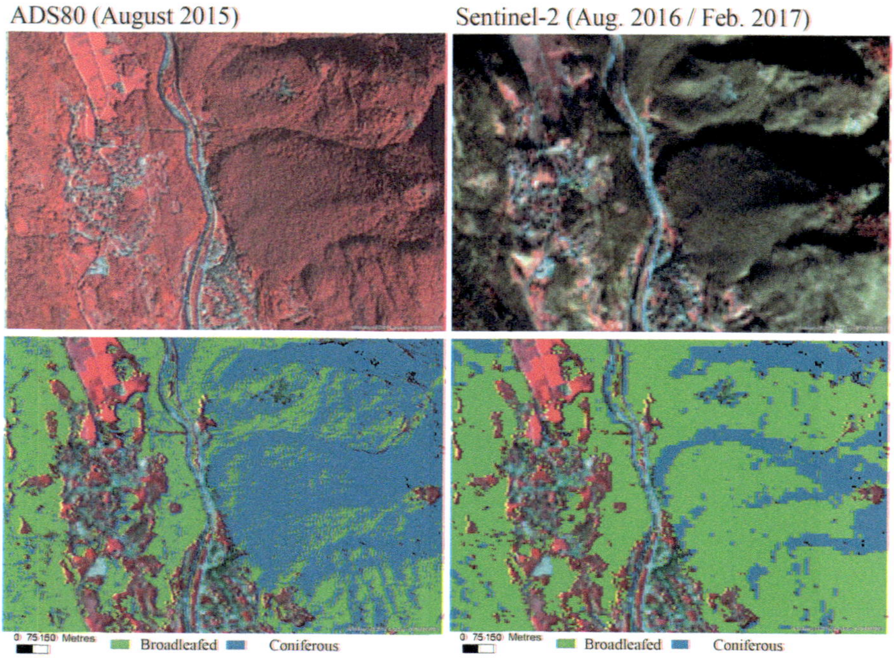

Fig. 7.8 Overestimation of the coniferous fraction in the tree type map using ADS summer images (left images) is reduced by using both summer and winter Sentinel-2 images (the winter image is shown in the upper right image)

reasonable computation time. Technical developments in remote sensing data and collection methods in the near future, and the increasing demand for long-term forest type maps, will further improve the quality of existing products and promote the development of new ones.

7.4 Growing Stock

Forest growing stock is considered a valuable indicator of forest health and function, providing information on forest resilience and the dynamics of forest stands (Debeljak et al. 2014). Estimates of growing stocks are therefore one of the most important variables in forest management and planning. In particular, the estimates are necessary for planning harvests and determining the sustainability of forest management practices, and they serve as the base components for the estimates of standing biomass and carbon required for international reporting. Growing stock is usually estimated in the form of means for certain domains based on data taken directly from national forest inventory assessments, for example the mean number of living trees with a diameter at breast height (dbh) >12 cm, with adjustments for

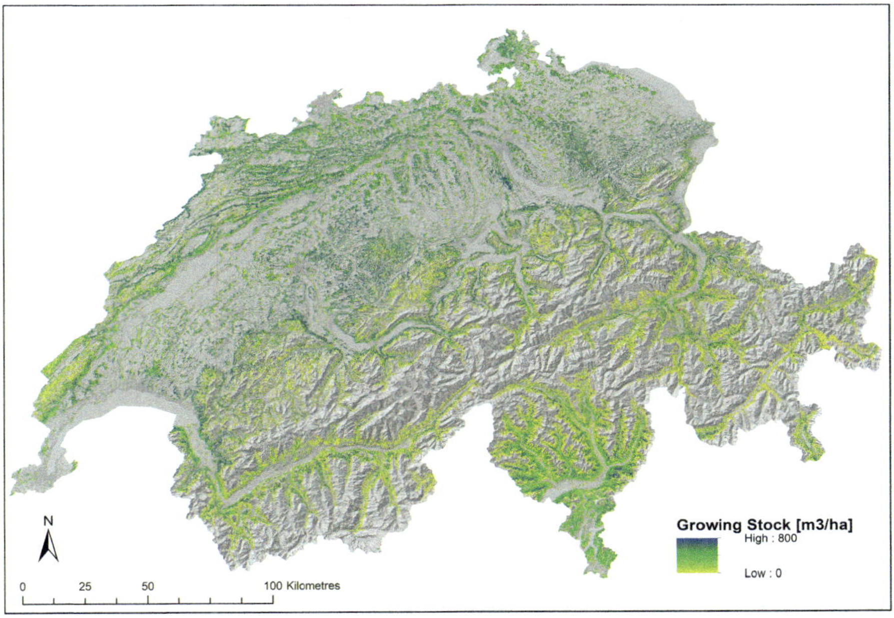

Fig. 7.9 Modelled growing stock (basis: forest cover map)

growth and removal to provide estimates for time periods between inventory years. In order to model area-wide growing stock across Switzerland, statistical models of the relationship between growing stock and measures of forest vegetation structure were constructed in which vegetation structure is derived from the vegetation height model (VHM) based on airborne stereo-images (ADS40/80). These models were then applied area-wide across the ADS data sets available for the whole country (Fig. 7.9).

Growing stock in m^3 per hectare, as estimated in NFI3 and NFI4 plots, is the response variable for the statistical modelling. The NFI reports growing stock for single trees on a plot and extrapolates these values to the hectare level. Plots located at forest and stand edges, and plots with <100% forest cover, are excluded from the modelling process to avoid errors when scaling up from individual tree measurements. As a result, a total of 3863 NFI3 and 2945 NFI4 plot-level measurements were used as modelling input.

The explanatory variables used for the statistical modelling include the following plot-level (500 m^2) metrics derived from point-cloud data from stereo ADS images: mean height of all points >3 m in height (avg); a variety of height percentiles (p50, p70, p80, p90, p95); standard deviation of the height of returns over 3 m (std); vegetation canopy cover derived from points (cov = the number of points >3 m in height divided by the number of all points); and percentage of points in the height classes 3–6 m (d0), 6–10 m (d1), 10–15 m (d2), 15–20 m (d3), 20–25 m (d4), 25–30 m (d5), 30–40 m (d6) and 40–60 m (d7). All metrics are derived using the

Table 7.2 Models for the prediction of growing stock according to elevation class

Elevation stratum (m a.s.l.)	Model	Model efficiency	Q-value	RMSE (%)
0–1000	566.45 + 12.94∗p90 + 1.82∗d3 + 1.6∗d4 + 1.92∗d5 + 2.43∗d6 – 20.82∗std + 0.32∗topo – 0.36∗temp	0.5	1.022	44.6
1000–1500	551.56 + 27.64∗p90 – 40.64∗std + 0.16∗topo – 0.49∗temp	0.59	1.045	43.4
>1500	688.98 – 2.75∗d0 – 4.68∗d1 – 3.07∗d2 – 1.22∗d3 + 24.9∗std + 4.99∗cov – 0.16∗Z – 0.38∗temp	0.59	1.338	66.0

LAStools lascanopy tool (Isenburg, German; http://rapidlasso.com/lastools) on normalised data. In addition, mean summer temperature (temp) from the Swiss baseline climatic variables (Zimmermann and Kienast 1999), elevation (Z) and topographical position index (topo) are included as explanatory variables. No explanatory variables that are highly correlated are included in the same models.

A number of modelling approaches were tested, including linear least-squares regression, non-linear least-squares regression and random forest models. Model performance was compared largely in terms of model efficiency (equivalent to R^2 in a linear model; e.g. Pinjuv et al. 2006), but also based on prediction error, which was calculated using a leave-one-out cross-validation algorithm and Q-Value (Holmgren et al. 2003). Q-value is defined as the ratio between the standard deviation from the cross-validation (prediction error) and the standard deviation of the regression (root mean square error), where a value closer to 1 indicates better model performance. Least-squares methods were used to calibrate the model (R functions *lm* and *nls* for linear and non-linear models, respectively). For linear models, the Akaike's Information Criterion (AIC) backward selection criterion was used to discard unnecessary variables. All the models were evaluated using a leave-one-out cross-validation algorithm. In general, the performance of the linear and non-linear (including random forest) models was very similar. As such, final implementation was based on the linear least-squares regression approach because it offers the easiest approach to implementation and interpretation without compromising performance.

Modelling was stratified by elevation into the classes 0–1000 m, 1000–1500 m and above 1500 m a.s.l. Stratification according to NFI production regions was also tested but not found to be meaningful.

The final set of variables which gave the best model performance consisted of: the 90th height percentile (p90), canopy cover (cov), standard deviation of height (std), height class percentages, topographic position index (topo), mean summer temperature (temp) and elevation (Z). Performance was improved by limiting the plots to those where NFI field data had been collected within 2 years of capturing the image. Final models per elevation stratum are given in Table 7.2.

Spatial accuracy of model performance was tested by comparing the model results to NFI predictions of mean growing-stock value for forest management districts. Out of 119 forest districts, the model predicted mean growing stock values

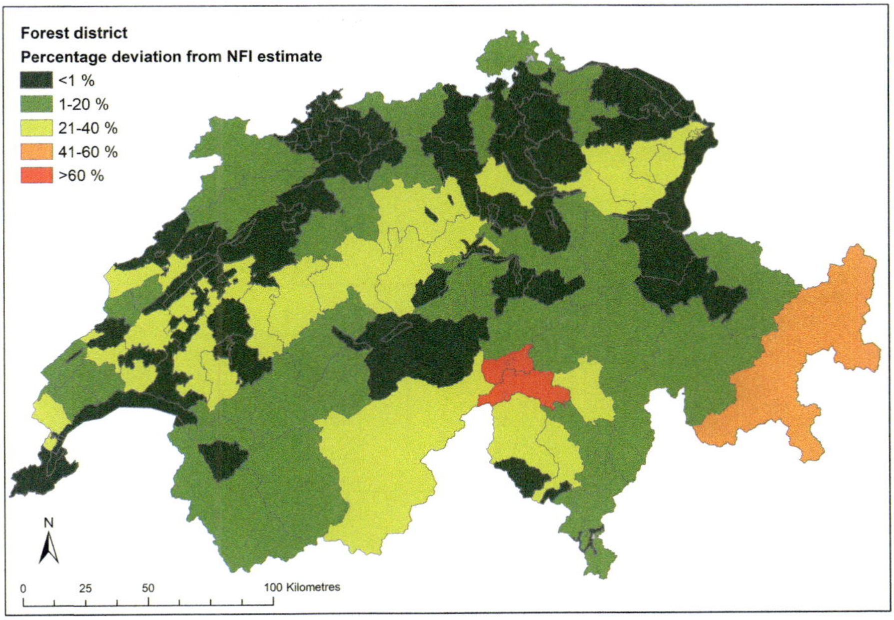

Fig. 7.10 Percentage deviation in mean growing stock per forest district of the growing stock model from the NFI estimates

within the NFI estimated range for 58 districts. The least accurate predictions occurred for forest districts within the cantons Grisons and Ticino (Fig. 7.10 values >40%).

The ADS data is expected to be available on a continuous cycle for the foreseeable future, making repeatable measurements possible and facilitating frequent model revisions and improvements, as well as potentially creating a cost-effective opportunity for monitoring changes in tree biomass and related carbon stocks (Ginzler and Hobi 2015). Closer date matching between field data and remotely sensed data could help improve the predictive ability of the models. The ongoing rolling update of ADS data, combined with the continuous NFI programme, should offer such an opportunity. The methodology also offers the potential to estimate growing stocks over areas and at time points not covered by the NFI field measurements, provided aerial stereo-images are available.

7.5 Changes in Forest Cover

One of the major tasks of national forest inventories is to record and assess data on forest resources and to provide authoritative information on forest change to support sustainable management of these resources. While changes in forest area, growing

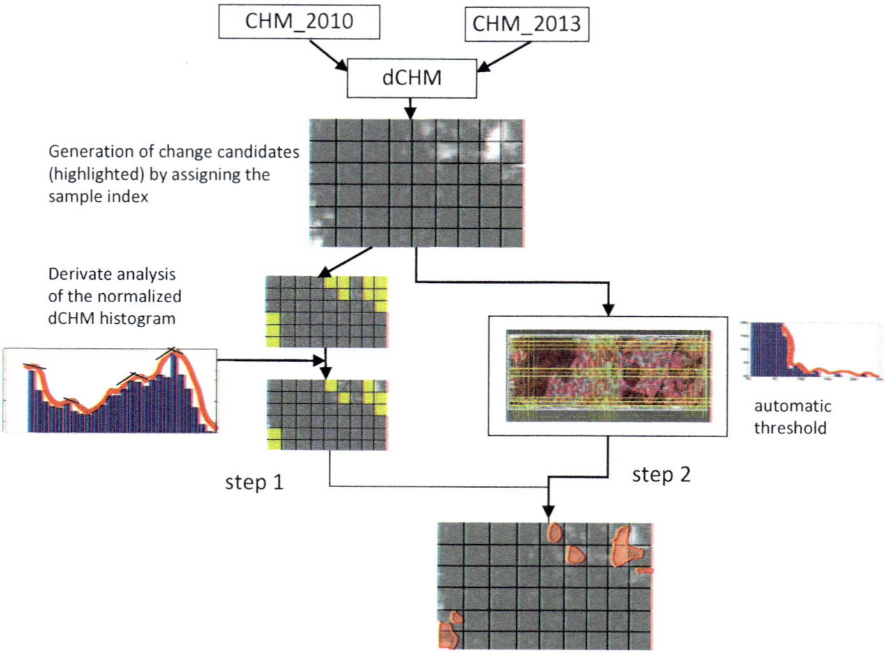

Fig. 7.11 Workflow of the approach to detecting forest-cover change developed by Wang et al. (2015)

stock and species composition between the inventories can be estimated with sample-based data, remotely sensed data offers ways to extend the specifically limited field-plot information to a continuous landscape-level representation.

An adaptive multi-scale approach to detecting forest-cover change with high spatial and temporal resolution was applied to two study areas in Switzerland by Wang et al. (2015). The challenge of this approach is to minimise height uncertainties in the **d**igital **s**urface **m**odel (DSM) and thus in the **v**egetation **h**eight **m**odel (VHM), which can influence the accuracy of the findings on forest-cover change. To achieve optimal results, a novel adaptive multi-scale approach was developed to capture real changes over a range of scales with a hierarchical strategy.

The approach consists of two steps (Fig. 7.11). In the first step, a *change index* parameter indicates the overall change status at a coarse scale. The tendency towards change was indicated by derivative analysis of the normalised histograms of the **d**ifference between the two **c**anopy **h**eight **m**odels (dCHMs) in different years (2010 and 2013). In the second step, detection of forest cover change at a refined scale was based on an automatic threshold and a moving window technique. Promising results were obtained and demonstrated that real changes in forest cover can be distinguished from falsely detected changes with a high degree of accuracy in managed mixed forests.

Using a threshold based on a priori knowledge enabled the identification of areas with erroneous height changes using a change index. Derivative analysis of the

Table 7.3 Accuracies of calculated forest changes in two study areas

	Producer's accuracy (change)	User's accuracy (change)	Producer's accuracy (non-change)	User's accuracy (non-change)	Kappa	Overall accuracy
Study area 1						
Proposed method	0.83	0.76	0.95	0.97	0.75	0.93
Threshold (= 15 m)	0.62	0.90	0.99	0.93	0.69	0.93
Threshold (= 5 m)	0.86	0.51	0.85	0.97	0.55	0.85
Study area 2						
Proposed method	0.53	0.42	0.99	0.99	0.46	0.98

normalised dCHM histograms, which tracked the tendency for change in each block, facilitated more accurate location of real changes. In this step, areas with erroneous changes were removed, resulting in substantially improved accuracy in detecting changes. The proposed automatic threshold selection for the refined-scale change detection made the method more robust than just using a predefined threshold.

In both study areas, producers' accuracies were higher than users' accuracies (Table 7.3) because more non-change areas referenced as change areas were detected than change areas referenced as non-change areas. This discrepancy could be related to the generation of DSMs and dCHMs, which was affected by the sun being at different elevations when the images were acquired, which could have led to there being shadows in the 2010 images and no shadows in the 2013 images, or vice versa. As the images were acquired by swisstopo during national flight cycles, the date and time of image acquisition could not be influenced and thus some shadows had to be accepted. Consequently, in future work, the shadow mask calculated for the forest-cover map will be used and these areas will be treated separately in the change-detection algorithm and excluded from the analysis.

A second reason for producers' accuracies being higher than users' accuracies is related to the oblique differences between the generated DSMs resulting from different flight angles of the images acquired in 2010 and 2013. This is an important issue for explaining lower accuracies in mountainous areas, such as study area 2 in this example. Therefore, using novel image-matching algorithms to further improve the accuracy of the generated DSMs is a research area worth exploring, as Mortensen et al. (2005), Choi and Kweon (2009), and Tack et al. (2012) have already suggested.

References

Barrett F, McRoberts RE, Tomppo E, Cienciala E, Waser LT (2016) A questionnaire-based review of the operational use of remotely sensed data by national forest inventories. Remote Sens Environ 174:279–289

Breiman L (2001) Random forests. Mach Learn 45:5–32

Choi O, Kweon IS (2009) Robust feature point matching by preserving local geometric consistency. Comput Vis Image Underst 113(6):726–742
Debeljak M, Poljanec A, Zenko B (2014) Modelling forest growing stock from inventory data: a datamining approach. Ecol Indic 41:30–39
DeVenecia K, Walker S, Zhang B (2007) New approaches to generating and processing high resolution elevation data with imagery. Photogrammetric Week 07:297–308
Gao J (2007) Towards accurate determination of surface height using modern geoinformatic methods: possibilities and limitations. Prog Phys Geogr 31:591–605
Ginzler C, Hobi M (2015) Countrywide stereo-image matching for updating digital surface models in the framework of the Swiss National Forest Inventory. Remote Sens 7:4343–4370
Ginzler C, Hobi ML (2016) Das aktuelle Vegetationshöhenmodell der Schweiz: spezifische Anwendungen im Waldbereich. Schweiz Z Forstwes 167:128–135
Holmgren J, Nilsson M, Olsson H (2003) Estimation of tree height and stem volume on plots-using airborne laser scanning. For Sci 49(3):419–428
Mathys L, Ginzler C, Zimmermann NE, Brassel P, Wildi O (2006) Sensitivity assessment on continuous landscape variables to classify a discrete forest area. For Ecol Manag 229(1–3):111–119
Mortensen EN, Hongli D, Shapiro L (2005) A SIFT descriptor with global context. In computer vision and pattern recognition, 2005. CVPR 2005. IEEE Comput Soc Conf 181:184–190
Pinjuv G, Mason E, Watt M (2006) Quantitative validation and comparison of a range of forest growth model types. For Ecol Manag 236(1):37–46
Price B, Gomez A, Mathys L, Gardi O, Schellenberger A, Ginzler C, Thürig E (2017) Tree biomass in the Swiss landscape: nationwide modelling for improved accounting for forest and non-forest trees. Environ Monit Assess 189:106
R Core Team (2017) R: a language and environment for statistical computing; R Foundation for Statistical Computing: Vienna, Austria. Available online: https://www.R-project.org/. Accessed 21 Sept 2017
Swisstopo (2017) The topographic landscape model TML. Available online: https://www.swisstopo.admin.ch/en/knowledge-facts/topographic-landscape-model.html. Accessed 14 Sept 2017
Tack F, Goossens R, Buyuksalih G (2012) Assessment of a photogrammetric approach for urban DSM extraction from tri-stereoscopic satellite imagery. Photogramm Rec 27(139):293–310
Tomppo E, Gschwantner T, Lawrence M, McRoberts RE (2010) National forest inventories-pathways for common reporting. Springer, Berlin, p 612
Wang Z, Ginzler C, Waser LT (2015) A novel method to assess short-term forest cover changes based on digital surface models from image-based point clouds. Forestry 88:429–440
Waser LT, Fischer C, Wang Z, Ginzler C (2015) Wall-to-wall forest mapping based on digital surface models from image-based point clouds and a NFI forest definition. Forests 6:4510–4528
Waser LT, Ginzler C, Rehush N (2017) Wall-to-wall tree type mapping from countrywide airborne remote sensing surveys. Remote Sens 9:766
Zellweger F, Braunisch V, Baltensweiler A, Bollmann K (2013) Remotely sensed forest structural complexity predicts multi species occurrence at the landscape scale. For Ecol Manag 307:303–312
Zellweger F, Morsdorf F, Purves R, Braunisch V, Bollmann K (2014) Improved methods for measuring forest landscape structure: LiDAR complements field-based habitat assessment. Biodivers Conserv 23:289–307
Zhang L, Gruen A (2006) Multi-image matching for DSM generation from IKONOS imagery. ISPRS J Photogramm Remote Sens 60:195–211
Zielewska-Büttner K, Adler P, Ehmann M, Braunisch V (2016) Automated detection of forest gaps in spruce dominated stands using canopy height models derived from stereo aerial imagery. Remote Sens 8:175
Zimmermann NE, Kienast F (1999) Predictive mapping of alpine grasslands in Switzerland: species versus community approach. J Veg Sci 10(4):469–482

Part IV
Terrestrial Inventory and Interview Survey

Chapter 8
Planning and Organisation of the Terrestrial Inventory and Interview Survey

Fabrizio Cioldi and Markus Keller

Abstract This Chapter describes the planning and organisation of the terrestrial inventory and the interview survey. The criteria for the selection of personnel and equipment are also given, and the instruction, training and safety standards are explained. The preparation and implementation of the fourth Swiss National Forest Inventory (NFI4) drew on the information and experience obtained during the previous inventories (NFI1–NFI3). Changing the survey into a continuous inventory over 9 years (2009-2017) had consequences for planning and organising both NFI4's field inventory and interview survey, and required redistributing the fieldwork over nine sub-grids. In the first three inventories (NFI1–NFI3), 10–12 field teams carried out the terrestrial survey over a period of 3 years, and the interview surveys were conducted immediately after completing the field surveys in each region. In NFI4, however, only three teams were involved in the terrestrial inventory, which included the assessment of about 820 sample plots each year. The NFI4 interview surveys were carried out in two stages, in 2013/2014 and in 2017/2018, also by field team members, but individually and no longer in pairs.

8.1 Introduction

The Swiss NFI (NFI) has three main data sources: interpretations of aerial photos, the terrestrial inventory and the interview survey. To determine the sample plots to be assessed by the field teams, the latest available aerial photos are interpreted and the forest/non-forest sample plots classified (Sect. 6.3). The field survey, which is the most important source of data for the NFI, can then start. The last step in the raw data collection is to interview foresters from the local forest services.

Why is the terrestrial inventory still so important considering the potential of aerial photography and laser scanning? Would it not be possible to rely more on remote sensing, which has made significant progress in recent years? No, this would

F. Cioldi (✉) · M. Keller
Scientific Service NFI, Swiss Federal Research Institute WSL, Birmensdorf, Switzerland
e-mail: fabrizio.cioldi@wsl.ch

C. Fischer, B. Traub (eds.), *Swiss National Forest Inventory – Methods and Models of the Fourth Assessment*, Managing Forest Ecosystems 35,
https://doi.org/10.1007/978-3-030-19293-8_8

not work primarily because the information needed for NFI requires high spatial and temporal resolutions. Given the limits of remote sensing, it is unlikely that it will be possible to do without the know-how and the data quality that only a proven expert can guarantee in the near future. The field data collected for the Swiss NFI must meet high precision and quality standards (Chap. 21), and should, ideally, be as close as possible to the "ground truth". Only a small number of field attributes (Chap. 9) could be assessed with remote sensing tools, and the rest require the expert knowledge of a field team to, e.g., judge whether a small plant has suffered browsing damage.

The preparation and implementation of NFI4 drew on the information and experience obtained during the previous inventories (NFI1–NFI3). Changing the survey into a continuous inventory over 9 years had consequences for planning and organising both NFI4's field inventory and interview survey, and required redistributing the fieldwork over nine sub-grids (Sect. 2.4.4). In the first three inventories (NFI1–NFI3), 10–12 field teams carried out the terrestrial survey over a period of 3 years, and the interview surveys were conducted immediately after completing the field surveys in each region. In NFI4, however, only three teams were involved in the field inventory (Sect. 8.2). The NFI4 interview surveys were carried out in two stages, in 2013/2014 and in 2017/2018, also by field team members, but individually and no longer in pairs (Sect. 8.3).

8.2 Terrestrial Inventory

8.2.1 Pilot Inventory

8.2.1.1 Design and Aim

In the pilot inventory conducted in 2008 to prepare for the NFI4 terrestrial inventory, some new attributes and methods were tested in terms of practicability, work processes and time expenditure. The pilot inventory was important for selecting the measuring instruments and tools, for detecting any technical and logical hard- and software problems, and for checking the consistency between the manual for the field survey and MAIRA, the software used for data collection. The experiences gained from the 2008 pilot inventory were very useful for improving the field survey instructions and for identifying insufficiently defined attributes and clarifying their definition in the field survey manual. Where necessary, the MAIRA software was adjusted, and in some cases the data-collection procedure was also modified.

8.2.1.2 Implementation

The pilot inventory was conducted in autumn 2008 over 30 days in four test areas representing the different topographical and climatic conditions in Switzerland.

In the test areas, a total of 136 sample plots were assessed, 24 of them twice by different field teams. The two field survey teams consisted of two people each, who were all forestry engineers and members of the in-house staff. Three of them had already worked in the NFI3 field survey, which meant the Project could count on highly qualified and experienced personnel.

The field tests focused on:

- Marking the sample-plot centres with the RECCO system
- Measuring the location of the sample-plot centres with GPS
- Assessing the forest boundary line (FBL)
- Estimating tree age
- Testing a new method to measure the regeneration
- Assessing ant mounds
- Taking photos around the sample-plot centres
- Measuring the tree diameter at a height of 7 m with the instrument Criterion400®©.
- Assessing the remains of sample trees after harvesting interventions (*logging residuals*).

8.2.1.3 Conclusions

Most of the new attributes or procedures tested in the pilot inventory were then introduced into NFI4's regular field inventory. They are described in detail in Chap. 9. It was decided, however, that *logging residuals* would not be assessed, as it was almost impossible to clearly identify such remains and assign them to the correct sample tree assessed in previous inventories.

Measurement of the tree diameter at a height of 7 m deserves a special mention. After the 2008 pilot inventory, the optical Dendrometer Criterion 400 was chosen as the new instrument for this measurement, as it is much lighter and easier to handle in rough and steep terrain than the Finnish calliper, with its telescopic aluminium pole, used in all inventories up to now. The data collected with it seemed to be reliable and promising. During the first major data analysis after the field seasons 2009–2011, however, biased results surfaced, especially with regards to the volume increment, which appeared to arise from the new method for measuring the upper diameter. It was therefore decided to switch back, as of 2012, to using the Finnish calliper to measure the tree diameter at this height.

8.2.2 Planning and Organisation

8.2.2.1 Annual Grid and Sample-Plot Numbers

In Switzerland the 1.4 × 1.4 km grid used in NFI4 contains a total of 20,638 sample plots, of which almost 1/3 are located in forested areas. This grid was divided into nine sub-grids, all of which are evenly spread across the whole country. Every year 1/9 of all the sample plots were selected to be visited in the field, based on the aerial-photo interpretation (Sect. 6.3). After 9 years, every sample plot in Switzerland had been assessed by the field teams. The number of sample plots to be assessed in NFI4 was estimated as follows:

Starting with the 6810 sample plots surveyed in NFI3, around 10% of new sample plots and 10% of sample plots for the repeat survey were added. This resulted in a total of approximately 900 sample plots per year (annual panel), which was used as the basis for planning the NFI4 field survey (personnel and equipment).

8.2.2.2 Working Regions and Time Schedule

To determine the number of field survey teams needed to complete one annual panel of 900 sample plots, it was assumed that the working period in the field lasts for 30 weeks (from April to November) and that, on average, two sample plots per day (or 10 per week) could be assessed. Three field teams were therefore considered sufficient.

Then, on the basis of the forest districts (state of 2007), three working regions were defined (Fig. 8.1) according to the following main criteria:

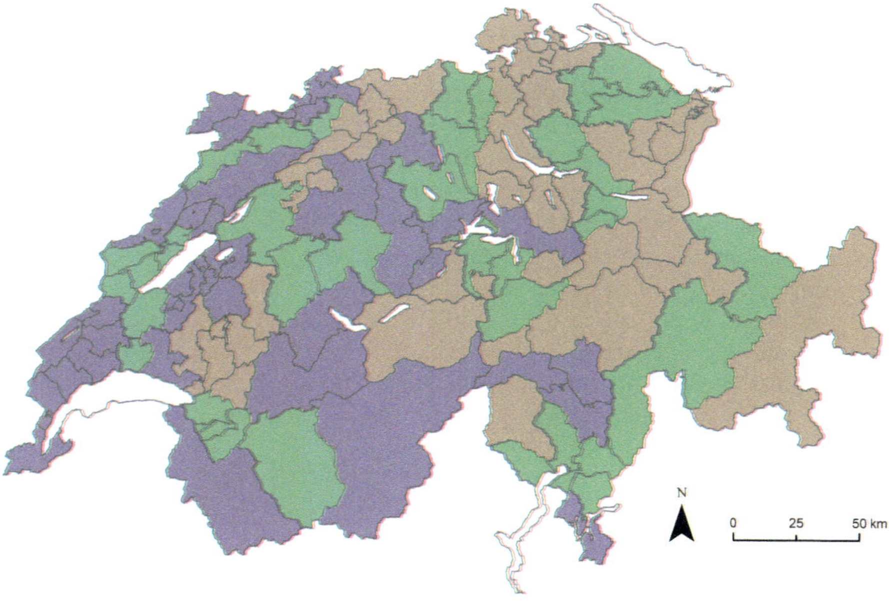

Fig. 8.1 Field survey regions in NFI4. Each colour corresponds to the work region of a field survey team

- The number of sample plots and their elevation distribution is the same in the three working regions
- The field teams work all over Switzerland, with their main focus on the western, the central or the eastern part of the country
- Larger cantons are assessed by at least two different teams
- The working regions of each field team are large enough for them to work there for at least a whole week to minimise time-consuming shifts and travel during the week

8.2.2.3 Personnel

As recruiting personnel in the previous inventories (NFI1–NFI3) had worked well, it was decided to rely on the same selection criteria for the field survey in NFI4, with a forest engineer and a forester or forest ranger in each field team. Being able to draw on the forest engineer's theoretical knowledge (university level) and the forester's or forest ranger's practical experience (apprenticeship and technical college) seems ideal for good work quality (Sect. 21.4.1).

In addition to the six people employed in the three field teams, three to four forest engineers or forest rangers were also contracted as replacements for the regular field team members in case of absence due to holidays, illness or accident. These collaborators also participated in the instruction and training courses. A key factor in recruiting the field staff was their physical fitness because the field teams often work in topographically difficult terrain in many parts of the country. We also aimed to have a good representation of the four official languages – German, French, Italian and Romansh – which was especially important for the interview survey.

Members of the field teams were paid by the hour with an annual contract at WSL for a period of 9 months. The advantage of having an hourly wage was that it gave the field teams considerable freedom and flexibility in organising their daily assignments to take into account the local conditions, e.g. the weather. In turn, this increased the number of sample plots they could assess each day. The team's expenses for food and accommodation were reimbursed on a per diem basis or, if this was insufficient, based on the actual costs incurred, e.g. hotel bills in expensive tourist regions. Travel expenses from their homes to the working region were reimbursed separately. WSL also covered the cost of professional mountain boots, as these are important for safety.

8.2.2.4 Preparation of the Annual Survey

From November to February each year, before the start of the new field season, the most recent aerial images available of the around 2300 sample plots in an annual panel (1/9 of the 1.4×1.4 km grid) were interpreted and the plots classified as 'forest' or 'non-forest' (Sect. 6.3). This information was used to determine which sample plots should be assessed by the field teams.

Sample plots were chosen for assessment in the field if they had been classified as:

- 'Forest' (including 'shrub forest') in the NFI1, NFI2 or NFI3
- 'Forest' (including 'shrub forest') according to the aerial-photo interpretation of NFI4
- 'Non-forest' with restrictions (only probably and not clearly 'non-forest') or unclassifiable in NFI4's aerial-photo interpretation because of e.g. shadows or clouds
- 'Non-forest' according to the NFI3 field survey

A few sample plots were removed 'manually' from the seasonal lists of the plots to be assessed each year if they were not interpretable in the aerial photo because of shadow or cloud coverage and could be considered clearly as 'non-forest' from other data sources such as maps or other aerial images. Examples are sample plots in rivers or lakes, or unproductive sites at high elevations such as glaciers. As a result, about 820 sample plots were chosen for the field survey each year (Fig. 8.2).

The dataflow from the data collection on the sample plots to the database is tested before the start of the field season, and if necessary, changes in the software are implemented or new attributes added. Five to six sample plots are assessed in a simulation in the office to test the procedure with different types of possible

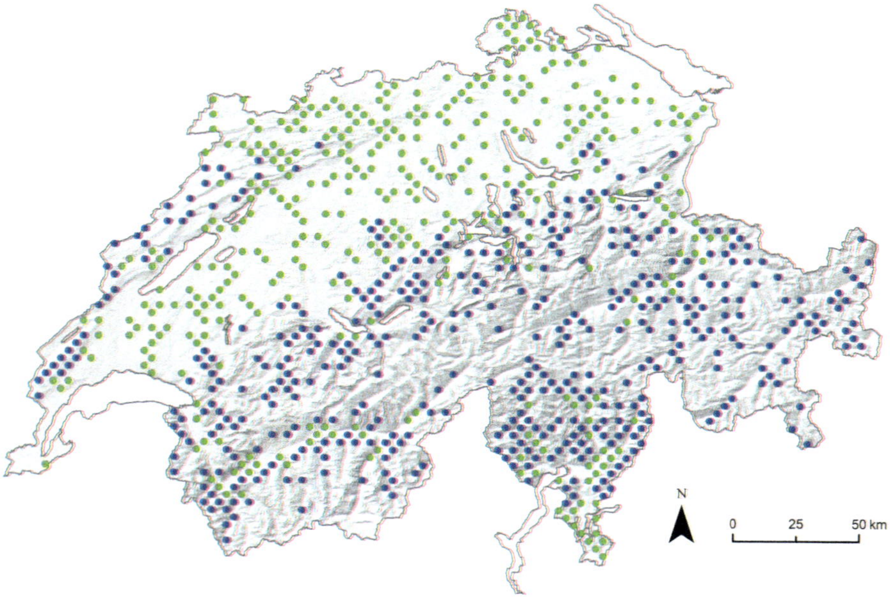

Fig. 8.2 Distribution of the sample plots for the terrestrial survey 2017, coloured green if at an elevation below 1000 m a.s.l. and blue if above 1000 m a. s. l.

situations, e.g. 'forest' with or without a forest-edge assessment, 'shrub forest' or 'non-forest'. The data entered in the program MAIRA is also recorded on paper. Once the data has been collected, it is sent to the database. The data entered in the program and recorded on paper is compared meticulously to the data actually stored in the database. Once this test run has been completed successfully, the field computers are loaded with the latest software version and the pre-determined data is made ready for the next survey.

In March all the cantonal forest services are informed about the inventory work plan and time schedule of the field survey. In every canton, one person (usually the person in charge of forest planning) is nominated as the contact person with NFI's headquarters. This person is updated periodically on the status and progress of the inventory work, and he/she in turn forwards the information from NFI's headquarters to the regional district foresters. Sometimes a forester interested in inventory techniques and the NFI accompanies a field team surveying a sample plot nearby.

From time to time in NFI4, NFI's headquarters informed the media in a press release about NFI's fieldwork. Quite often journalists accompanied a survey team and observed their work in surveying a sample plot, and later reported about their visit in national or local media (newspapers, radio or television). In addition, the field teams were trained to handle meetings with forest owners and any other people they might encounter during their work. It was stressed that they should take enough time to explain the objectives of the inventory and why they were in the forest. Each field team has a set of printed flyers in German, French, Italian and English with information about the NFI's goals and objectives.

Every field survey team is responsible for the following tasks: planning the details of the field survey in their regions (daily and weekly plan), conducting the sample-plot assessments and uploading the collected data periodically (several times per week), maintaining vehicles and equipment, and writing the work reports. The survey teams are free to plan and organise the work themselves, although obviously they have to comply with certain general rules: in spring and autumn the field surveys must be carried out in the low-elevation regions. Summer is reserved for sample plots at high elevations. This concerns all the sample plots above 1500 m a.s.l., which must be assessed before the end of September to avoid problems with any seasonal snowfall.

8.2.2.5 Training and Safety Standards

The field season begins with an instruction course lasting 1 week, although in the first field season, at the start of NFI4 in 2009, the instruction course lasted for 2 weeks. In the course the participants are introduced to all the theoretical and practical aspects relevant for the NFI method. In particular, the manual for the field survey (Düggelin and Keller 2017) is studied in detail, and the field team members have time to become familiar with the software MAIRA and to participate in special training courses providing specific information, e.g. on identifying woody species, fungi and ants. Furthermore, each field team receives the full equipment needed to start the work. The general objective of the course is for each team at the end of the week to be able to make a complete survey of a sample plot. In addition, the planning of the work and the administrative issues involved should be clear to everyone.

During the first days of the regular field surveys immediately after the instruction course, each team is accompanied for several days by a staff member from NFI's headquarters to ensure good data quality and to clarify any remaining uncertainties directly in the sample plot (Sect. 21.4.3). In addition, the field teams can contact headquarters by phone via a special hotline number throughout the season should there be unexpected problems or incidents, or if special situations need to be discussed and dealt with.

In addition to the initial instruction course, two additional training days are organised in NFI4 every year (usually in June and September), where some typical aspects of the survey in a particular region are studied in depth with specific exercises on, e.g. neophytes in the Southern Alps. More details about the instruction are described in Sect. 21.4.3.

In NFI special attention is paid to safety. Often the survey teams have to work in rather inaccessible terrain under dangerous conditions. In addition, they frequently have to travel by vehicle for long distances on tracks that are sometimes almost impassable. Therefore, periodic driving courses and first-aid courses (basic life support) are part of the basic training of the field teams.

An important NFI safety rule is to never work alone in the forest. Not only must personal equipment (such as mountain boots) be checked and maintained regularly, but so too must the instruments and the vehicles, including brakes and tyres.

Should a field team encounter a dangerous situation in the course of fieldwork, they are always advised not to take any unnecessary risks. If one team member does not feel happy about continuing up to a very steep sample plot, for example, another access route must be found. If the second attempt fails as well, the sample plot is declared inaccessible. Thanks to the emphasis placed on safety aspects, no serious accidents have occurred since the beginning of the first NFI in 1983.

8.2.2.6 Control and Repeat Survey

During the first months of fieldwork, especially in spring, staff members from NFI's headquarters carry out control surveys to detect any errors, particularly systematic ones (Sect. 21.5). In contrast, the repeat survey is carried out three times per season for 1 week each (normally in June, September and November). During a repeat survey week, each team assesses on average eight to ten sample plots that have already been assessed by another field team in the course of the regular survey (Sect. 21.5). An additional organisational measure for maintaining good work quality is to shift field team members between field teams two to three times a year so that they can experience how their colleagues work and deal with difficult situations. The aim of this exchange is to standardise working procedures as much as possible among the different field teams (Sect. 21.4).

8.2.2.7 Equipment and Material

WSL provides the survey teams with all the necessary instruments and equipment, including the vehicles (Sect. 9.2). Important factors for the selection of the measuring instruments are their reliability, robustness and simplicity, but their precision and weight are also considered because the amount of equipment used in the course of the inventories has increased, especially the number of electronic instruments. However, the equipment for the NFI4 field survey was largely the same as in NFI3 with a few exceptions, e.g. the Trimble GPS and the REGA-transceiver radio for emergencies.

At the end of each season, the teams store all their equipment in the depot at WSL, where it is maintained and, if necessary, calibrated. Furthermore, many of the important documents, such as forms or tree maps of the sample plots, are prepared at WSL in winter before the start of the field season.

8.3 Interview Survey

8.3.1 Testing of the Interview Survey

Changing from the earlier periodic survey to a continuous inventory over 9 years in NFI4 also affected the implementation of the interview survey with the local forest services. In the previous inventories, the interview surveys were carried out immediately after completing the field surveys in a given region, but in NFI4 it was obviously not possible to interview the same foresters every year about the few sample plots assessed in their district during the last season. It was therefore decided to synchronise the interview survey with the publication of the inventory results. The main products of each NFI cycle are the publication of intermediate results based on data from the first 5 years, i.e. from 5/9 of the sample plots, followed by the publication of the final results at the end of the 9-year period using data from all the sample plots. The interview survey for NFI4 took place in two stages: the first one in 2013/2014 and the second one in 2017/2018.

The interview survey in NFI4 consisted of two main parts: information collected from the forest services about the sample plots assessed in the forest, and a forest road survey for the entire country. The first part was the same as in the interview surveys from previous NFIs (Sect. 10.2) with exactly the same attributes as in the NFI3 interview survey. The manual therefore remained unchanged and no special tests or pilot interviews were necessary before starting the NFI4 interview survey in 2013. Interviews concerned all sample plots classified as ‘forest’ or ‘shrub forest’ in the field survey, and all inaccessible plots classified as ‘forest’ or ‘shrub forest’ in the aerial-photo interpretation. About 50 attributes were assessed on the following topics:

- Forest management and planning tools
- Forest use and history
- Timber harvesting technique

The second part of the interview survey, the forest road assessment, was modified and extended, in particular at the request of the Federal Office of the Environment (FOEN), which wanted to have better information on forest road dimensions and categories (Sect. 10.3). A pilot study with an extended forest road assessment was tested in some regions of Cantons Zurich and Grisons in 2012. The results of the pilot study showed that:

- Information about forest roads already available in the cantons could not be integrated into the NFI system, mainly because different categories for forest roads were used
- The procedure for the forest road assessment during the interview with the forester had to be modified and optimised
- Several attributes or attribute categories, such as road width and carrying capacity, had to be corrected

After the pilot test, the final changes in the interview survey manual were made and the operative implementation of the NFI4 interview surveys was planned.

8.3.2 *Planning and Organisation*

The interviews were always conducted after the field assessments because they required some information collected by the field survey teams on the sample plots about, for example, the last silvicultural intervention and the photos. The first stage of the interview survey took place in 2013/2014 with a focus on the sample plots assessed in the first 5 years (5/9 of the entire terrestrial survey) and the assessment/mapping of all the forest roads in the country. The second stage of the interview survey was carried out in 2017/2018 and related to the data on the remaining sample plots assessed between 2014 and 2017 (4/9 of the grid). The forest road assessment was not, however, updated in the second stage.

The following description of the planning relates only to the interview survey conducted in 2013/2014.

As in the previous interview surveys (NFI1–NFI3), all 800 foresters in Switzerland were questioned. The content of the NFI4 interview survey was almost identical to that of NFI3 and refers to all sample plots assessed terrestrially in the years 2009–2013. The maps from the forest road survey NFI3 were used as basic information for the forest road survey NFI4, but unlike in NFI3, the forest roads were classified according to their carrying capacity (total weight) and width.

8.3.2.1 Personnel

In the earlier inventories, the interview survey was carried out by the field team (two people) immediately after the field assessment of all sample plots in the forest district. In NFI4, however, the interviews were carried out by one surveyor, a member of the field staff. The extra effort of sending two people to the different regions to carry out the interviews would not have been justified. To decide who should work in which region (Fig. 8.3) the following criteria were considered:

- Ability to speak the local language fluently
- Place of residence
- Participation in the field survey of the region in question
- Knowledge of the specific region

The same system as in the field surveys was used for reimbursing expenses for food, accommodation and travel of the people who conducted the interviews (Sect. 8.2.2).

8.3.2.2 Survey Regions and Time Schedule

The workload of the forest rangers was taken into account in the general planning of the interview survey. In the lowlands, timber harvesting is mainly carried out during the winter months. In mountain regions, forest rangers are more likely to be available in the office during winter months when snow blocks the roads and makes working in the forest almost impossible. For these reasons, it was decided to carry out the interviews during the field season (April to June) in the Jura and Plateau and during the winter (November to February) in the Pre-Alps, Alps and Southern Alps. This means the workload is heavy for staff at headquarters and for field team members,

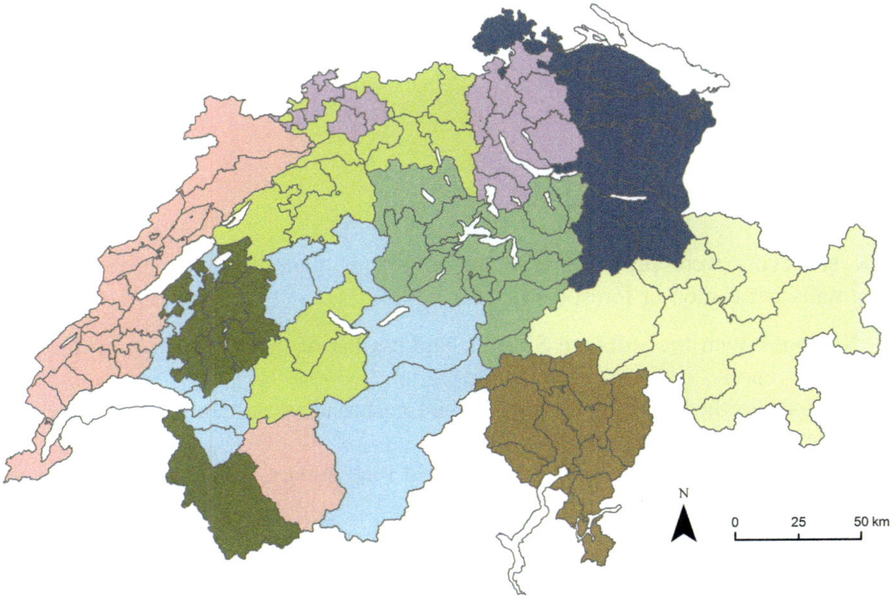

Fig. 8.3 Interview survey regions in NFI4. Each colour corresponds to the work region of one surveyor

especially in spring when some of them are involved in both the field and the interview survey.

During their regular meetings the contact person of each canton informed the district foresters about the beginning of the interview survey (April 2013) and encouraged them to reserve the time needed for the meeting with the NFI surveyors. The interview survey was carried out by nine people, all of whom were members of the NFI field staff. Each one was responsible for specific regions (normally a canton or forest district). The foresters were contacted by phone to fix an appointment and supplied with the necessary information and basic data, such as the coordinates of the sample plots concerned and the manuals for the interview survey and forest road assessment, for preparing the interview. The foresters were asked to prepare all the documents they have about forest planning, ownership and functions, as well as intervention maps.

8.3.2.3 Training

The instruction courses for the NFI4 interview survey took place during the general course prior to the start of the field season in 2013. Two days of this one-week course were dedicated exclusively to the interviews with the local forest services, including the forest road assessment. In particular, the manual for the interview survey (Keller 2013) was explained in detail and the field staff had time to get familiar with the software SILVIS (Sect. 23.2.2). In addition to the specific training for the interviews according to the NFI manual, the surveyors participated in a specific course on interview techniques and received the equipment and documentation they needed to start the interviews. In November, after the end of the field survey, a second course took place before the beginning of the interviews in the Pre-Alps, Alps and Southern Alps.

8.3.2.4 Equipment and Material

The surveyors were provided with the following equipment per survey region (normally per canton or forest district):

- Notebook with the software SILVIS, address list of the foresters, form for the hourly report
- Manuals for the interview survey and the forest road assessment (Chap. 10) and a leaflet on the correct interview procedure
- Forest road maps from NFI3 for the forest road survey, including a large-scale map of the entire survey region
- Folder with sample-plot documentation: documentation form from NFI4, sample-tree information from NFI3 and NFI4, selected list of collected data from the terrestrial survey NFI4, and photos from NFI4
- Documents for public relations (NFI flyer, NFI3 poster set, NFI3 report on results, and DVD with film on NFI)

During the first interview days, people from NFI's headquarters accompanied the surveyors to ensure the data was of good quality and to clarify any uncertainties. Throughout the interview period, the surveyors could contact the headquarters via the hotline if they needed help.

Unlike with the field surveys, no repeat survey for the interview survey was carried out, as forest rangers would probably not appreciate being visited twice. Making them spend time and answering the same questions again would not promote NFI.

References

Düggelin C, Keller M (2017) Schweizerisches Landesforstinventar, Feldaufnahme-Anleitung 2017. Eidg. Forschungsanstalt für Wald, Schnee und Landschaft, Birmensdorf

Keller M (2013) Schweizerisches Landesforstinventar, Anleitung für die Umfrage- und Erschliessungserhebung 2013. Eidg. Forschungsanstalt für Wald, Schnee und Landschaft, Birmensdorf

Chapter 9
Field Assessment

Christoph Düggelin, Markus Keller, and Fabrizio Cioldi

Abstract The field assessment is the most important source of data for the Swiss National Forest Inventory (NFI). In NFI4 the regular terrestrial survey contained 7455 sample plots, of which 6357 were accessible and classified as 'forest' or 'shrub forest'. On each forest or shrub-forest sample plot, around 280 different attributes were assessed. The individual work steps for assessing a sample plot as well as the equipment and tools used, are explained in detail.

9.1 Introduction

The terrestrial inventory is the most important source of data for the Swiss NFI (NFI). In NFI4 the regular terrestrial survey contained 7455 sample plots, of which 6357 were accessible and classified as 'forest' or 'shrub forest'. On each accessible forest or shrub-forest sample plot, around 280 different attributes were assessed. These attributes were also the basis for hundreds of further variables, the so-called derived data (Sect. 20.3.1). In the following, the individual work steps for assessing a sample plot (Fig. 9.1) and the equipment and tools used are explained in detail. Most of the assessed attributes are described in this section or listed in Table 9.5 (Appendix). A detailed description of all attributes assessed in the field can be found in the field survey manual (Düggelin and Keller 2017).

C. Düggelin (✉) · M. Keller · F. Cioldi
Scientific Service NFI, Swiss Federal Research Institute WSL, Birmensdorf, Switzerland
e-mail: christoph.dueggelin@wsl.ch

C. Fischer, B. Traub (eds.), *Swiss National Forest Inventory – Methods and Models of the Fourth Assessment*, Managing Forest Ecosystems 35,
https://doi.org/10.1007/978-3-030-19293-8_9

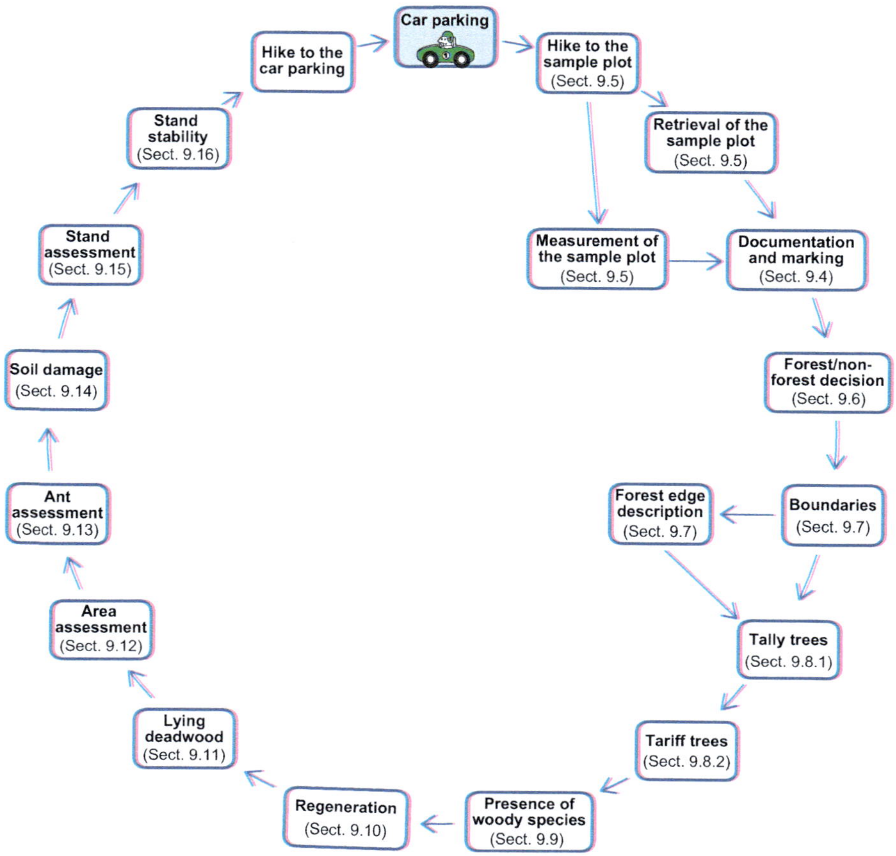

Fig. 9.1 Workflow on a sample plot. A detailed description of each work step can be found in the section indicated

9.2 Field Survey Equipment

Since the field-survey teams work independently in the Swiss forests for weeks at a time, carefully selected equipment is essential for them to be able to collect high quality data efficiently. Surveying the plots is physically demanding and can involve long treks and difficult terrain. The equipment should therefore be as robust, weather-resistant and light as possible. Moreover, the measuring instruments should be as precise as possible to ensure high data quality and electronic devices should be up to date. At the same time, any changes in the measuring instruments used should not lead to any methodological deviations from previous inventories where older measuring instrument were used. A list and a photo of the complete field equipment can be found in Fig. 9.7 (Appendix).

For the data collection in NFI4, the software application MAIRA developed at WSL was used. It is in the form of modules and assists the field teams in the process of data assessment with elaborate checks of data correctness and consistency. In addition, it allows the assessed data to be transferred to the central database using the mobile network. A detailed description of the MAIRA field application can be found in Chap. 23.

To increase work safety and work comfort, every member of the field staff received an annual allowance for work clothes and optimal footwear. Each field team had a first-aid kit for minor injuries or an emergency. In case of emergencies in areas without mobile-phone reception, every team member had a REGA-transceiver radio to organise help if needed.

Each field survey team had a four-wheel-drive Toyota HiAce van – or later a VW T6 van – at their disposal so they could drive with all the equipment to the work region and get as close as possible to the sample plots. This often involved driving on narrow and sometimes slippery forest roads.

9.3 Field Survey Manual

The field survey manual (field manual) describes the procedures used in the NFI terrestrial survey and contains binding definitions of the individual attributes with assessment instructions. The field manual is divided into the following chapters: introduction, preparation of sample-plot assessment, sample plot, 'forest/non-forest' decision, boundaries and forest edge assessment, single tree assessment, regeneration assessment, assessment of lying deadwood, assessment of area data, stand assessment and stand stability. A list of woody plant species recorded in the various field assessment phases is given in Table 9.6 (Appendix).

The field manual NFI4 (Düggelin and Keller 2017) is based on the field manuals NFI1 (Zingg and Bachofen 1988), NFI2 (Stierlin et al. 1994) and NFI3 (Keller 2005), supplemented with attributes newly introduced in NFI4. In NFI4, every year a new version of the field manual was issued with mostly only minor adjustments between issues to refine, for example, attribute definitions or assessment techniques.

9.4 Documentation and Permanent Marking of the Sample Plot

The main reason for marking and documenting a sample plot permanently is to be able to find it in a future inventory. Apart from this, documentation can help clarify ambiguities, for example, to check that a certain attribute is reasonable.

In NFI1–3, each sample-plot centre was inconspicuously marked with a 20 cm long aluminium bar sunk in the ground and three eye-catching marks were painted in blue on trees, stones or rocks near the plot centre. The polar coordinates of the blue marks were measured with respect to the sample-plot centre. In addition to the

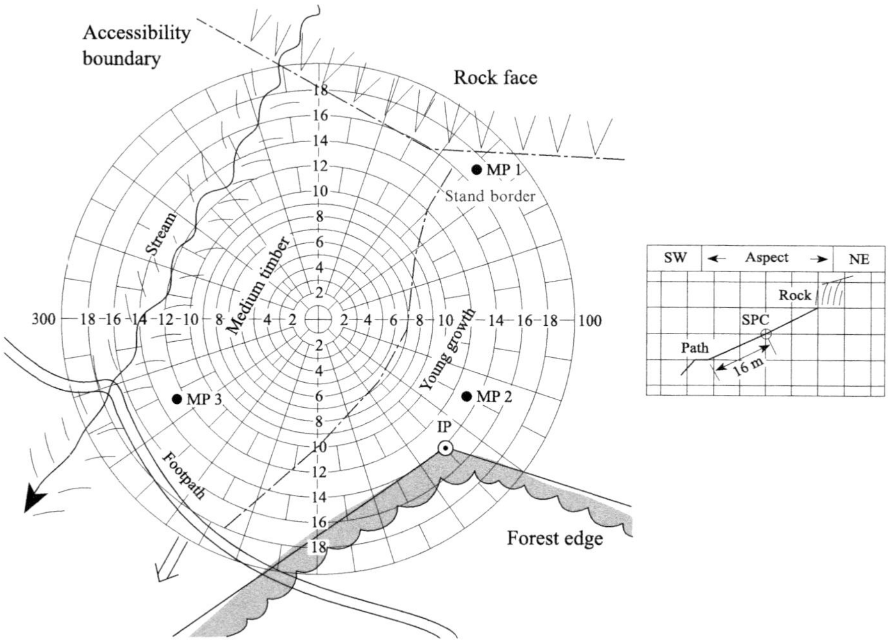

Fig. 9.2 Sketch (left-hand side: plan view; right-hand side: front view) of the sample plot as part of the documentation form where IP is the inflexion point of the stocking boundary (forest edge); MP the marking point and SPC the sample-plot centre

permanent marking, a documentation form with a sketch of the sample plot (Fig. 9.2) and information about how to access the sample plot was generated.

Marking the sample plot with eye-catching colours visible for everyone could be problematic as the forest on the plot might receive special treatment because people know it is surveyed. The inventory must be based on representative sample plots in order to correctly describe the reality in the Swiss forest. After NFI2, a special analysis detected no clear influence of the permanent marking of the sample plot (Traub 2001), but further investigations were recommended.

It was decided in NFI4 to mark the sample plots less visibly. Unfortunately, accurate position measurements using satellites were only possible on some of the sample plots when NFI4 started since only the American system (GPS) was available. The sample plots were therefore equipped with two RECCO-reflectors, with one in the ground next to the aluminium bar (maximum location distance approximately 5 m) and one on, for example, a branch at about 1.5 m aboveground within the 500-m^2 circle (maximum location distance approximately 40 m). RECCO is a technology developed to rescue people caught in avalanches and has the advantage that the reflector reacts passively. No batteries are needed and it can be expected to last a long time. One disadvantage is that the RECCO-reflectors could only be installed on 77% of the sample plots outside a possible avalanche perimeter (modelled with the avalanche simulation program AVAL-2D) to avoid endangering any rescue operations. From 2013 onwards, only the RECCO-reflector in the ground next to the aluminium bar was installed as, by then, more precise coordinate measurements of the sample-plot centres nationwide had become possible. The three conspicuous blue points near the sample-plot centre from NFI1–3 were painted over with an olive-grey camouflage colour. In addition to the RECCO-reflectors, two very small inconspicuous blue points were painted near the sample-plot centre and one larger blue point approximately 30 m away. The centre itself was marked, as in NFI1–3, with an aluminium bar (L-shaped) and a documentation form was created.

A new feature in NFI4 was the inclusion of five photos taken with a 24 mm focal length of the sample plot as additional documentation. Four were taken from the centre towards the North, East, South and West, and one towards the sample-plot centre (Fig. 9.3).

Another change introduced in NFI4 was to use a precise GNSS-receiver to measure the coordinates of the sample-plot centres. From 2013 onwards, the GLONASS-satellites could be used in addition to the GPS-satellites. The data was collected in the field and, after post-processing, the coordinates could be determined for 91.9% of the sample-plot centres. The mean value of the horizontal precision indicated by the software was 0.56 m. In NFI1–3, the newly determined sample-plot centres were measured with a measuring tape and a compass from a known reference point defined in an aerial photo or on a topographic map. In NFI4, however, the reference point was defined in most cases near the desired sample-plot centre using the GNSS-receiver. Table 9.1 shows the horizontal distances of the sample-plot centres from the desired coordinates measured in NFI1–3 with the conventional method and in NFI4 with the new method. While it shows, on the one hand, how exactly people measured in NFI13, on the other hand, it also shows the advantages of using a precise GNSS-receiver to define a reference point.

9.5 Measurement and Retrieval of the Sample-Plot Centre

To reach a sample-plot centre, the field team drove as close as it could to the sample plot. The teams have permission from the Swiss federal government to drive on all forest roads. In just a few cases, additional means of transport, such as aerial cableways or helicopters, were used to get closer to the sample plot. After this, the team went on foot using mainly the Swiss topographic map 1:25,000 to locate the plot, plus the documentation form from the previous inventory and a handheld GNSS-receiver. In the last field season in 2017, the field teams were also able to use a handheld GNSS-receiver combined with topographic maps and aerial images. Once they reached the sample plot, the field team searched for the blue marking points. The sample-plot centre itself could normally be localised using the known polar coordinates of the blue marking points, and if necessary, by referring to a tree map of the sample plot from the previous inventory with information about species, diameter-class and position of the tally trees (Fig. 9.4). In 92.8% of the sample plots already measured in a previous inventory, the field teams were able to find the

Fig. 9.3 Five photos of a sample plot as examples of the new documentation added in NFI4

aluminium bar at the centre. In 6.8% of the sample plots, the centre could be reconstructed precisely. Only in 0.4% of the sample plots was it necessary to take measurements to determine a new centre.

Table 9.1 Horizontal distance of newly determined sample-plot centres from the desired coordinates, classified after inventory

	Horizontal distance from the desired coordinate		
	NFI1–4 (n = 5845)	NFI1–3 (n = 5738)	NFI4 (n = 107)
Mean value (m)	6.23	6.31	1.86
Standard deviation (m)	7.19	7.23	1.91
Minimum (m)	0.05	0.05	0.17
First quartile (m)	2.69	2.77	0.92
Median (m)	4.56	4.62	1.43
Third quartile (m)	7.23	7.3	2.12
Maximum (m)	129.52	129.52	15.18

In those sample plots that were assessed for the first time and those already established that could not be found or reconstructed, the centre had to be determined. Usually three reference points from the aerial-image interpretation were available. Additionally, in NFI4 it was possible, for the first time, to define reference points in the field with a precise GNSS-receiver. This was the procedure chosen in most cases. The sample-plot centre was determined with the help of a compass, a measuring tape, a clinometer and the MAIRA field application on the tablet PC by measuring from a selected reference point.

9.6 'Forest/Non-forest' Decision

The decision whether a sample plot is located in a 'forest', 'shrub forest' or 'non-forest' is a point decision taken with reference to the sample-plot centre. The 'forest/non-forest' decision is mainly based on three inputs: the width of the stocked area, the degree of cover and the top height. To classify a forest as 'shrub forest', more than two-thirds of the reference stand cover must consist of green alder (*Alnus viridis*), shrubby mountain pine (*Pinus mugo prostrata*) or shrubs at least 3 m high. The complete flowchart for making a 'forest/non-forest' decision in NFI is illustrated in Fig. 9.5. See the field manual (Düggelin and Keller 2017), for a more detailed description.

If the 'forest/non-forest' decision was close, the field team assessed the forest boundary line with the help of the MAIRA field application software for documentation purposes. On newly forested areas, the mean sample-plot slope was determined from up- and downslope measurements immediately after the 'forest/non-forest' decision, and the sample-plot radii derived to obtain one horizontal plot 200 m^2 and one 500 m^2 in area. For sample plots classified as 'forest' or 'shrub forest' in an earlier inventory, the slope and the radii were given and could not be changed.

Sample plot in the inventory 310			650000 / 251000
Plot identifier: 16722	Inventory: 310	Survey date: 4th June 2004	m.a.s.l.: 360
Slope: 1 %	Sample Plot: Accessible	Centre: O-profile (NFI1)	Decision: Forest or Shrub-Forest
Stand development: Mixed	Stand Border: No	Forest edge assessment: No	Radius R2 and R5: 7.98 m/12.62 m

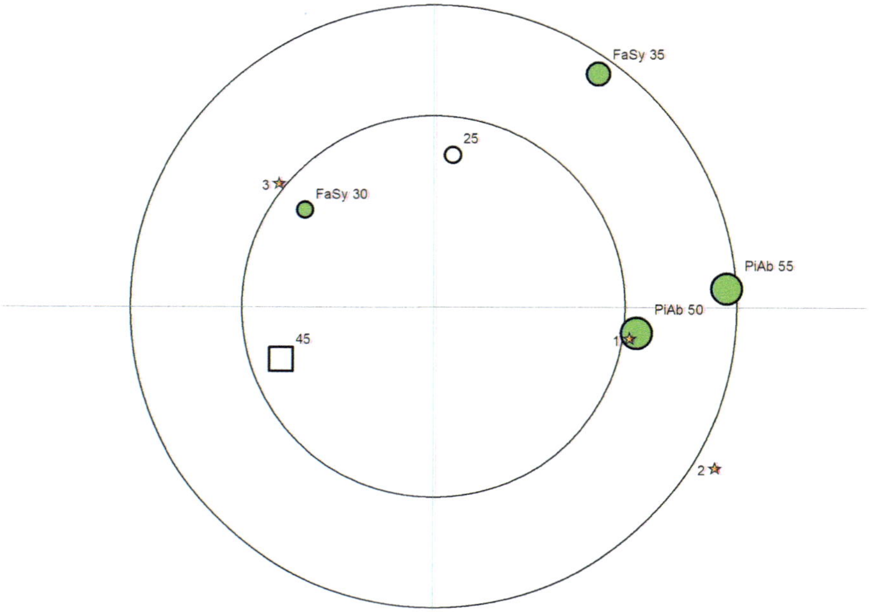

Tree species	DBH-Class	Distance [m]	Azimuth [°]	Comments
	25	6.4	8	Standing dead
Fagus sylvatica	35	11.88	39	Standing living
Picea abies	55	12.2	96	Standing living
Picea abies	50	8.5	108	Standing living
	45	6.7	279	Lying dead
Fagus sylvatica	30	6.7	341	Standing living

Mark. point	Description	Distance [m]	Azimuth [°]
1	PiAB dbh 48	8.25	110
2	AbAl dbh 29	13.47	133
3	FaSy dbh 48	8.25	343

Fig. 9.4 Example of a tree map of a sample plot

9.7 Boundaries and Forest Edge

Within the 500-m^2 circle, two potential boundaries, the accessibility boundary and the reduction line, could be assessed to delimit the accessible and stocked parts of a sample plot. They were used in calculating different attributes in the 500-m^2 circle.

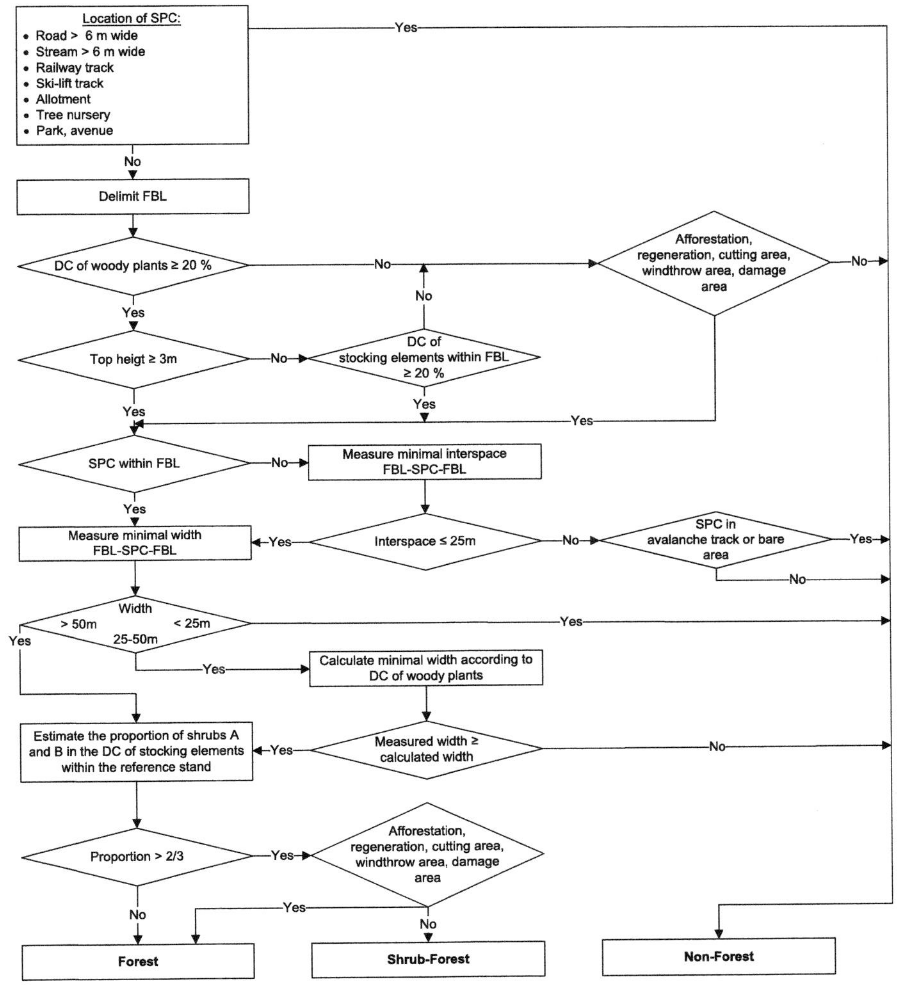

Fig. 9.5 Flowchart for the 'forest/non-forest' decision (*DC* Degree of Cover, *FBL* Forest Boundary Line, *SPC* Sample-Plot Centre)

No assessment was then allowed outside the boundaries. To assess the boundaries, an inflexion point and two azimuths were determined. The accessibility boundary delimits inaccessible parts of the 500-m^2 circle. The accessibility boundaries were usually adopted from the previous inventory. The reduction line reduces the sample plot inside the 500-m^2 circle to just that area which is stocked. It is defined as the line connecting the stems of the outermost trees and shrubs that have a minimum height of 3 m and that form the forest border. Unlike the accessibility boundary, the course of the reduction line may well change naturally or anthropogenically, and is therefore not adopted from the previous inventory. The reduction line is defined and measured each time independently.

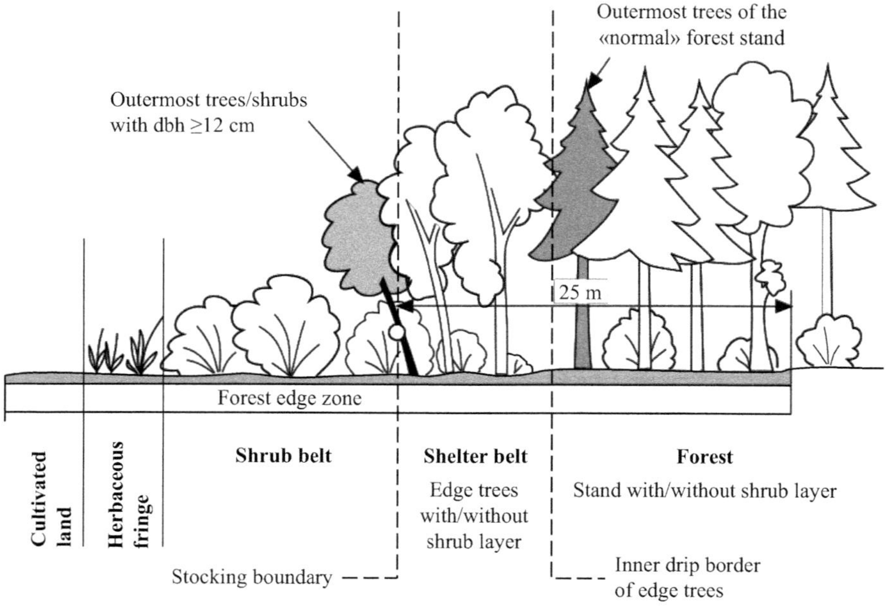

Fig. 9.6 Cross-section of a structured forest edge

Existing stand borders were identified in each 500-m^2 circle. Stand borders separate the reference stand, i.e. the stand containing the sample-plot centre, from the other stands present in the 500-m^2 circle. The stand borders were not measured, but if an assessed tally tree was part of the reference stand, this was recorded.

The *forest edge* is the transition zone between forest and open land. Structured forest edges often contain many ecologically valuable habitats and are very species-rich. Since NFI2, the forest edge has been assessed in detail on sample plots with a stocking boundary within the 25-m circle. The stocking boundary is the tangent connecting the outermost trees and shrubs with a dbh of at least 12 cm that form the forest border. To assess the stocking boundary, an inflexion point and two azimuths were measured. The forest edge was assessed on each side of the inflexion point 25 m along the azimuths, resulting in a so-called assessment line 50 m in length. If the stocking boundaries had been determined in previous inventories and were still meaningful, they were adopted to ensure comparability.

The forest edge is assessed along the assessment line according to: aspect, shape, immediate neighbourhood type, delimiting barriers, vertical structure and vegetation density. The widths of the different forest-edge zones, the herbaceous fringe, shrub belt and shelter belt are measured (Fig. 9.6). All woody plants visible from the outside are recorded, and their share of the vertical projection estimated. In addition, visible human-made interventions and influences are assessed.

In NFI4, 929 forest edges in total were assessed (on 15% of the forest and shrub-forest sample plots).

9.8 Single Trees

9.8.1 Assessment of Tally Trees

Tally trees are woody plants according to the NFI species list (Table 9.6 in Appendix) on the sample plot that fulfil the following criteria:

- Those inside the 200-m^2 circle must have a dbh $\geq$12 cm.
- Those outside the 200-m^2 circle but inside the 500-m^2 circle must have a dbh $\geq$36 cm.

Tally trees also include lying and dead trees and are identified by their polar coordinates (distance and azimuth from the sample-plot centre).

To facilitate the assessment of the tally trees already assessed in the previous inventory, their polar coordinates and tree species are pre-determined in the MAIRA field application. In NFI4, a total of 72,316 tally trees were measured and assessed. Additionally, 10,030 of the tally trees from the previous inventory were no longer recorded as tally trees because they did not meet the necessary conditions, for example if only the stump was left or the whole tree had disappeared or could not be clearly identified (Table 9.2).

The attributes assessed for each tally tree depended on the tree's status and type, i.e. whether it was living, dead, standing or lying (Table 9.5 in Appendix provides a detailed list of attributes). For example, for standing living sample trees, the attributes assessed were: species, diameter at breast height, special growth habits, crown

Table 9.2 Individual trees according to their status and type

TREE STATUS	Standing living	Standing dead	Lying living	Lying dead	Number
Tally tree in earlier NFIs (A)	54,148	3393	307	3035	60,883
New in NFI4 (B)	10,642	147	98	546	11,433
Total no. assessed	64,790	3540	405	3581	72,316
Tariff trees (subset of the total no. assessed)	13,507				13,507
No assessment (C)					2219
New stump (D)					6975
Nothing present (E)					830
Not identified (F)					6
Total no. of trees considered					82,346

length, vertical stand position, estimated tree age and kind of damage if any. For standing or lying dead tally trees, the attributes assessed were: diameter at breast height, stage of decomposition according to different attributes, ecological features such as woodpecker cavities, the presence of specific fungus species and colonisation by moss, lichens, young trees or shrubs. Their year of death was also estimated. For stumps, their height and diameter were measured and the year of cutting estimated. Where possible, tree rings were counted.

9.8.2 Measurement of Tariff Trees

To derive tariff functions to estimate individual tree volumes using only one measurement variable (the dbh in cm), a subset of the standing living trees was defined as tariff trees (Sect. 2.3.4). For tariff trees, the attributes tree height, height of the crown base and the upper stem diameter at a height of 7 m (measured with the *Finnish calliper*) were also assessed. Prerequisites for tariff trees are that they must not be bent, tilted or pollarded and must not have a broken stem. The probability of a tally tree being selected as a tariff tree depends on its dbh and is calculated according to Eqs. 9.1, 9.2, and 9.3:

$$p_i = 5.625 \times 10^{-6}\ (DBH_i)^{2.00}\ 50\ \mathit{if}\ 12\ cm \le DBH_i \le 35\ cm \tag{9.1}$$

$$p_i = 1.125 \times 10^{-8}\ (DBH_i)^{3.75}\ 20\ \mathit{if}\ 36\ cm \le DBH_i \le 59\ cm \tag{9.2}$$

$$p_i = 1\ \mathit{if}\ DBH_i \ge 60\ cm \tag{9.3}$$

The tariff trees were selected within the sample circles in circular sectors starting in the North (0 gon) with an opening angle in a clockwise direction. A tally tree was selected as a tariff tree if it was within its specific circular sector. The opening angle of each specific circular sector depends on the dbh of the tally tree and is defined such that the ratio of the area of the circular sector to the area of the sample circle is equal to p_i. Thus trees selected as tariff trees will also be considered tariff trees in future inventories provided they still meet all the tariff tree prerequisites.

9.9 Presence of Woody Species

The presence of woody species is assessed mainly in order to have a basis for assessing the diversity of woody species and their geographical and ecological distribution. Within the 200-m^2 circle, all trees, shrubs and some woody climbing plants with a minimum height of 40 cm and a maximum dbh of 11.9 cm were assessed (see the NFI species list in Table 9.6 in Appendix). The plants were classified as either: (a) plants 40–129 cm in height or (b) plants ≥130 cm height and a dbh <12 cm. Individuals were not counted, and only the presence of a species in the respective class was recorded.

9.10 Regeneration

Since regeneration is the basis of future forest stands, a detailed description is essential. Thus information about the frequency, spatial distribution and condition of the species regenerated was obtained. Any protective measures taken were assessed, as was general information about what facilitates or hinders regeneration.

In NFI4, the regeneration was assessed on 98.6% of the accessible forest and shrub-forest sample plots. The assessment usually took place on a subplot 10 m westwards of the sample-plot centre. The statistical method is described in detail in Chap. 2. The species registered were all tree species, as well as the seven stand-building shrub species: *Alnus viridis*, *Corylus avellana*, *Ilex aquifolium*, *Juniperus communis*, *Laburnum anagyroides*, *Pinus mugo prostrata* and *Prunus padus*. Four regeneration classes were distinguished:

- Regeneration class 1 (plants with 0.10–0.39 m length)
- Regeneration class 2 (plants with 0.40–1.29 m length)
- Regeneration class 3 (plants with a dbh of 0.1–3.9 cm)
- Regeneration class 4 (plants with a dbh of 4.0–11.9 cm)

In a first step, for each class, the plant closest to the subplot centre within a given maximum radius (class 1: 2.5 m; classes 2–4: 5.0 m) was described in terms of: horizontal distance from the subplot centre, species, diameter at breast height (only for classes 3 and 4), origin and various types of possible damage such as browsing of the previous year's leading shoot.

In a second step, all individuals of the registered species were assessed within a given radius of a horizontal circle adjusted for slope according to the four regeneration classes: class 1 within a horizontal circle 0.9 m in radius, class 2 within a horizontal circle 1.5 m in radius, class 3 within a horizontal circle 2.5 m in radius, and class 4 within a horizontal circle 4.0 m in radius. For each individual in classes 1 and 2, browsing of the previous year's leading shoot was assessed. For each individual in classes 3 and 4, the diameter at breast height was measured and for each individual in all classes, species and coppice growth was recorded. In NFI4, 11.7 individuals on average were assessed on each subplot (maximum 310), with 3.6 on average in class 1 (maximum 301), 2.6 in class 2 (maximum 150), 3.3 in class 3 (maximum 153) and 2.2 in class 4 (maximum 32).

In a third step, factors facilitating or hindering regeneration were assessed for the subplot area within a horizontal circle 4.0 m in radius. In Table 9.5 (Appendix) a detailed list of the attributes assessed during the three steps of the regeneration assessment is provided.

In addition to the quantitative assessment of the regeneration on the subplots, in the reference stands (Sect. 9.15) the degree of cover of the regeneration (plants with a minimum height of 0.10 m and with a dbh up to 11.9 cm) and the established regeneration (plants with a minimum height of 1.3 m and with a dbh up to 11.9 cm) were estimated. The proportions of the main tree species relative to the established regeneration were also estimated.

9.11 Lying Deadwood

Deadwood provides nourishment and habitats for many different species and fixes greenhouse-gas-relevant carbon. It is also an important structural element of the forest and can play a decisive role in forest regeneration. In NFI4, the quantity, dimensions and degrees of decomposition of lying deadwood were recorded along three 10-m-long transect lines oriented at 35 gon, 170 gon and 300 gon around the sample-plot centre according to the Line Intersect Sampling method. For a piece of wood to be assessed, it had to have a minimum cross-diameter of 7 cm at the point of intersection with the transect line – see Table 9.5 in Appendix for a detailed list of the attributes assessed on the recorded deadwood. The statistical methods are described in Sect. 2.3.5. In NFI4, 1.8 pieces of deadwood on average were assessed per sample plot. The data for 1.8% of the recorded pieces is based on rather 'daring' estimates where it was assumed that mixed-up heaps of branches contained such pieces but it was not possible to identify and measure them exactly.

9.12 Area Assessment

For the survey of the interpretation area (50 × 50 m), a range of site factors were recorded and any traces of natural hazards, such as rockfall or landslides, and of pasturing were noted. The ecological aspects assessed included the presence of heaps of branches, root plates and geomorphological objects. The proportion of the area with obstacles and other relevant constraints that could interfere with timber harvesting were estimated – see Table 9.5 in Appendix for a detailed list of the attributes assessed on the interpretation area.

9.13 Ant Assessment

Red wood ants (*Formica rufa*-group), which build mounds, are keystone species and relatively rare in Switzerland, especially at lower elevations. To obtain more information on their distribution and habitat requirements, in NFI4 all inhabited and abandoned ant mounds within the 500-m^2 circle were recorded and their polar coordinates, height and diameter measured. Ant samples from each inhabited ant mound were taken so that experts could determine the species later. Normally, each ant mound was photographed. In addition to recording the ant mounds, the first standing living tally tree (smallest azimuth) was also searched for ants and, where possible, samples were taken.

9.14 Soil Damage

Soil damage caused by logging influences the natural development of the soil, the hydrological balance, the availability of nutrients and, consequently, the sustainability of forest development. In NFI4, skidding tracks and the percentage of cover with vehicle tracks within the 200-m^2 circle were assessed. Where vehicle tracks were clearly visible, the track closest to the sample-plot centre was described in terms of its length within the 200-m^2 circle and its age, the width of the vehicle track, the width and depth of the wheel track, the presence of puddles in the wheel track and the severity of soil damage.

9.15 Stand Assessment

Once a stand has been assessed in detail, conclusions can be drawn about the forest's management, protection function and habitats. In NFI, a stand is defined as a community of trees/shrubs that differs clearly from other plant communities with respect to species composition, age and/or structure. The minimum area of a stand is 500 m^2. In NFI, the area within the interpretation area (50 × 50 m) of the stand containing the sample-plot centre, the so-called reference stand, is assessed. The main attributes used in describing a stand concern its horizontal and vertical structure, the tree species in the upper layer, the stand's age and stage of development, and the degree of regeneration cover.

Different attributes were assessed according to the type of forest use. In Table 9.5 (Appendix) a detailed list of the attributes assessed in the reference stands in NFI4 for the different forest-use categories is provided. The type of forest use is determined at the sample-plot centre (Table 9.3).

In preparation for the interview with the local forest service (Chap. 10) the field team carried out a preliminary assessment of the sample plot according to attributes concerning the origin of the stand, recent and appropriate future silvicultural interventions, areal damage since the previous inventory and the type and intensity of current recreational use – see Table 9.5 in Appendix for a detailed list of the attributes assessed.

Table 9.3 Forest-use categories and their frequency in NFI4

Forest-use category	Number of sample plots	Frequency (%)
Road	56	0.9
Timber yard	3	0
Recreation area	4	0.1
Nursery	0	0
Stream	9	0.1
Unstocked, channel/track	14	0.2
Meadow, cropland	76	1.2
Other bare patches	10	0.2
Swath	16	0.3
Road shoulder	21	0.3
Stand	6050	95.2
Cutting area	55	0.9
Windthrow area	27	0.4
Damage area	16	0.3
Total	6357	100

9.16 Stand Stability

Stand stability describes how resilient to disturbing influences the reference stand is expected to be during the next 10 years in the Plateau, Jura and Pre-Alps and during the next 20 years in the Alps and Southern Alps. In the interpretation area, the magnitude of the impact of potential damage sources, such as snow load, wind, landslide, rockfall, snow movement, fire, pasturing, game or humans, is first estimated on a scale of 1–10. Second, the resilience of the following reference stand characteristics is estimated by considering the relevant loads on a scale of 1–10: species, slenderness ratio, crown length, crown shape, skewness, stage of stand development, degree of closure/gaps, stand structure/steep edges, vitality, damage and diseases.

On the basis of these two ratings, the probability that the reference stand will be seriously damaged or even collapse within the next 10 or 20 years is then estimated.

Appendix

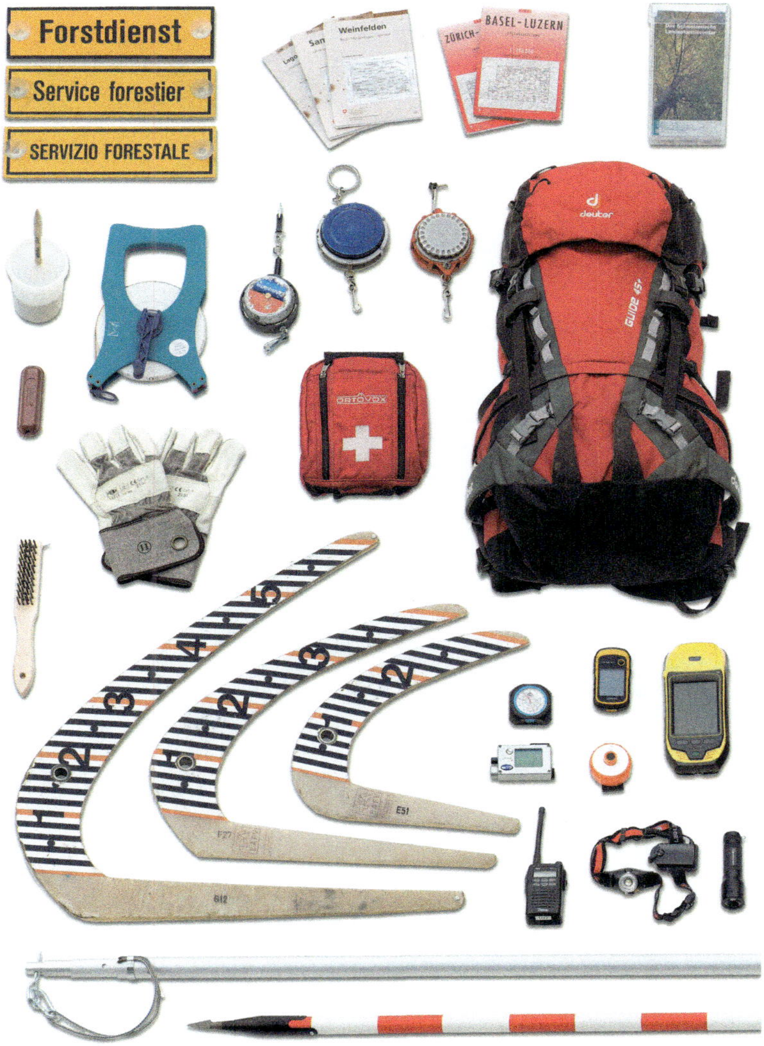

Fig. 9.7 Photos of the complete equipment – see the list in Table 9.4

Fig. 9.7 (continued)

Table 9.4 List of equipment for fieldwork in NFI4

Documents and books
Field manual NFI4
Complete set of topographic maps 1:25,000 and 1:100,000
Map and list of sample plots to be assessed[a]
Site diagram NFI3, tree diagram NFI3 and aerial photo of sample plots to be assessed[a]
Botanical field guides
Information box with NFI leaflets
Equipment and tools
Backpack (DEUTER, Guide 45+)
First-aid kit
Radio for REGA
Mobile phone (NOKIA, 3720 classic)
Smartphone with stored topographic maps, aerial images and sample plots to be assessed (MOTO G5 Plus, navigation software OruxMaps)
GNSS-receiver (GARMIN, eTrex 10)
Altimeter (THOMMEN TX22)
Tool vest
Work gloves
Billhook
Folding saw
Pocket knife (Victorinox: SwissTool)
Files (round and flat)
Brick hammer
Wire brush
Paint-tins with brushes and blue (RAL 5010) and olive grey (RAL 7002) paint
RECCO reflector
Aluminium profile to mark the sample-plot centre permanently
Torch
Headlamp
Extension cable with multiple socket
Magnifying glass (thread counter)
Umbrella
Cleaning material and maintenance equipment
Surveying material
Hand compass (SUUNTO, KB-14, 400 g)
Clinometer (SUUNTO, PM-5, 400PC)
Compass (WYSSEN, MI-4007)
Tripod for compass (Gitzo GT2531LVL)
Dendrometer to measure distances and heights (VERTEX IV)
Binoculars (NIKON, 8 × 25)
Digital camera (LUMIX, DMC-LX7)
GNSS-receiver (TRIMBLE, Geo7X)
Long ranging pole (2 m)

(continued)

Table 9.4 (continued)

Measuring tape 50 m (fibreglass)
Measuring tape 25 m (steel)
Measuring tape 20 m (steel)
Circumference measuring tape 10 m (steel)
Folding metre rule
Pocket knife (Victorinox: Picknicker)
Marking out hook
Calliper 60 cm
Calliper 30 cm
Telescopic pole (7 m)
Finnish calliper (30, 40 and 60 cm)
Data collection and data transfer
Tablet PC and accessories (PANASONIC, Toughpad FZ-G1)
USB Memory Stick 16 GB[a]
Vehicle
TOYOTA, HI-ACE 4 WD 2.7 SWB or VW, T6 Kombi RS[a]
Forest service signs
Car accessories[a]

[a]Not illustrated

Table 9.5 List of selected attributes assessed in NFI4

1.3.8. Single trees
Assessment of a pre-determined tally tree or of a new tally tree (A+B)
Azimuth of tree measured from sample-plot centre (for identified tally trees data from previous inventory is pre-determined)
Distance of tree from plot centre (for identified tally trees data from previous inventory is pre-determined)
Species (for identified tally trees data from previous inventory is shown)
Growth habit if the genus is Salix, to distinguish between a dendriform and a shrublike growth
Comments about any special features of the tally tree
Reason for new tally tree (only for new tally trees)
Diameter at breast height (for trees with a diameter ≤60 cm)
Circumference (for trees with a diameter >60 cm)
Stand affiliation (only if a stand border crosses the 500-m^2 circle)
In case of standing living trees
Inclination of tally tree to sample-plot centre
Crown length
Stand layer to which the tally tree belongs
Tree age
Method for determining the tree age
Type of most important damage
Place of most important damage
Cause of most important damage

(continued)

Table 9.5 (continued)

Type of the second most important damage
Place of the second most important damage
Cause of the second most important damage
Stem height of broken tally tree (where the damage is 'Bole broken' or 'Stem broken')
Remain of the tree top (where the damage is 'Bole broken' or 'Stem broken')
In case of lying living trees
Tree age
Method for determining the tree age
In case of standing and lying dead trees
Inclination of tally tree to sample-plot centre if the dead tree is still standing (i.e. a snag)
Coniferous or broadleaved tree if the species is no longer identifiable
Stand layer to which the tally tree belongs if the dead tree is still standing
Estimated year of death of the tally tree
Status of stump to describe if the tally tree is still in place in one piece if the dead tree is lying
Broken stem or cut stem
Stem height of broken snag if the dead tree is standing with a broken or cut stem
Length of dead tally tree if it is lying
Presence of twigs <3 cm
Bark cover
Stage of deadwood decomposition
Ground contact if the dead tree is lying
Species of long-living fungal fruiting-bodies if present
Presence of woodpecker cavities
Degree of moss cover
Degree of lichen cover
Number of shrubs regenerating on dead tally tree
Number of trees regenerating on dead tally tree
Assessment of an existing pre-determined tally tree that is no longer considered a tally tree (C)
Azimuth of tree measured from sample-plot centre (data from previous inventory is pre-determined)
Distance of tree from plot centre (data from previous inventory is pre-determined)
Species (data from previous inventory is shown)
Growth habit if the genus is Salix to distinguish between a dendriform and a shrublike growth
Reason for no longer assessing the tree
State of the former tally tree (alive/dead, standing/lying)
Assessment if only the stump of a predetermined tally tree is present (D)
Species (data from previous inventory is shown)
Growth habit if the genus is Salix to distinguish between a dendriform and a shrublike growth
Reason why the tally tree is missing
Number of tree rings if countable
Estimated year of leaving the population if the tally tree left the population
Estimated year of death of the tally tree
Assessment if nothing at all is left of a pre-determined tally tree (E)
Reason why the pre-determined tally tree is missing

(continued)

Table 9.5 (continued)

Estimated year of disappearance if the tally tree is no longer part of the tally tree population
Assessment of an unidentified tree (F)
Tally trees which cannot be identified clearly by means of the pre-determined data lose their own ID-number and are assessed as new tally trees (and thus get a new ID-number)
1.3.10 Regeneration
Assessment of single plants
Horizontal distance subplot centre to plant (relevant measuring point: regeneration classes 1 and 2 = base point; regeneration classes 3 and 4 = dbh point)
Species
Dbh (only for regeneration classes 3 and 4)
Indication whether it is a coppice regrowth
Indication whether it is the main coppice if it is coppice regrowth
Distinction between natural and artificial regeneration
Distinction between generative and vegetative reproduction
Individual plant protection
Various types of damage (dry top, rubbed trees or stripped bark, disease, damage caused by logging, other damage)
Browsing of the leading shoot of the previous year (only for regeneration classes 1 and 2)
Length of the leading shoot of the previous year (only for regeneration classes 1 and 2)
Substrate the plant is growing on
Silvicultural value
Counting the young plants
Species
Indication whether it is a coppice regrowth
Indication whether it is the main coppice if it is coppice regrowth
Indications which coppices belong to the same individual if it is coppice regrowth
Dbh (only for plants of the regeneration classes 3 and 4)
Browsing of the leading shoot of the previous year (only for regeneration classes 1 and 2)
Areal data referring to subplot
Proportion of area with naturally poor growing conditions
Proportion of area with anthropogenically poor growing conditions
Proportion of area with competing vegetation
Group of competing species where competing vegetation is present
Main type of topsoil
Degree of shading of the subplot centre at heights of 0.40 m and 1.30 m
1.3.11. Lying deadwood
Cross diameter at the point of intersection with the transect line
Method for measuring the cross diameter (measured or estimated)
Angle to the horizontal at the point of intersection with the transect line
Broadleaved or coniferous wood
Degree of decomposition at the point of intersection with the transect line
Indication whether the piece is an element of the NFI tree population
Indication whether the piece is an uprooted trunk
Indication whether the piece belongs to a confusing heap of branches

(continued)

Table 9.5 (continued)

Indication whether the intersected piece includes a section at least 1 m in length that is larger than 10 cm in diameter
1.3.12 Area assessment
General information
Determinability of the aspect (only on new forest plots, otherwise given)
Azimuth of the aspect (only if the aspect is determinable; only on new forest plots, otherwise given)
Type of relief (only on new forest plots, otherwise given)
Traces of landslides
Traces of water erosion
Traces of rockfall
Assessment of the largest three boulders (size and shape; only if at least one boulder is present; assessment within the 500-m^2 circle)
Traces of slow snowpack movements
Traces of avalanches
Traces of forest fires
Traces of pasturing (only surveyed within the stocking boundaries)
Proportion of area which could interfere with timber harvesting due to obstacles
Presence and most relevant type of constraints on logging
Presence of root plates (distinguished as large, small and former)
Presence of heaps of branches
Presence of stumps
Presence of snags
Presence of dry stonewalls and heap of stones
Presence and most relevant type of excessive ecological pressure and disturbances
Types of recreational facilities
Presence and most relevant type of gap (only surveyed within the stocking boundaries)
Presence and most relevant type of geomorphological objects or micro-reliefs
Presence and type of azonal site
Number of mould cavities on standing tally trees $\geq$36 cm (distinguished as 'broadleaved', 'coniferous' and 'undefined wood'; only within the 500-m^2 circle)
1.3.13. Stand assessment
General information
Presence of an inner edge on forest land
Stand size
Distinction between 'shrub forest', 'scattered stocking' and 'closed forest' (not assessed if the forest-use category = B)
Distinction between 'high forest', 'coppice forest', 'coppice with standards', 'selva' and 'plantation' (not assessed if the forest-use category = B)
Top height of the stand (not assessed if the forest-use category = B)
Degree of cover of stand layer (not assessed if the forest-use category = B)
Type of closure of stand layer (not assessed if the forest-use category = B)
Vertical stand structure (not assessed if the forest-use category = B)
Stage of stand development (depending on the dominant dbh; not assessed if the forest-use category = B)

(continued)

Table 9.5 (continued)

Stand age (not assessed if the stage of stand development is 'mixed' or the forest-use category = B)
Method for determining the stand age (not assessed if the stage of stand development is "mixed" or the forest-use category = B)
Proportion of basal area of conifers and deciduous trees (not assessed if the forest-use category = B)
Crown closure (not assessed if the forest-use category = B)
Species in the upper layer (all tree species as well as shrubs A; not assessed if the forest-use category = B)
Relative species proportion of the degree of cover in the upper layer (not assessed if the forest-use category = B)
Degree of cover of regeneration (all trees higher than 0.1 m and with a dbh <12 cm)
Degree of cover of established regeneration (all trees at least 1.3 m high with a dbh <12 cm)
Relative main tree species proportion of the degree of cover of established regeneration (not assessed if the degree of cover of established regeneration is <1%)
Type of regeneration ('natural', 'planted' or 'mixed')
Protection method of regeneration
Degree of cover of shrub layer (all tree and shrub parts from 0.5 m to 3.0 m in height)
Degree of cover of berry bushes (Rubus and Vaccinium species)
Main species of berry bush (Rubus and Vaccinium species) (not assessed if the degree of cover of berry bushes is <1%)
Attributes required to prepare the interview survey
Type of silvicultural treatment to be done next
Urgency of silvicultural treatment to be done next (not assessed if under type of silvicultural treatment 'no treatment' is proposed)
Type of first areal damage (refers to the whole interpretation area)
Year of first areal damage
Type of second areal damage (refers to the whole interpretation area)
Year of second areal damage
Extent of areal damage (only where areal damage has occurred, refers to the whole interpretation area)
State of clearing of the areal damage (only where areal damage has occurred, refers to the whole interpretation area)
Type of origin of the forest (data from previous inventory is shown)
Year of afforestation (data from previous inventory is shown) (only assessed if the type of origin of the forest is 'artificial reforestation' or 'mixed reforestation')
Type of stand origin (data from previous inventory is shown)
Years since the last silvicultural treatment (data from previous inventory is shown)
Type of last silvicultural treatment since the previous inventory (only assessed if there was a silvicultural treatment since the previous inventory)
Intensity of current recreational use (refers to the forest area within a radius of 100 m around the sample-plot centre)
Type of current recreational use (refers to the forest area within a radius of 100 m around the sample-plot centre)

Table 9.6 Species list for NFI4

Conifers		**Code**	**A**	**B**	**C**	**D**
Abies alba	Silver Fir	**11**	x	x	x	x
Larix decidua	European Larch	**146**	x	x	x	x
Larix kaempferi	Japanese Larch	**147**	x	x	x	x
Picea abies	Norway Spruce	**10**	x	x	x	x
Pinus cembra	Swiss Pine	**19**	x	x	x	x
Pinus mugo arborea	Mountain Pine	**18**	x	x	x	x
Pinus nigra	Black Pine	**16**	x	x	x	x
Pinus strobus	Eastern White Pine	**17**	x	x	x	x
Pinus sylvestris	Scots Pine	**15**	x	x	x	x
Pseudotsuga menziesii	Douglas Fir	**22**	x	x	x	x
Taxus baccata	Common Yew	**25**	x	x	x	x
Exotic conifers						
Abies sp.	Firs	**30**	x	x	x	x
Cedrus sp.	Cedars	**31**	x	x	x	x
Chamaecyparis sp.	Chamaecyparis	**32**	x	x	x	x
Cryptomeria sp.	Cryptomeria	**33**	x	x	x	x
Metasequoia	Dawn Redwood	**34**	x	x	x	x
Picea sp.	Firs	**35**	x	x	x	x
Pinus sp.	Pines	**36**	x	x	x	x
Sequoiadendron	Giant Sequoia	**37**	x	x	x	x
Thuja sp.	Thuja	**38**	x	x	x	x
Tsuga sp.	Hemlock	**39**	x	x	x	x
	Other conifers	**49**	x	x	x	x
Broadleaved trees						
Acer campestre	Field Maple	**56**	x	x	x	x
Acer opalus	Italian Maple	**59**	x	x	x	x
Acer platanoides	Norway Maple	**57**	x	x	x	x
Acer pseudoplatanus	Sycamore Maple	**58**	x	x	x	x
Aesculus hippocastanum	Horse Chestnut	**89**	x	x	x	x
Ailanthus altissima	Tree of Heaven	**161**	x	x	x	x
Alnus glutinosa	Black Alder	**63**	x	x	x	x
Alnus incana	Grey Alder	**64**	x	x	x	x
Betula pendula	Silver Birch	**65**	x	x	x	x
Betula pubescens	Downy Birch	**66**	x	x	x	x
Carpinus betulus	Common Hornbeam	**67**	x	x	x	x
Castanea sativa	Sweet Chestnut	**62**	x	x	x	x
Celtis australis	European Hackberry	**68**	x	x	x	x
Cinnamomum camphora	Camphor Tree	**162**	x	x	x	x
Fagus sylvatica	Common Beech	**50**	x	x	x	x
Fraxinus excelsior	Common Ash	**60**	x	x	x	x
Fraxinus ornus	Manna Ash	**61**	x	x	x	x
Juglans regia	Walnut	**69**	x	x	x	x
Liriodendron tulipifera	Yellow Poplar	**90**	x	x	x	x
Malus sylvestris s.l.	European Wild Apple	**71**	x	x	x	x
Ostrya carpinifolia	Hop Hornbeam	**70**	x	x	x	x
Platanus sp.	Plane Tree	**163**	x	x	x	x

(continued)

Table 9.6 (continued)

Broadleaved trees		**Code**	**A**	**B**	**C**	**D**
Populus alba	Silver Poplar	**150**	x	x	x	x
Populus x canescens	Grey Poplar	**149**	x	x	x	x
Populus nigra s. l.	Black Poplar incl. hybrids	**74**	x	x	x	x
Populus sp.	Other Poplars	**76**	x	x	x	x
Populus tremula	Aspen	**75**	x	x	x	x
Prunus avium	Sweet Cherry	**77**	x	x	x	x
Pyrus communis & P. pyraster	European Pear	**72**	x	x	x	x
Quercus cerris	Turkey Oak	**54**	x	x	x	x
Quercus petraea	Sessile Oak	**52**	x	x	x	x
Quercus pubescens	Downy Oak	**53**	x	x	x	x
Quercus robur	English Oak	**51**	x	x	x	x
Quercus rubra	Northern Red Oak	**55**	x	x	x	x
Rhus typhina	Staghorn Sumac	**191**	x	x	x	x
Robinia pseudacacia	Locust	**78**	x	x	x	x
Salix alba	White Willow	**79**	x	x	x	x
Salix caprea	Goat Willow	**152**	x	x	x	x
Salix sp.	Willow	**80**	x	x	x	x
Sorbus aria	Common Whitebeam	**81**	x	x	x	x
Sorbus aucuparia	Rowan	**82**	x	x	x	x
Sorbus domestica	True Service Tree	**83**	x	x	x	x
Sorbus latifolia s.l.	Service Tree of Fontainebleau	**159**	x	x	x	x
Sorbus mougeotii	Mougeots-Whitebeam	**158**	x	x	x	x
Sorbus torminalis	Wild Service Tree	**84**	x	x	x	x
Tilia cordata	Small-leaved Linden	**85**	x	x	x	x
Tilia platyphyllos	Large-leaved Linden	**86**	x	x	x	x
Ulmus glabra	Wych Elm	**88**	x	x	x	x
Ulmus laevis	European White Elm	**160**	x	x	x	x
Ulmus minor	Field Elm	**87**	x	x	x	x
	Other broadleaved trees	**99**	x	x	x	x
Shrubs A						
Alnus viridis	Green Alder	**5**	x	x	x	x
Corylus avellana	Hazel	**106**	x	x	x	x
Ilex aquifolium	Holly	**6**	x	x	x	x
Juniperus communis	Common Juniper	**2**	x	x	x	x
Laburnum anagyroides	Golden Chain tree	**7**	x	x	x	x
Pinus mugo prostrata	Mountain Pine	**1**	x	x	x	x
Prunus padus	Bird Cherry	**8**	x	x	x	x
Shrubs B						
Amelanchier ovalis	Snowy Mespilus	**102**	x	x	x	
Berberis vulgaris	European Barberry	**100**	x	x	x	
Buddleia sp.	Butterfly Bush	**173**	x	x	x	
Buxus sempervirens	Common Box	**101**	x	x	x	
Chamaerops humilis	European Fan Palm	**189**	x	x	x	
Clematis alpina	Alpine Clematis	**192**	x	x	x	
Colutea arborescens	Bladder Senna	**194**	x	x	x	
Cornus mas	European Cornel	**105**	x	x	x	
Cornus sanguinea	Common Dogwood	**104**	x	x	x	

(continued)

Table 9.6 (continued)

Shrubs B		Code	A	B	C	D
Coronilla emerus	Scorpion Senna	195	x	x	x	
Cotinus coggygria	Smoke tree	209	x	x	x	
Cotoneaster integerrima	Common Cotoneaster	196	x	x	x	
Cotoneaster tomentosa	Woolly Cotoneaster	197	x	x	x	
Crataegus monogyna	Common Hawthorn	171	x	x	x	
Crataegus oxyacantha	Midland Hawthorn	172	x	x	x	
Cytisus scoparius	Common Broom	198	x	x	x	
Daphne alpina	Alpine Daphne	199	x	x	x	
Daphne laureola	Spurge Laurel	200	x	x	x	
Daphne mezereum	Mezereon	201	x	x	x	
Euonymus europaeus	European Spindle	169	x	x	x	
Euonymus latifolius	Large-leaved Spindle	170	x	x	x	
Ficus carica	Common Fig	174	x	x	x	
Hippophae rhamnoides	Sallow Thorn	113	x	x	x	
Juniperus com-munis ssp alpina	Alpine Juniper	202	x	x	x	
Juniperus sabina	Savin Juniper	203	x	x	x	
Laburnum alpinum	Scottish Laburnum	175	x	x	x	
Laurus nobilis	Bay Laurel	176	x	x	x	
Ligustrum vulgare	Privet	110	x	x	x	
Lonicera alpigena	Alpine Honeysuckle	165	x	x	x	
Lonicera caerulea	Blue-berried Honeysuckle	167	x	x	x	
Lonicera nigra	Black-berried Honeysuckle	166	x	x	x	
Lonicera periclymenum	Common Honeysuckle	168	x	x	x	
Lonicera xylosteum	Fly Honeysuckle	164	x	x	x	
Mespilus germanica	Common Medlar	177	x	x	x	
Myricaria germanica	False Tamarisk	204	x	x	x	
Prunus cerasus	Sour Cherry	205	x	x	x	
Prunus laurocerasus	Cherry Laurel	178	x	x	x	
Prunus mahaleb	St. Lucie Cherry	117	x	x	x	
Prunus serotina	Black Cherry	179	x	x	x	
Prunus spinosa	Blackthorn	116	x	x	x	
Rhamnus alpina	Alpine Buckthorn	180	x	x	x	
Rhamnus cathartica	Purging Buckthorn	109	x	x	x	
Rhamnus frangula	Alder Buckthorn	112	x	x	x	
Rhamnus saxatilis	Rock Buckthorn	181	x	x	x	
Rhododendron hirsutum	Hairy Alpenrose	207	x	x	x	
Rhododendron ferrugineum	Rusty-leaved Alpenrose	206	x	x	x	
Rhododendron ferrugineum	Rusty-leaved Alpenrose	206	x	x	x	
Ribes alpinum	Alpine Currant	182	x	x	x	
Ribes nigrum	Blackcurrant	183	x	x	x	
Ribes petraeum	Rock Currant	184	x	x	x	
Ribes rubrum	Red Currant	185	x	x	x	
Ribes uva-crispa	Gooseberry	186	x	x	x	
Rosa sp.	Dog Rose	122	x	x	x	
Ruscus aculeatus	Butcher's Broom	208	x	x	x	
Sambucus nigra	Black Elder	107	x	x	x	
Sambucus racemosa	Red Elderberry	108	x	x	x	

(continued)

Table 9.6 (continued)

Shrubs B		**Code**	**A**	**B**	**C**	**D**
Sorbus chamaemespilus	Dwarf Whitebeam	**187**	x	x	x	
Staphylea pinnata	Bladdernut	**188**	x	x	x	
Trachycarpus fortunei	Chusan Palm	**190**	x	x	x	
Viburnum lantana	Wayfaring tree	**114**	x	x	x	
Viburnum opulus	Guelder Rose, Snowball tree	**115**	x	x	x	
	Other shrubs	**9**	x	x	x	
Shrubs C: assessment of shrubs, tall forbs and creepers						
Clematis vitalba	Clematis	**124**		x	x	
Hedera helix	Ivy	**123**		x	x	
Rubus fruticosus	Blackberry	**120**		x	x	
Rubus idaeus	Raspberry	**121**		x	x	
Vaccinium myrtillus	Blueberry	**125**		x	x	
Vaccinium oxycoccus	Cranberry	**128**		x	x	
Vaccinium uliginosum	Bog Bilberry	**127**		x	x	
Vaccinium vitis-idaea	Cowberry	**126**		x	x	
Species indeterminable, dead trees, shrubs or stumps		**999**	x			

The columns marked A-D indicate which species are assessed in different phases of the survey:

A - Assessment of sample trees

B - Presence of woody species

C - Assessment of the forest edge

D - Assessment of the regeneration

References

Düggelin C, Keller M (2017) Schweizerisches Landesforstinventar, Feldaufnahme-Anleitung 2017. Eidg. Forschungsanstalt für Wald, Schnee und Landschaft, Birmensdorf

Keller M (2005) Schweizerisches Landesforstinventar, Anleitung für die Feldaufnahmen der Erhebung 2004–2007. Eidg. Forschungsanstalt für Wald, Schnee und Landschaft, Birmensdorf

Stierlin H-R, Brändli U-B, Herold A, Zinggeler J (1994) Schweizerisches Landesforstinventar, Anleitung für die Feldaufnahmen der Erhebung 1993–1995. Eidg. Forschungsanstalt für Wald, Schnee und Landschaft, Birmensdorf

Traub B (2001) Representativeness of the sample grid. In: Brassel P, Lischke H (eds) Swiss national forest inventory: methods and models of the second assessment. WSL Swiss Federal Research Institute, Birmensdorf, pp 132–149

Zingg A, Bachofen H (1988) Schweizerisches Landesforstinventar, Anleitung für die Erstaufnahme 1982–1986. Eidg. Anst. forstl. Versuchswes, Birmensdorf

Chapter 10
Interview Survey with the Local Forest Service

Christoph Fischer and Marielle Fraefel

Abstract In the Swiss National Forest Inventory (NFI), data on about 6400 sample plots is collected directly in the field, but not all information of interest can be assessed in this way. During NFI1, an interview survey was therefore designed to complement the terrestrial inventory. All local Swiss foresters (about 800) responsible for the forest districts and representatives of the forest authorities (forest service) in Switzerland were interviewed. The interviews were conducted twice in NFI4: the first for the first five NFI annual panels and the second for the remaining four annual panels (Sect. 8.3). This procedure resulted in one interview for each sample plot for every NFI cycle. The interview survey consists of two main parts: interviews to obtain information about the assessed sample plots, conducted twice during NFI4, followed by interviews conducted only once about forest roads throughout the entire country (independent of the sample plots). The interviews in NFI4 covered about 50 different attributes.

10.1 Introduction

In the Swiss NFI (NFI), data on about 6400 sample plots is collected directly in the field, but not all information of interest can be assessed in this way. During NFI1, an interview survey was therefore designed to complement the terrestrial inventory. All local Swiss foresters (about 800) responsible for the forest districts and representatives of the forest authorities (forest service) in Switzerland were interviewed. The interviews were conducted twice in NFI4: the first for the first five NFI annual panels and the second for the remaining four annual panels (Sect. 8.3). This procedure resulted in one interview for each sample plot for every NFI cycle. The interviews focused on all sample plots that were 'forest' or 'shrub forest' according to the field

C. Fischer (✉)
Scientific Service NFI, Swiss Federal Research Institute WSL, Birmensdorf, Switzerland
e-mail: christoph.fischer@wsl.ch

M. Fraefel
GIS, Swiss Federal Research Institute WSL, Birmensdorf, Switzerland

C. Fischer, B. Traub (eds.), *Swiss National Forest Inventory – Methods and Models of the Fourth Assessment*, Managing Forest Ecosystems 35,
https://doi.org/10.1007/978-3-030-19293-8_10

survey, as well as on all plots that were terrestrially inaccessible, but 'forest' or 'shrub forest' according to the aerial-photo interpretation.

The interview survey consists of two main parts: interviews to obtain information about the assessed sample plots, conducted twice during NFI4, followed by interviews conducted only once about forest roads throughout the entire country (independent of the sample plots). The interviews in NFI4 about the sample plots covered about 50 different attributes. Since the attributes were identical to those in the previous interview survey, no changes were made to the manual and no special pilot interviews were necessary before the start of this part of NFI4. In contrast, the NFI4 forest road survey was slightly different from that in NFI3. The changes are explained in detail in Sect. 10.3.

For the first part of the interview survey, the specifically developed software application SILVIS was used to lead the surveyer through the interview and record the complete data. It supported them in making plausibility checks, and provided them with information recorded in the field during NFI4 and during previous inventories (see Part VIII for a detailed description of SILVIS).

In addition to the interview survey with the local foresters, the cantonal forest services are periodically asked to provide updated digital maps of their forest districts and to report on the status of the regional forest-planning tools (forest development plans) in written form.

10.2 Interview Survey Concerning the Sample Plots

10.2.1 General Information

Prior to interviewing the local forest-service staff, the field teams receive extensive instructions (Sect. 8.3). These focus mostly on the variables to be assessed during the interview, and, to a lesser extent, on how best to ask questions in an interview, e.g. how to avoid asking suggestive questions. See the manual of the forest service survey (Keller 2013) for detailed descriptions of the variables covered in the interview.

10.2.2 Conducting the Interviews About the Sample Plots

Once the survey areas have been defined as described in Sect. 8.3.2 the field teams are given an address list with the contact information for the local foresters they need to interview so that they can arrange the meetings with them independently. A control list with all plots to be assessed is also given to each field-team member.

For the interview survey, the most important piece of equipment is a laptop with the custom-built software SILVIS installed (Part VIII). The software guides the field crew through the interview and ensures it is complete. As an additional help, each field-team member receives a data sheet for each sample plot covered during the field survey, which is produced prior to the interview. It includes information from the latest field assessment about the sample plot with a documentation form, a tree map, information on the forest stand and pictures, as well as information from the previous interview survey (Sect. 8.3.2).

It is not only the field-team members who need to prepare for the survey, but also the local foresters themselves. They are sent the plot coordinates in the survey so that they know where the plots are and provided with a copy of the field manual. To ensure the interviews are as efficient as possible, the foresters are asked to have various maps and other documents at hand including the forest management map, forest stand maps and maps with information on silvicultural treatments and drinking-water sources, as well as documents concerning land ownership.

Once the survey with a forester has been completed, a backup of the assessed data is mandatory. Data from the plot-related interview survey is backed-up daily on USB sticks, and the forest road survey maps are photographed weekly. All data is transferred to the NAFIDAS raw data base (Sect. 20.3.1) as soon as an Internet connection is available, at least once per week.

10.2.3 Topics Covered in the Interviews

The topics are subdivided into three major thematic fields, covering a total of about 50 attributes: management and planning, forest use and its history, and timber harvesting techniques. *Management and planning* attributes are mostly used for describing the organisation of the forest enterprise and the higher-level forest management and silvicultural framework. They include information on forest ownership, forest functions, the date of the forest management plan, the existence of a regional forest plan, the size of the management unit, as well as information on certification and the corresponding label. Information on forest ownership is particularly helpful as management processes in private forests are usually not as well documented or accessible as those in public forests. The NFI interview survey provides insights into forest management in private forests.

Forest use and history covers a wide range of attributes on the forest stand and forest management intentions, including the type of forest establishment, year of afforestation, year and type of silvicultural treatments since the last NFI, year of last pasturing, type and urgency of the next silvicultural treatment, type and year of forest damage, the terms for the next treatment, as well as information on recreational use and its intensity. These attributes provide insights into the current forest management approach, and are used for reporting on the *naturalness* of the forest and evaluating the sustainability of forest management.

Timber harvesting technique attributes focus on the timber harvesting technique applied, assortments and who carries out the harvest. Where no timber harvesting has taken place, the most probable timber harvesting technique is indicated. The attributes include: the timber harvesting technique, whether the timber was short or long and harvested by forest service staff or contractors, the skidding phases and corresponding distances, the means of transport and pre-haulage distances. The timber harvesting techniques distinguished range from felling with a chainsaw to highly mechanised systems using harvesters. The potential timber harvesting costs can then be calculated (Chap. 16). These are relevant not only for national reports, but also for modelling future forest development (Chap. 17).

10.3 Forest Road Assessment

This subchapter describes the Swiss forest road data set and how it was produced. The information about the roads was obtained by interviewing local forest service staff and letting them write the road dimensions on maps. These were then added to the road geometries of the national Topographic Landscape Model swissTLM^{3D} (swisstopo 2012). As in the previous surveys, roads were included in the survey if they fulfilled the following criteria: (a) they are situated in the forest or along the forest edge; (b) they have a minimum transport capacity of a 10-ton axle load; and (c) they are at least 2.5 m wide (Keller 2013). In addition, in NFI4, information about the roads' suitability for specific vehicle categories was collected for the first time, and the connections between the forest truck roads and the higher-level road network were also mapped.

10.3.1 General Information

The forest roads were mapped for the entire Swiss forest, unlike in the plot-based interview survey (Sect. 10.1) in which only information related to the sample plots was collected. In order to limit the amount of work required for mapping and digitisation, only changes since the previous survey were mapped.

The same definition of an NFI forest road has been used since 1983 to ensure the data in the time series is comparable. As further information was required, it was decided to implement a complementary forest road survey in addition to the original NFI road survey to assess the suitability of the roads for specific vehicle categories. The connections between the forest roads and the higher-level road network outside the forest were also included. For a road to be included in this assessment, it must be at least 3.0 m wide and trafficable with a 26-ton 3-axle truck (Keller 2013).

For clarity, we refer to the roads in the original NFI forest road survey as *forest truck roads* and use the term *forest roads* to refer to either the *forest truck roads* or the roads that fulfil the criteria for the complementary forest road survey.

10.3.2 Data Preparation for the Forest Road Survey

The forest truck road data from NFI3 was available as a polyline data set. Each of the forest road segments (polyline segments) has attribute fields containing information about the surface material and changes since NFI2. The underlying vector data set is the swisstopo Vector25 data set. For NFI4, a more accurate and up-to-date vector data set (swissTLM^{3D} 2012) from swisstopo (swisstopo 2012) was used.

To allow comparisons between NFI3 and NFI4, all attributes from NFI3's forest truck road survey had to be transferred from the Vector25 data set to the swissTLM3D geometries by: (a) automatically matching road segments from the old and the new data set using buffers, (b) manually allocating all other roads if an unambiguous, one-to-one (or one-to-many) match could not be found, and (c) copying the relevant attributes to the new road data set.

The connection from the forest roads assessed in the complementary survey to the higher-level road network was also mapped. The higher-level road network was defined as main roads that can be used by a 40-ton vehicle at any time. Potential higher-level roads were marked on the maps using the TomTom/Multinet Functional Road Classes *Major Roads of High Importance*, *Other Major Roads*, *Secondary Roads* and *Local Connection Roads* (TomTom 2013). Both the data from the NFI3 forest truck road survey and the higher-level roads were printed on top of the most recent swisstopo 1:25,000 topographic maps using an A1 format corresponding to the format in which the 1:25,000 map sheets are published. These maps were used as the basis for the NFI4 forest road survey.

10.3.3 Conducting the Forest Road Survey

The interviews with the local forest services were planned as described in Sect. 8.3.2 with first the plot-related survey and then the forest road survey. Each forester was requested to evaluate only their own forest district. All foresters in Switzerland were interviewed, which meant that a wall-to-wall survey of the Swiss forest roads was carried out.

Each forester first marked their forest district on the map and any changes in the road segments fulfilling the NFI's requirements for a *forest truck road* since the previous inventory were recorded and classified according to the categories listed in Table 10.1. In addition to any changes to the roads themselves, corrections to the data set could be made if roads had been omitted or incorrectly recorded (amendment), or had been mistakenly recorded during the previous survey. Three types of surface material were distinguished: gravel, asphalt and concrete. The changes were mapped in writing in different colours and coded according to the type of change (Table 10.1).

Table 10.1 Changes in road material and road use since the previous survey

Road type	Surface material
New surface material	Gravel
	Asphalt
	Concrete
Newly constructed road	Gravel
	Asphalt
	Concrete
Upgraded road	Gravel
	Asphalt
	Concrete
Amendment of omitted road or incorrect surface material	Gravel
	Asphalt
	Concrete
Renaturalised road	
Abandoned/no longer maintained	
Mistakenly recorded	
In need of reconstruction	
Used for other purposes	

Table 10.2 Minimum criteria for forest roads to be assessed

Criterion	Forest truck roads	Complementary forest truck road survey
Minimum width	2.5 m	3.0 m
Minimum weight allowed	10 t axle load (20 t total weight)	26 t total weight
Location	Forest and forest edge	Forest, forest edge, most important connecting routes outside the forest

Once the forest truck road survey had been completed, the complementary forest truck road survey was started for a subset of the forest truck road data set. The subset included only those forest truck roads that were at least 3 m wide and trafficable with a 26-ton 3-axle vehicle. All roads inside the forest that met these criteria and could be used at least 10–11 months per year were included (Table 10.2). For roads outside of the forest, only the most important connecting routes to the nearest large and permanently trafficable road were included.

The complementary forest truck road survey describes the roads in more detail than the forest truck road survey. For each road segment, the road width, the largest vehicle type allowed to drive on the road, obstacles and piggyback transport were assessed (Table 10.3). The attribute *road width* is categorised as 3.0–3.49 m or ≥3.5 m. *vehicle type* specifies the largest truck category allowed on the road. *Obstacle* indicates road sections with limited trafficability due to structural barriers (e.g. underpass) or official restrictions (e.g. authorisation required). *Piggyback* signals dead-end sections that can be entered with a piggyback trailer and left with a six-axle vehicle.

Table 10.3 Variables assessed in the forest road surveys

Forest truck road survey		Complementary forest truck road survey	
Surface material	Gravel	Road width	3.0–3.49 m
	Asphalt		≥3.5 m
	Concrete	Largest vehicle allowed	3-axle, 26 t vehicle
Type of change	New surface material, new road, etc. (Table 10.1)		4-axle, 28 t or 32 t vehicle
			5- or 6-axle, 40 or 44 t vehicle
		Obstacle	Undersized road or structural barrier
			Regulatory restriction
		Piggyback	Piggyback transport required

During the survey, a code indicating the truck category and road width was written next to the roads on the map (Fig. 10.1). An additional code for obstacles or piggyback system was added where appropriate, e.g. '*A4+*' for a 3 m-wide road that can be used by a piggyback 26-ton six-wheeler.

All variables and categories for the attributes of the complementary forest truck road assessment were defined in collaboration with the Federal Office of Environment and tested for their applicability in a pilot study prior to implementation in NFI (Müller et al. 2016; Sect. 22.2).

After completing the forest road assessment, each map was checked for consistency and completeness. In unclear cases, e.g. border roads between two forest districts or missing road segments, the surveyor resolved the situation together with the foresters concerned. Once a map sheet had been completed and had passed the consistency and completeness checks, it was taken to the WSL office for digitisation.

10.3.4 Forest Road Digitisation

By the end of the NFI4 survey, approximately 500 paper maps had been returned to the WSL office, where they were scanned and georeferenced. They were then used as background layers in a Geographic Information System (GIS), and the mapped information was transferred to the road geometries (heads-up digitisation) of the swissTLM3D by a team of digitising staff, who visually checked all the map drawings on screen, selecting the appropriate road line and allocating the correct code. Predefined attribute domains and specially developed software tools helped reduce data-entry errors (Sect. 22.5). Where needed, lines were split in order to correctly assign the appropriate attributes. The digitisation approach was devised during the design of the road survey to ensure an efficient workflow and clear terminology.

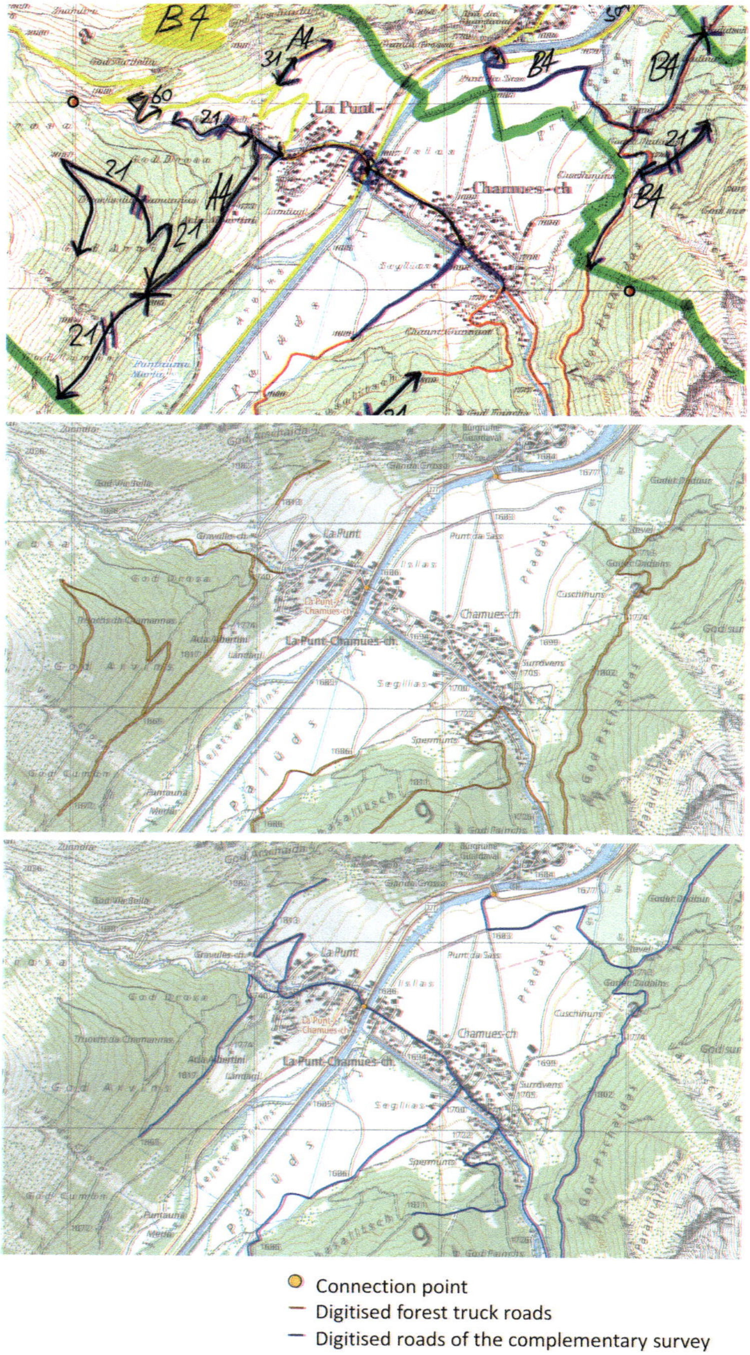

Fig. 10.1 (**a**) Example of mapping showing the codes and changes to the forest-road system selected on the maps during the NFI4 survey. (**b**) Resulting data set for the forest truck roads after the digitisation. (**c**) Resulting road data set from the complementary survey

After digitisation, the data was checked for errors, such as omissions or attribute inconsistencies, and corrected as necessary. The measures taken to ensure good data quality are described in Chap. 22.

The forest definitions used in the forest road survey and in the NFI field survey should match as well as possible. A polygon data set was developed for the forest-cover map (Sect. 7.2) to classify all roads as: (a) 'forest', (b) 'non-forest', (c) 'forest edge', or (d) 'open forest'. The road data set was overlaid with the forest cover map so that every road segment could be assigned the correct information regarding its location. The original forest truck road data set was then transferred into the NFI database together with the derived data sets.

References

Cioldi F, Speich S (2008) Aufwand Feldaufnahmen LFI3. Eidg Forschungsanstalt WSL 1–4

Düggelin C, Keller M (2017) Schweizerisches Landesforstinventar, Feldaufnahme-Anleitung 2017. Eidg. Forschungsanstalt für Wald, Schnee und Landschaft, Birmensdorf

Keller M (2013) Schweizerisches Landesforstinventar, Anleitung für die Umfrage- und Erschliessungserhebung 2013. Eidg. Forschungsanstalt für Wald, Schnee und Landschaft, Birmensdorf

Keller M (2005) Schweizerisches Landesforstinventar, Anleitung für die Feldaufnahmen der Erhebung 2004–2007. Eidg. Forschungsanstalt für Wald, Schnee und Landschaft, Birmensdorf

Müller K, Fraefel M, Cioldi F et al (2016) Der Datensatz «Walderschliessungsstrassen 2013» des Schweizerischen Landesforstinventars. Schweiz Z Forstwes 167:136–142. https://doi.org/10.3188/szf.2016.0136

Stierlin H-R, Brändli U-B, Herold A, Zinggeler J (1994) Schweizerisches Landesforstinventar, Anleitung für die Feldaufnahmen der Erhebung 1993–1995. Eidg. Forschungsanstalt für Wald, Schnee und Landschaft, Birmensdorf

swisstopo B für L (2012) swissTLM3D Version 1.1 Ausgabe 2012 – Nachführungsinformation

TomTom (2013) Multinet user guide v.1.1.5. https://files.be.ch/bve/agi/geoportal/geo/lpi/TELEDAT_2013_01_LANG_DE.pdf. Accessed 23 Jan 2018

Traub B (2001) Representativeness of the sample grid. In: Brassel P, Lischke H (eds) Swiss national forest inventory: methods and models of the second assessment. WSL Swiss Federal Research Institute, Birmensdorf, pp 132–149

Zingg A, Bachofen H (1988) Schweizerisches Landesforstinventar, Anleitung für die Erstaufnahme 1982–1986. Eidg. Anst. forstl. Versuchswes, Birmensdorf

Zinggeler J, Herold A (1997) Zeitlicher und finanzieller Aufwand für die terrestrische Probenahme im zweiten Schweizerischen Landesforstinventar (LFI). Schweiz Z Forstwes 148:939–959

Chapter 11
Time Study of the Terrestrial Inventory and Interview Survey

Fabrizio Cioldi

Abstract The time required for the fieldwork for large-scale forest inventories is often not known, or only partially known. Knowing, however, how much time is needed for the main working stages is indispensable for planning and conducting an inventory. This Chapter describes the time needed for the different phases in the fourth Swiss National Forest Inventory (NFI4). Up to the NFI3 (2004–2006), the terrestrial inventory and the interview survey were two interconnected processes that took place immediately after each other. In NFI4, however, the field surveys (2009–2017) and the interview surveys (2013/2014 and 2017/2018) were carried out as completely separate processes. The average total time required in NFI4 to assess one sample plot is 4.6 hours (including driving time and training), with large differences from region to region. The total time expenditure for the interview survey 2013/2014 amounted to 4360 person-hours, with a mean of 5.5 hours spent on each interview. These data were valuable for the operational and financial planning of NFI5, which began in 2018.

11.1 Introduction

The time required for the fieldwork for large-scale forest inventories is often not known, or only partially known. Knowing, however, how much time is needed for the main working stages is indispensable for planning and conducting an inventory. Up to the third Swiss NFI (NFI3) (2004–2006), the terrestrial inventory and the interview survey were two interconnected processes that took place immediately after each other. In NFI2 (1993–1995), the time and costs involved in conducting the interview survey were not assessed separately from the field survey (Zinggeler and Herold 1997). Information about the time expenditures in the NFI3 can be found in Cioldi and Speich (2008). In NFI4 the time required for the terrestrial survey (2009–2017) and the interview survey (2013/2014) is described in Sects. 11.2 and 11.3. These data were valuable for the operational and financial planning of NFI5, which began in 2018.

F. Cioldi (✉)
Scientific Service NFI, Swiss Federal Research Institute WSL, Birmensdorf, Switzerland
e-mail: fabrizio.cioldi@wsl.ch

C. Fischer, B. Traub (eds.), *Swiss National Forest Inventory – Methods and Models of the Fourth Assessment*, Managing Forest Ecosystems 35,
https://doi.org/10.1007/978-3-030-19293-8_11

11.2 Terrestrial Inventory

The two main sources for data about the time spent on the terrestrial inventory and the duration of the different work steps are: (a) the exact time information per attribute group registered in the software MAIRA (Table 11.1 phases A+B) and (b) the working reports of the field teams (German: *Turnusberichte*). A working report is usually handed in after ten working days (phases C+D and total time). Phase E (other work), refers to the difference between the total time and the phases A–D.

The time expenditures presented in Table 11.2 relate only to the work of the field teams for the regular sample-plot survey and the repeat survey. The time spent at the WSL office on planning, organising and supporting the field teams is not included.

From 2009 to 2017, a total of 7455 sample plots were assessed in NFI4. Of these, 485 sample plots were assessed twice by two different teams in the course of the repeat survey (Table 11.3). The total time needed to survey the in effect 7940 sample plots, including training courses, amounted to 73,500 person-hours. The average total time required to assess one sample plot is 4.6 h (Table 11.2), with large differences from region to region. In the Jura and the Plateau, a mean of just over

Table 11.1 Definitions of the different work phases

Phase of work	Definition of the activity
A = Sample-plot assessment	Data collection on the sample plot
B = Hike	Hike from the vehicle to the sample plot and back, or from the helicopter landing place/arrival point of cable car to the sample plot and back
C = Driving	Drive from the place of residence or from WSL to the survey region and back. In addition, all driving within a survey region, e.g. driving from one sample plot to another
D = Training	Instruction course and special training during the field season (e.g. driving courses and basic live support)
E = Other work	Preparatory work (daily/weekly plan), data transfer, maintenance of the equipment, public relations. Inactive time due to bad weather or technical problems, accident/illness

Table 11.2 Average time spent per survey team in minutes (hours) and sample plot, according to work phase and region

	Time expenditure per work phase						
Region	A	B	C	D	E	Total time (minutes)	Total time (hours)
Jura	88	24	66	18	59	254	4.2
Plateau	87	19	66	18	59	249	4.1
Pre-Alps	90	36	66	18	59	269	4.5
Alps	87	57	82	18	59	303	5.0
Southern Alps	92	88	82	18	59	339	5.7
Switzerland	**88**	**42**	**71**	**18**	**59**	**278**	**4.6**

Table 11.3 Average time spent on assessing different types of sample plot (phase A) and reaching them (phase B)

Type of sample plot	Number of plots	Time expenditure (in minutes)
Forest (without shrub forest)	6042	134
Shrub forest	315	183
Non-forest	780	74
Inaccessible sample plots	318	128
Regular field survey	**7455**	**130**
Repeat survey	485	118
Total sample-plot field survey	**7940**	**129**

4 h was needed per sample plot, whereas in the Southern Alps the average was 5.7 h, i.e. almost 40% longer.

The main reasons for this difference are the long hiking distances to the sample plots due to the rugged topography and the low density of forest roads in the Southern Alps, where the field teams often have to hike for many hours to reach the sample plot, and thus cannot assess the minimum of 2–3 sample plots per day. To reach particularly remote and isolated sites, normally around 20 sample plots per year, a helicopter is used to fly the teams. The time required for data collection on only the sample plot is approximately the same in all regions and is, on average, 88 min (about one third of the total time needed). Since the distances between the sample plots in NFI4 were at least 4.2 km (Sect. 2.4), i.e. greater than in the previous inventories, the time taken to reach them is also longer than in the past, amounting to about 70 min. Around 6.5% of the total time spent was used for field team instruction and training (phase D). The phase E (other work) took up around 1 h per sample plot, and was also about the same in all the regions.

For all types of sample plot, the average time spent on phases A and B was 130 min. 'Shrub forest' was not surveyed in NFI1 and NFI2, but since NFI3 it has been assessed according to the same procedures as for the other forest types. During NFI4, a total of 315 shrub-forest sample plots were assessed, i.e. about 4% of all the 7455 sample plots surveyed, taking on average 183 min for each assessment. This is 37% longer than for the other forest plots, which require on average 134 min, of which 31% was spent on hiking and localising the plot's centre and 69% on data collection alone. In 'shrub forest', on the other hand, only 36% of the time was spent on collecting data and 64% (117 min on average) on hiking to and from the sample plot and finding it.

The average time taken to assess the 780 'non-forest' plots was 74 min, with approximately 40 min spent on walking and measurement or localisation of the sample plot, and the rest (34 min) on marking the sample-plot centre and completing the 'forest/non-forest' decision. For the clarification of inaccessible sample plots, a mean of 128 min was needed.

The repeat survey is an independent survey of selected sample plots (7% or 485 of the total) already assessed by another field team (Sect. 21.5) excluding inaccessible areas, sample plots that were clearly 'non-forest', and plots reached by helicopter.

Table 11.4 Average time spent measuring each attribute group in phase A

Attribute group	Time expenditure (in minutes)
Marking of the sample-plot centre	10.4
Basic decisions (e.g. 'forest/non-forest') and boundaries	6.0
Forest edge[a]	2.4
Single tree measurements	23.2
Tariff sample trees	7.8
Presence of woody species	5.4
Regeneration	11.3
Lying deadwood	4.8
Assessment of area data	5.1
Stand assessment including stand stability	11.7
Total sample-plot assessment	**88.1**

[a]Only on 15% of the sample plots

The time spent on assessing the plots in the repeat survey was about 12 min less than on assessing the same plots in the regular survey because some work steps (e.g. localisation) were not needed. However, the sample plots in the repeat survey are much further away from each other, which meant that more driving time was needed.

The working time needed for all attribute groups (phase A in Table 11.1) to assess all sample plots in the forest and shrub forest was systematically recorded using the field survey software (MAIRA, Sect. 23.2.1). Not all sample plots have all features, e.g. a forest edge was assessed only on 930 sample plots, taking on average about 16 min each time. Extrapolated for all sample plots, the forest edge assessment adds on average 2.4 min per sample plot (Table 11.4).

The permanent marking of the sample-plot centre took on average 10 min. Measuring and assessing single trees was the most time-consuming activity (23 min on average), but also the most important for the whole inventory. Measuring the tree height and diameter at 7 m of tariff trees took, on average, 8 min. The stand assessment and the assessment of the regeneration took between 11 and 12 min.

Conclusions

The time spent on collecting data (including hiking time) in NFI4 was about 20 min less than in NFI3 (2004–2006), largely because the regeneration was assessed on one subplot only in NFI4 and not on two as in NFI3. However, the time saved in NFI4 was spent on additional travelling by car between the sample plots because of the greater distance between them.

Thirty-two percent of the total working time was spent on collecting data on about 280 attributes (phase A). Eliminating attributes from the data catalogue thus reduces the total cost only slightly. On the other hand, the data catalogue should not be increased much to ensure the assessment time of about 150 min as it would otherwise not be possible to assess 2–3 sample plots per day, especially in the Alps and

Southern Alps. In general, little can be done to make the inventory more efficient because many of the costs are fixed. For example, the hiking time depends on the topography.

11.3 Interview Survey

The field teams' working reports contain data on the time spent on the interviews during all the phases, A to F (Table 11.5), and are usually handed in after ten working days.

The times given in Table 11.6 relate exclusively to the work of the surveyors for the interview survey 2013/2014, including instruction and driving time. This does not include the expenditures for the planning, organisation and support of the interview survey provided by NFI's headquarters.

Information about a total of 6617 sample plots was collected during the interviews in NFI4 between 2013 and 2014 and later between 2017 and 2018. For this, nearly 800 interviews were necessary. Organising the interviews and arranging meetings with all the foresters within a relatively short period was very demanding and time-consuming.

Collecting the data during the interviews with the local foresters took around 50% of the time needed for the interview survey, with the rest of the time spent on organisation, driving, instruction and other work related to the interviews (Table 11.6). The total time expenditure for the interview survey 2013/2014 (3695 plots), including training time, amounted to 4360 person-hours, with a mean of 5.5 h spent on each interview and considerable differences between the regions

Table 11.5 Definitions of the work phases in the interview survey

Phase of work	Definition of the activity
A = Sample-plot survey (interview)	Data collection during the interview
B = Forest road assessment (interview)	Assessment of the forest roads on maps (scale 1:25,000)
C = Driving	Drive from the place of residence or from WSL to the forester's office and back. In addition, all driving within an interview survey region
D = Preparation/ organisation	Preparatory work (daily/weekly plan). Documentation, e.g. maps for the forest road assessment. Contacting foresters to organise meetings and sending them information and data (e.g. list of the sample plots, attribute catalogue) so that they could prepare for the interview
E = Training	Instruction course
F = Other work	Quality control of the data, completeness check, data transfer, public relations. Inactive time due to technical problems or postponing/cancelling an interview, accident/illness

Table 11.6 Average time expenditure in minutes and hours per interview, plus the total time spent per sample plot according to work phase and region

	Time expenditure per work phase						Total time (minutes)	Total time (hours)	Total time (hours)
Region	A	B	C	D	E	F	per interview	per interview	per sample plot
Jura/Plateau	50.7	80.7	70.9	46.3	18.1	22.5	289.3	4.8	1.9
Pre-Alps/Alps/ Southern Alps	103.4	90.0	67.9	52.0	12.8	40.6	366.7	6.1	1.0
Switzerland	**80.1**	**84.0**	**67.9**	**48.5**	**15.0**	**33.0**	**328.5**	**5.5**	**1.2**

(Table 11.6). In the Jura and Plateau, slightly less than 5 h per interview were required, but in the other regions the total time was just over 6 h, i.e. about 25% more, mainly because the forest districts in the Pre-Alps, Alps and Southern Alps are much larger with many more sample plots. In the Jura and Plateau, the ratio of the time spent during the interview on the sample-plot survey and the forest road assessment was approx. 39:61%, whereas in the Pre-Alps, Alps and Southern Alps proportionally less time was spent on the forest road survey (the ratio was 53:47%) because the forest road network there is much less dense than in the Jura and Plateau, which require more time for the forest road assessment.

Driving to the interviews took on average more than 1 h, or 20% of the total time required per interview. Organising the interviews (weekly and daily plans) took a further 50 min (15% of the total time needed), and included preparing the equipment and documentation (e.g. maps for the forest road assessment), telephoning to making appointments with the local forest service and sending information and documentation (e.g. lists of the sample plots and attribute catalogue) to the foresters so they could prepare for the interviews. Instructing the surveyors took up 5% of the total time spent, and phase F (other work) around 30 min per sample plot. Phase F included checking for completeness after the interviews, transferring the data to the WSL office and any 'inactive times' due to technical problems or the interview being postponed.

The digital processing of the raw road-survey data was very time-consuming (Sect. 10.3.4). It would probably therefore make sense to see whether it would be possible in the next interview survey in 2022/2023 to carry out the forest road assessment in digital form directly during the interview with the local forester.

References

Cioldi F, Speich S (2008) Aufwand Feldaufnahmen LFI3. Eidg Forschungsanstalt WSL 1–4

Zinggeler J, Herold A (1997) Zeitlicher und finanzieller Aufwand für die terrestrische Probenahme im zweiten Schweizerischen Landesforstinventar (LFI). Schweiz Z Forstwes 148:939–959

Part V
Modelling of Forest Resources

Chapter 12
State and Change of Forest Resources

Anne Herold, Jürgen Zell, Brigitte Rohner, Markus Didion, Esther Thürig, and Erik Rösler

Abstract In the Swiss National Forest Inventory (NFI), wood volume and changes in wood volume are estimated based on the stem volume of individual trees using various models: stem volume models and tariff models, models estimating volumes of large and small branches and growth models. Many of the models applied in the fourth NFI were described in previous publications, but some of them have subsequently been completed or adjusted based on methodological developments. This chapter mainly updates descriptions published previously.

12.1 Introduction

In the Swiss NFI (NFI), wood volume and changes in wood volume are estimated based on the stem volume of individual trees using various models. Many of the models applied currently were described by Kaufmann (2001), and this earlier text is thus referred to repeatedly below. However, some of them have subsequently been completed or adjusted based on methodological developments, and some of the descriptions provided by Kaufmann (2001) are not comprehensive. Therefore, this chapter mainly updates descriptions published previously, but also recapitulates fundamental information when necessary for readability and completeness.

Stem volume models estimate single stem volume using three measured tree dimensions: diameter at breast height ($d_{1.3}$), upper diameter at 7 m height (d_7) and tree height (h). The stem volume models currently implemented are the same as those described by Kaufmann (2001). Section 12.2 therefore provides complementary information and a summary of the essential information, such as equations and coefficients. Tariff models are the most general and estimate single stem volume as a function of $d_{1.3}$ as the only measured tree dimension. The current tariff models have

A. Herold (✉) · J. Zell · B. Rohner · M. Didion · E. Thürig · E. Rösler
Resource Analysis, Swiss Federal Research Institute WSL, Birmensdorf, Switzerland
e-mail: anne.herold@wsl.ch

C. Fischer, B. Traub (eds.), *Swiss National Forest Inventory – Methods and Models of the Fourth Assessment*, Managing Forest Ecosystems 35,
https://doi.org/10.1007/978-3-030-19293-8_12

been slightly modified from the descriptions given by Kaufmann (2001). Section 12.3 provides updated descriptions.

In addition to stem volumes, volumes of large and small branches are estimated, mainly as inputs for biomass estimation. The ratio of the volume of large branches to stem volume over bark is estimated using a logit model described by Kaufmann (2001). The corresponding model for the ratio of small branches is very similar but has not been published previously. These two models and their estimated coefficients are described in Sect. 12.4.

As the NFI is based on permanent inventory plots, changes in resources in general can be estimated as the difference between consecutive NFIs. In particular changes in standing volume such as growth, harvest and mortality can be estimated directly as the difference in tree volume. However, in some cases growth models are needed. The models and coefficients for basal area increment and volume increment are described in Sect. 12.5.

12.2 Stem Volume Models

The stem volume models map $d_{1.3}$, d_7 and tree height to total stem volume over bark of individual trees, including the tree top and the aboveground part of the stump but excluding branches. They are able to explain nearly all variance and are more precise than the tariff models (Sect. 12.3). However, d_7 and tree height are time-consuming measurements and are therefore only assessed on a subsample, namely the tariff trees.

The method for selecting tariff trees is explained in Sect. 2.3.4.5. The estimated volume of the tariff trees is used for two purposes. First, it is used as the dependent variable when fitting tariff models (Sect. 12.3). Second, it is used to improve stem volume predictions with the tariff models, which use $d_{1.3}$ as the only explanatory variable measured in the field (Sect. 12.3). For the tariff trees, stem volume can be estimated using both the more precise stem volume model and the generally applicable tariff model. The difference between the two models is scaled up (by the inverse of the selection probability) and added to the tariff volumes to correct for the difference to the stem volume. Assuming that the stem volume is known and is unbiased for the tariff trees, this procedure results in unbiased estimates of standing and total volume even in small sampling units, i.e. for rare tree species or small regions (Mandallaz 1991, 1997). A full description of this two-stage estimation procedure and the reasons to adopt it were presented by Kaufmann (2001, page 176 ff.).

The stem volume functions of the NFI were developed in 1991. Nine different functions were fitted: seven were developed for the main tree species Norway spruce (*Picea abies*), silver fir (*Abies alba*), pine (*Pinus sylvestris* and *P. nigra*), European larch (*Larix decidua*), Douglas fir (*Pseudotsuga menziesii*), European beech (*Fagus sylvatica*) and oak (*Quercus* spp.), and the remaining two were developed for coniferous and broadleaved species (Kaufmann 2001; page 162 ff.). For the sake of completeness, the equations and the coefficients (Table 12.1) are presented here.

Norway spruce (*Picea abies*):

Table 12.1 Coefficients of the stem volume models used in the NFI

	b_0	b_1	b_2	b_3	b_4	b_5
Norway spruce	0.029504	0.46756	2.43885	−5.74664	−0.0018265	–[a]
Silver fir	0.039594	0.35832	−0.39142	3.75195	−0.013314	1.62E-07
Pine	0.055349	0.40341	−0.63535	4.84573	−0.10114	–[a]
European larch	−0.0173	0.36366	2.49123	0.000107	–[a]	–[a]
Douglas fir	0.013166	0.35079	2.67531	−2.95083	0.0010962	–[a]
European beech	0.0025428	0.39446	2.56612	−3.67034	0.03567	–[a]
Oak	−0.026759	0.31686	5.01484	−7.71408	0.19704	–[a]
Coniferous (all species)	0.0084865	0.5436	2.8898	−1.94043	−4.93601	−1.33E-05
Broadleaved (all species)	−0.021786	0.39992	0.28036	2.30656	−1.20368	–[a]

[a]Variable not part of the model

$$V = b_0 + b_1 d_7^2 h + b_2 d_{1.3}^2 + b_3 d_7^3 + b_4 h \tag{12.1}$$

Silver fir (*Abies alba*):

$$V = b_0 + b_1 d_7^2 h + b_2 d_{1.3} + b_3 d_{1.3}^2 + b_4 d_{1.3}^3 h + b_5 h^4 \tag{12.2}$$

Pine (*Pinus sylvestris* and *P. nigra*):

$$V = b_0 + b_1 d_7^2 h + b_2 d_{1.3} + b_3 d_{1.3}^2 + b_4 d_{1.3}^3 h \tag{12.3}$$

European larch (*Larix decidua*):

$$V = b_0 + b_1 d_7^2 h + b_2 d_{1.3}^2 + b_3 h^2 \tag{12.4}$$

Douglas fir (*Pseudotsuga menziesii*):

$$V = b_0 + b_1 d_7^2 h + b_2 d_{1.3}^2 + b_3 d_{1.3}^3 + b_4 d_{1.3}^2 h^2 \tag{12.5}$$

European beech (*Fagus sylvatica*):

$$V = b_0 + b_1 d_7^2 h + b_2 d_{1.3}^2 + b_3 d_7^3 + b_4 d_{1.3}^3 h \tag{12.6}$$

Oak (*Quercus* spp.):

$$V = b_0 + b_1 d_7^2 h + b_2 d_{1.3}^2 + b_3 d_{1.3}^3 + b_4 d_{1.3}^3 h \tag{12.7}$$

Coniferous (all species):

$$V = b_0 + b_1 d_7^2 h + b_2 d_{1.3}^2 + b_3 d_7^2 + b_4 d_7^3 + b_5 d_{1.3} h^3 \tag{12.8}$$

Broadleaved (all species):

$$V = b_0 + b_1 d_7^2 h + b_2 d_{1.3} + b_3 d_{1.3}^2 + b_4 d_7^2 \tag{12.9}$$

where V is the stem volume over bark (in m^3), $d_{1.3}$ is the diameter at breast height (in m), d_7 is the upper diameter measured at 7 m height (in m), and h is the total tree height (in m).

As the stem volume models are fundamental to the NFI, this section includes information about their derivation and the data used to fit them.

Model fitting was based on a data set including long-term growth and yield data from approximately 38,000 tree stems from the permanent plot network of the Experimental Forest Management (EFM) trials conducted at WSL. These stems were measured (lying) in 2-m sections as the trees were harvested, between 1888 and 1974. Thus, their precise volume is known. Table 12.2 gives an overview of the number of available stem measurements for each species and the regional distribution of the measured stems in Switzerland.

The NFI tariff sample trees and the trees from the EFM trials were assessed for very different objectives: the NFI aimed to infer the state of forests representatively over the whole of Switzerland and its production regions, whilst the EFM experiments aimed to answer various research questions (e.g. Zell 2018; Peck et al. 2014) but are not representative of Switzerland's forests. In fact, the tree population from the EFM trials is clearly different from that of the corresponding species in Switzerland's forests: trees of the regions Plateau and Pre-Alps are generally over-represented in the EFM data set, whereas trees of the Alps region are under-represented.

The two data sets actually represent tree populations with different characteristics. For example, the stem forms of EFM and NFI sample trees are clearly different for all species and species groups. More specifically, the slenderness ratio ($h/d_{1.3}$) of trees measured in EFM plots is systematically larger than that of NFI sample trees, as shown for Norway spruce in Fig. 12.1

Stem volume models are assumed to be valid for all of Switzerland, but this is not so the tariff models, which respect regional differences. Therefore, the stem forms were shown for the regions with respect to their fitting data set (i.e. the EFM data) compared to their prediction data set (i.e. the NFI data). The stem form of trees in the Swiss Plateau is similar to that of EFM sample trees, whereas trees in the Jura and the Pre-Alps show a lower slenderness ratio compared to trees from the EFM trials. This difference even increases in the regions Alps and Southern Alps. Figures 12.2 and 12.3 show examples of these differences for Norway spruce and European beech, respectively.

Table 12.2 Number of measured stems from the Experimental Forest Management data set for each species corresponding to a fitted model, along with the regional distribution of the trees used for measurements

	N stems	Jura	Plateau	Pre-Alps	Alps		Southern Alps
					Under	Over	
					1500 m a.s.l.		
Coniferous (all species)	27,799	15%	39%	36%	6%	4%	0%
		(14%)	(16%)	(22%)	(19%)	(22%)	(7%)
Norway spruce (Picea abies)	15,292	15%	46%	36%	0%	3%	0%
		(12%)	(17%)	(24%)	(20%)	(22%)	(6%)
Silver fir (Abies alba)	7,222	22%	28%	49%	1%	0%	0%
		(29%)	(22%)	(33%)	(12%)	(1%)	(3%)
Pine (Pinus sylvestris and Pinus nigra)	1,751	9%	43%	2%	46%	0%	0%
		(19%)	(16%)	(3%)	(53%)	(5%)	(4%)
European larch (Larix decidua)	1,599	0%	34%	1%	41%	24%	0%
		(1%)	(3%)	(1%)	(16%)	(51%)	(28%)
Douglas fir (Pseudotsuga menziesii)	601	9%	5%	86%	0%	0%	0%
		(23%)	(71%)	(5%)	(0%)	(0%)	(1%)
Broadleaved (all species)	10,903	23%	52%	25%	0%	0%	0%
		(21%)	(22%)	(16%)	(18%)	(1%)	(22%)
European beech (Fagus sylvatica)	8,234	24%	45%	31%	0%	0%	0%
		(29%)	(25%)	(20%)	(13%)	(0%)	(14%)
Oak (Quercus spp.)	1,805	32%	68%	0%	0%	0%	0%
		(27%)	(33%)	(3%)	(14%)	(0%)	(23%)

In brackets is the corresponding regional distribution of the species in Switzerland's forests according to NFI3

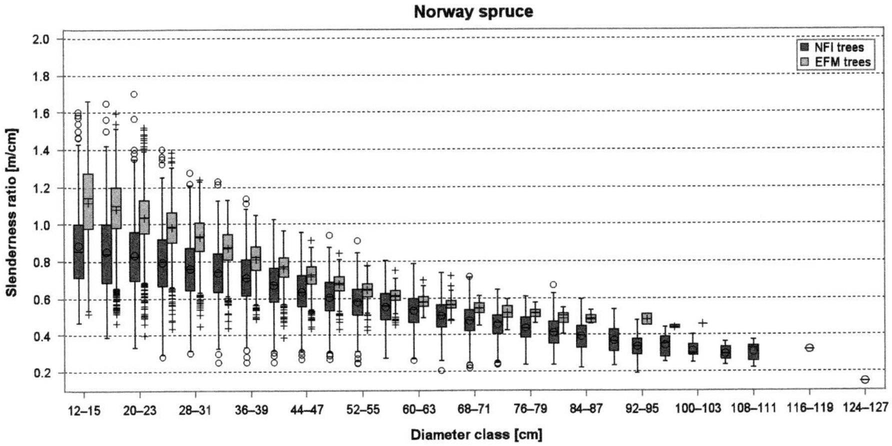

Fig. 12.1 Slenderness ratio ($h/d_{1.3}$) per diameter class of Norway spruce (*Picea abies*) sample trees measured in the NFI (measurements on standing trees) and as part of the EFM Project (measurements on lying trees) over all Switzerland

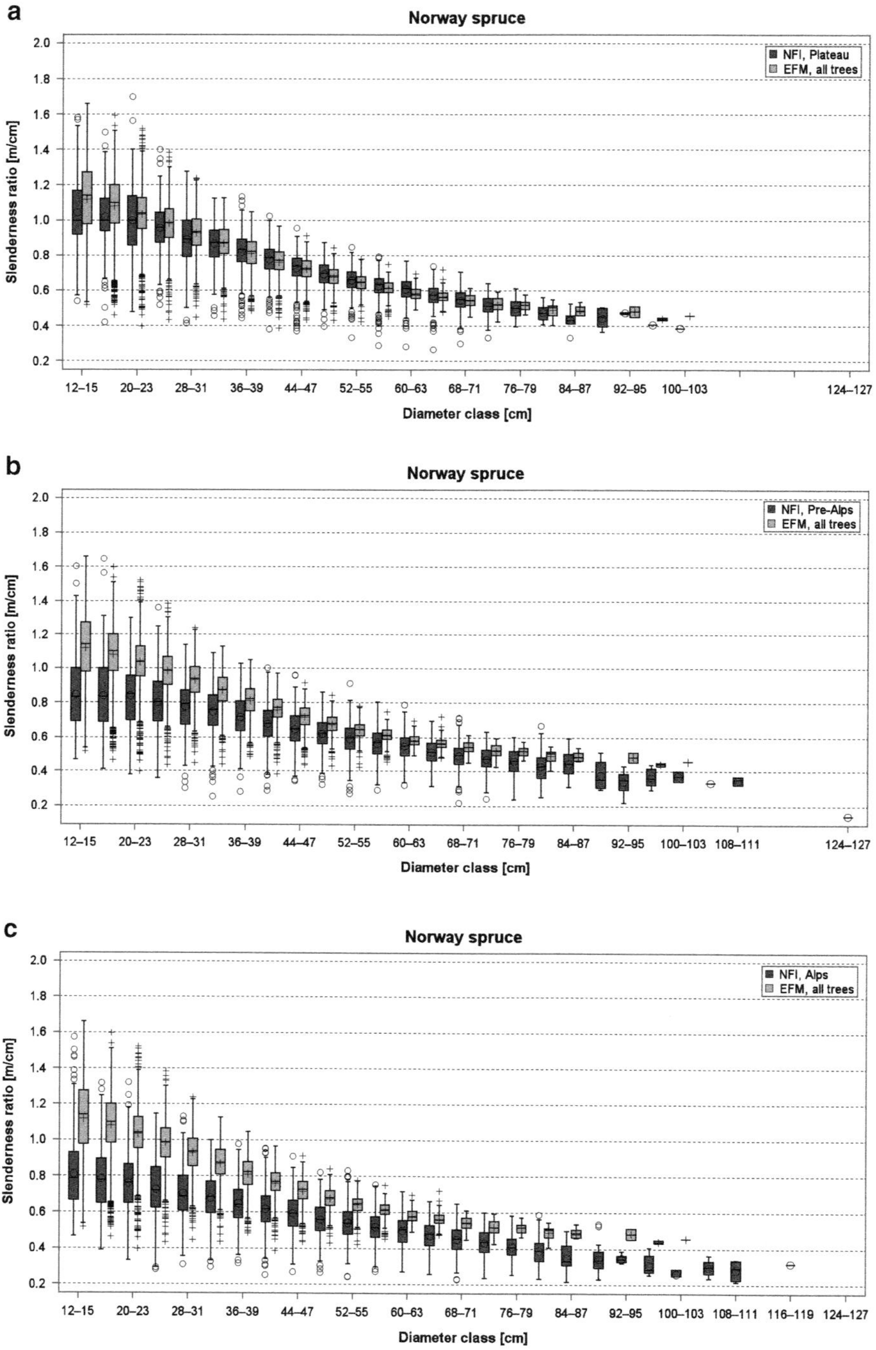

Fig. 12.2 Slenderness ratio ($h/d_{1.3}$) per diameter class of Norway spruce (*Picea abies*) sample trees measured in the NFI (measurements on standing trees) in the regions Plateau (**a**), Pre-Alps (**b**) and Alps (**c**) and as part of the EFM Project (measurements on lying trees) over all Switzerland

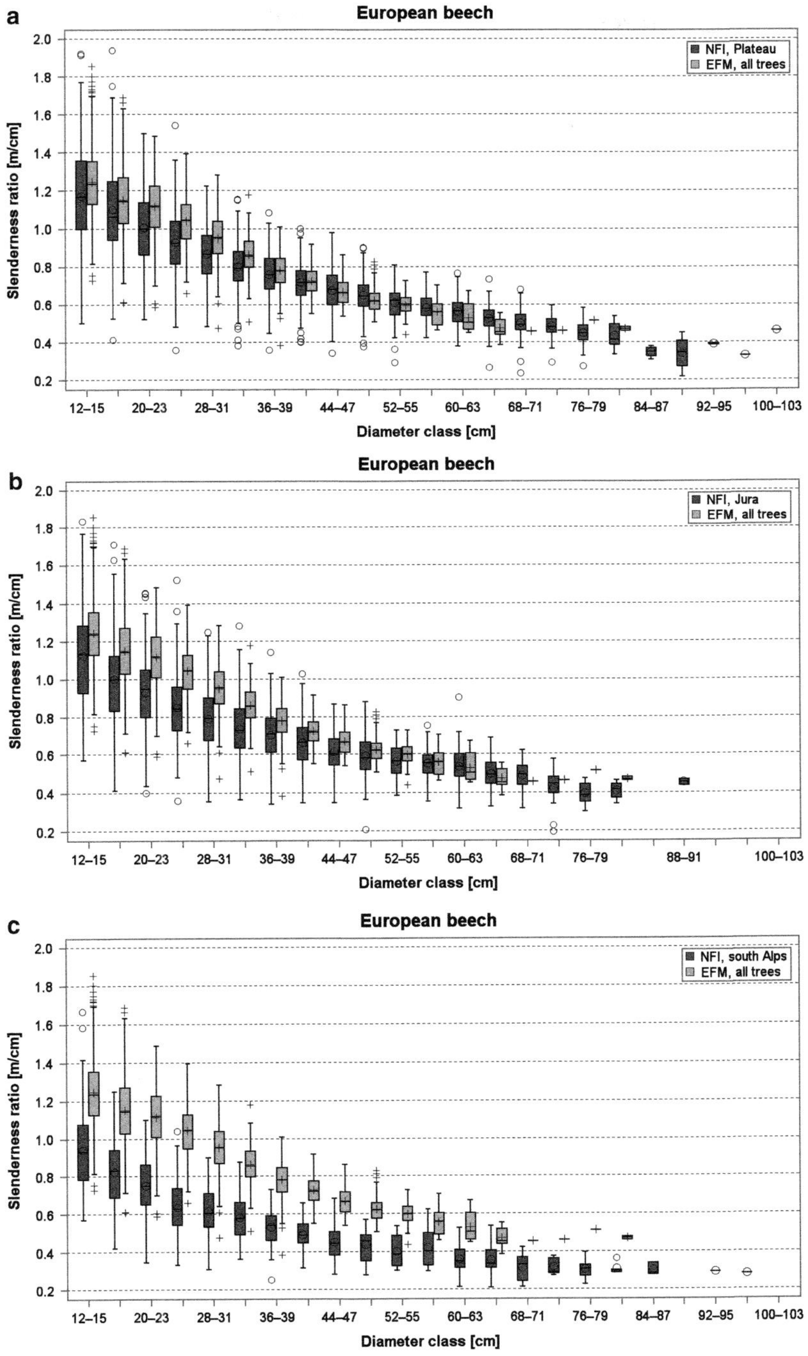

Fig. 12.3 Slenderness ratio ($h/d_{1.3}$) per diameter class of European beech (*Fagus sylvatica*) sample trees measured in the NFI (measurements on standing trees) in the regions Plateau (**a**), Jura (**b**) and Southern Alps (**c**) and as part of the EFM Project (measurements on lying trees) over all Switzerland

To compensate for these differences and to complete the EFM data set over the full range of slenderness ratios, approximately 500 additional trees were sampled in 1990 and 1992. They were selected with the criteria of a low slenderness ratio and a large diameter at breast height (spruce: 199, fir: 11, larch: 89, pine: 20, beech: 174 and oak: 10). The diameters at 10%, 30%, 50%, 70% and 90% of the total tree height, as well as the diameter at breast height ($d_{1.3}$), the upper diameter at 7 m height (d_7) and the tree height (h), were measured on these standing trees. The stem volumes were then calculated using a method similar to that described in Sect. 13.2. Cubic interpolation splines were fitted between the measured diameters to achieve continuous stem profiles (Sect. 13.2 step 2), and the stem volumes were then calculated with rotational integrals (Kaufmann 1993, unpublished internal report; Sect. 13.2 step 7). This data set was then merged with the EFM data set to fit the stem volume functions.

Another difference between the NFI and EFM sample trees is the diameter distribution. In the EFM data set, small diameters are over-represented while middle and large diameters are under-represented compared to the NFI tariff sample trees (see Fig. 12.4 for examples with Norway spruce and European beech). To increase the comparability of the data, only a subsample of the full EFM data set was used to fit the models: stems <12 cm in breast height diameter ($d_{1.3}$) were not considered and stems 12–35 cm in $d_{1.3}$ were selected randomly and proportionally to their $d_{1.3}$. Stems with diameters >35 cm were all included.

Coefficients of the Eqs. 12.1, 12.2, 12.3, 12.4, 12.5, 12.6, 12.7, 12.8 and 12.9 were estimated with weighted linear regressions. The increasing variance over predicted values (heteroscedasticity) was respected by weighting the residuals with the term $1/\left(d_7^3h\right)$, i.e. assuming that the variance is proportional to $\left(d_7^3h\right)$. To reduce bias, a grouped jackknife (Efron and Stein 1981) was applied instead of using only

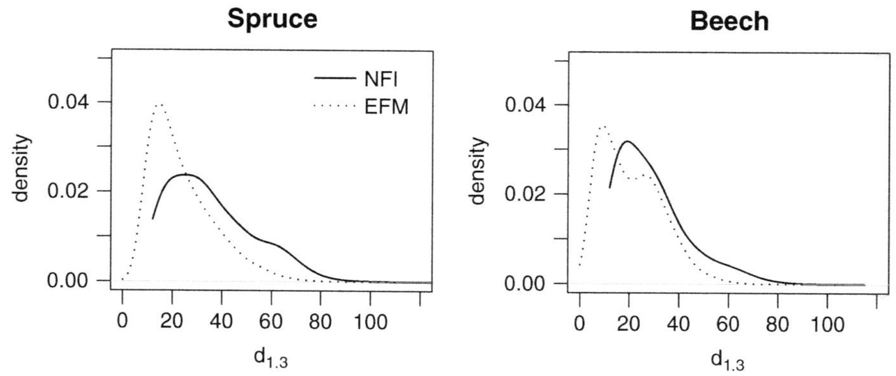

Fig. 12.4 Densities of the diameter distributions of Norway spruce (left) and European beech (right) sample trees measured in the NFI and in the EFM Project

one estimator. The data were randomly split into 20 groups. An estimator was calculated for each group, as well as for the whole data set.

This fitting methodology has the following advantages. First, the data set used to fit the coefficients is closer to observations made in the NFI, owing to subsampling and complementing of the data with extreme forms. Second, by weighting the residuals, the inherent increasing variance is incorporated into the prediction. Third, a smaller bias in the estimated coefficients can be assumed, owing to the grouped jackknife. Fourth, as the models may result in negative predictions, a simple imputation for these trees was established using geometrically reasonable volume estimates. Fifth, the standardised residuals are unbiased over $d_7/d_{1.3}$ and $h/d_{1.3}$, as illustrated in Fig. 12.5.

To analyse the predictive behaviour of the models using the NFI data, a sample from the EFM data (the source of information for volumes) was selected proportionally to the three-dimensional density of $d_{1.3}$, h and d_7 in the NFI. The stem volume models were applied to this sample data set and residuals were calculated. As the residuals are representative of the NFI data, they could be used to show over- or underestimation of the stem volume models. Figure 12.5 shows the results of this analysis for examples with Norway spruce and European beech. For the largest trees (with an estimated volume >5 m^3), an overestimation of about 5–10% was found, whereas the form factor and the slenderness showed better residual distributions.

Some critical points of the stem volume functions are: (a) the model selection was not fully described; (b) biased predictions were obtained over the size of trees (see Fig. 12.5 for examples with Norway spruce and European beech); and (c) reproduction of the coefficients was not entirely possible, and discrepancies were large for some species (see Fig. 12.6 for examples with Norway spruce and European beech).

Possible Further Developments A re-analysis of the entire system for estimating volume is planned. Special attention will be given to unbiased predictions over the range of observations. Therefore, different prediction methods will be tested. The weighted residuals technique will be compared with a model-based approach in which all explanatory variables of the stem form functions are predicted (by simple or multiple imputation). Further, the stem volume functions will be analysed with respect to model selection, model formulation and heteroscedasticity. The overall aim will be to develop a prediction system capable of reproducing the observed (i.e. the representatively resampled) volumes.

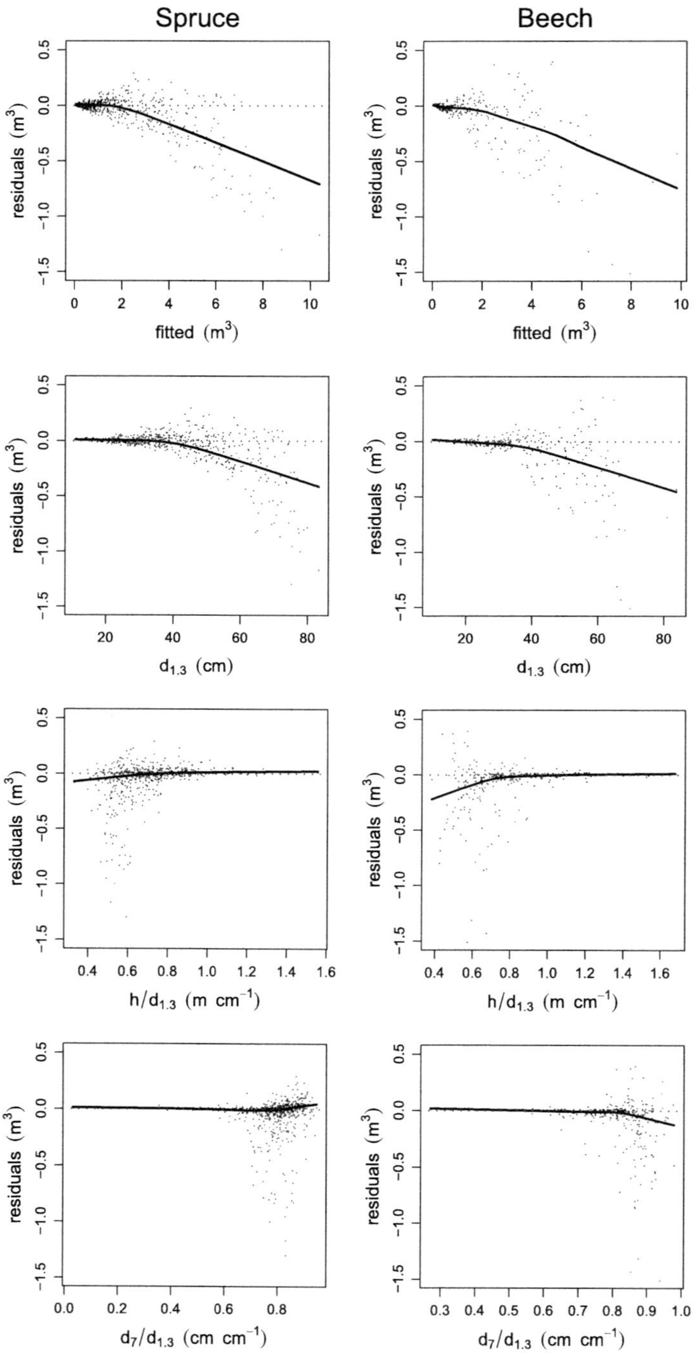

Fig. 12.5 Residual analysis of the stem volume functions for Norway spruce (*left*) and European beech (*right*)

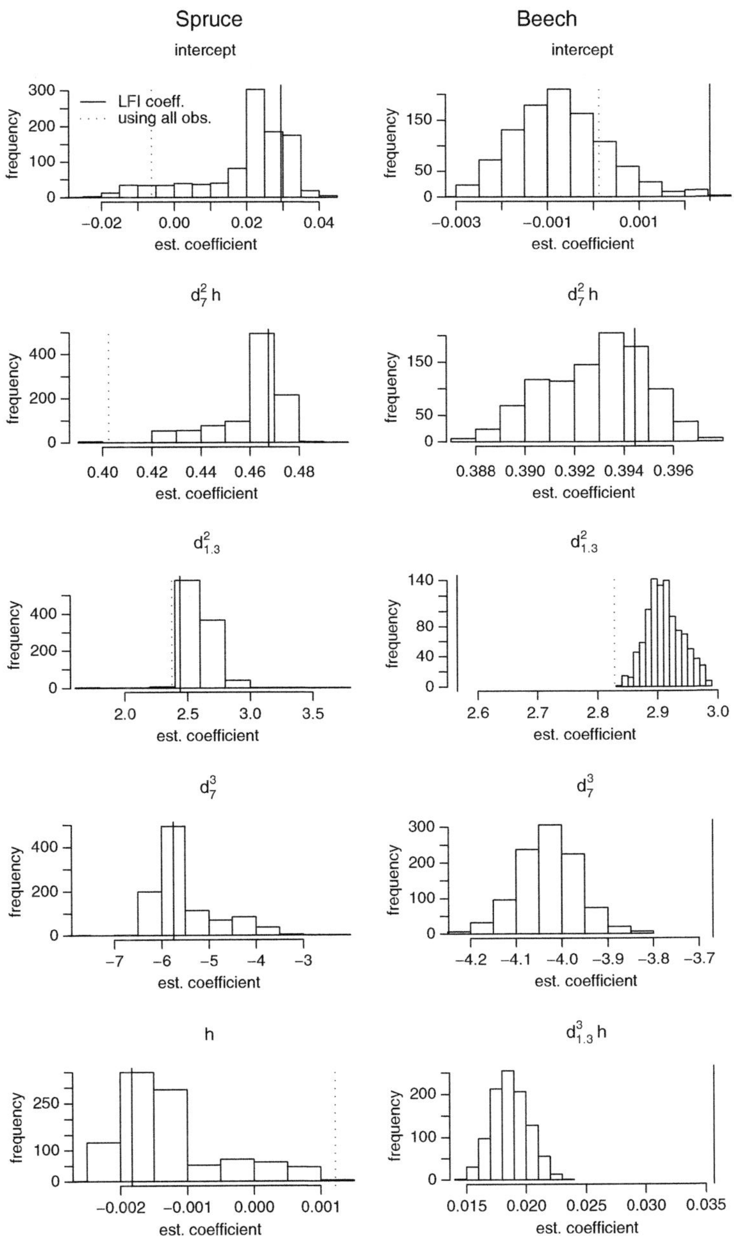

Fig. 12.6 Distribution of 1000 refitted coefficients for Norway spruce (*left*) and European beech (*right*), based on the fitting procedure described in this section (subsampling proportional to $d_{1.3}$, grouped jackknife). Continuous lines indicate the coefficients reported in the NFI and the dotted lines indicate coefficients based on all observations and jackknifing)

12.3 Tariff Models

The stem volume models described in Sect. 12.2 can only be applied to trees with measured $d_{1.3}$, d_7 and h, namely the tariff trees (Sect. 2.3.4). However, for the majority of NFI sample trees only $d_{1.3}$ is measured. For these trees, stem volume over bark, including the tree top and the aboveground part of the stump, is estimated with a tariff model that only depends on measured $d_{1.3}$ and on stand and site attributes as auxiliary variables. A total of 30 different tariff models were fitted for different combinations of tree species and production regions (Table 12.3).

The tariff models have largely remained the same since 1992, and the model descriptions and analysis documented by Kaufmann (2001; pages 166–169) are mostly still valid. This section therefore focuses on minor changes.

Tariff models were fitted based on the stem volume of the tariff sample trees (Sect. 12.2). The models that were described by Kaufmann (2001) had an independent set of coefficients for each inventory cycle. However, this caused artificial jumps in the volume prediction between inventories and therefore resulted in implausible estimations of growth in some cases, for example in small sampling units or for rare tree species. For this reason, the tariff models were refitted in a slightly different form so that all information from NFI1–3 was used simultaneously. The different inventories were considered through two additional coefficients (b_8 and b_9). Table 12.3 shows the number of observations used to fit each model. Stem volumes in NFI4 are currently estimated using coefficients fitted for NFI3.

$$\hat{Y} = exp\left(b_0 + b_1 \ln\left(d_{1.3}\right) + b_2 \ln^4\left(d_{1.3}\right) + \sum_{i=3}^{9} b_i B_i\right) \qquad (12.10)$$

where $\hat{Y}$ is the tariff volume over bark fitted for each tariff number (201, 202, . . ., 230; Table 12.3). The index i corresponds to the additional single tree and sample-plot attributes (3,. . ., 9), and $d_{1.3}$is the measured diameter at breast height. B_3 to B_9 are the following additional single tree and sample-plot attributes:

- B_3 TMI: site quality expressed as the maximum of the total mean increment from stand establishment until the age of 50 years, in kg dry weight ha^{-1} $year^{-1}$ (Sect. 15.5)
- B_4 d_{dom}: dominant diameter, i.e. mean diameter of the 100 thickest trees per hectare (derived from the diameters of the sample trees in the plot)
- B_5 bifurcation of the stem (0 = no bifurcation, 1 = bifurcation) based on field observations
- B_6 elevation (m a.s.l.), taken from the digital elevation model with a 25 m grid
- B_7 stand layer to which the single tree belongs (0 = upper layer, 1 = understorey) based on field observations

B_8 *(inv2)* and B_9 *(inv3)* together indicate the inventory cycle(s) in which the tree was measured:

inv2=0 and inv3=0: tree measured in NFI1
inv2=1 and inv3=0: tree measured in NFI2
inv2=0 and inv3=1: tree measured in NFI3

Table 12.3 Combinations of tree species and production regions along with the number of stems used to fit each tariff model

Tariff number	Tree species	Production region	N of stems
201	Norway spruce (Picea abies)	Jura	3509
202		Plateau	5854
203		Pre-Alps	8230
204		Alps	10,774
205		Southern Alps	1476
206	Silver fir (Abies alba)	Jura	2374
207		Plateau	2247
208		Pre-Alps	3149
209		Alps/Southern Alps	1227
210	Pine (Pinus sylvestris, P. nigra, P. mugo arborea)	Jura	423
211		Plateau	591
212		Pre-Alps/Alps/Southern Alps	1287
213	Larch (Larix decidua, L. kaempferi)	Jura/Plateau/Pre-Alps/Alps	2836
214		Southern Alps	744
215	Other conifers	All	595
216	European beech (Fagus sylvatica)	Jura	3751
217		Plateau	3303
218		Pre-Alps	2101
219		Alps	1219
220		Southern Alps	874
221	Oak (all species) (Quercus spp.)	Plateau	721
222		Jura/Pre-Alps/Alps/Southern Alps	607
223	Sycamore and Norway maple (Acer pseudoplatanus, A. platanoides)	Jura/Plateau	899
224		Pre-Alps/Alps/Southern Alps	821
225	Ash (Fraxinus excelsior, F. ornus)	Plateau	786
226		Jura/Pre-Alps/Alps/Southern Alps	1099
227	Chestnut (Castanea sativa)	All	802
228	Other broadleaved	Jura/Plateau	1043
229		Pre-Alps/Alps	762
230		Southern Alps	721

The model coefficients b_0 to b_9 are presented in Table 12.7.

Depending on the tariff, some explanatory variables of the general model do not contribute substantially to improving the model. Variables with a p-value >0.05 were sequentially deleted using a backward model selection procedure. These deleted variables are marked with a hyphen in Table 12.7.

Discussion of the Tariff Models The tariff models are used to predict single tree volumes, where the only measured biometric variable is $d_{1.3}$. All other variables represent auxiliary information and reflect changes in growth conditions that influence tree height and stem form. In this sense, the tariff models use information beyond $d_{1.3}$. However, it is unclear how precise volumes predicted by tariff models are compared to observed volumes. Since stem volumes are not measured in the NFI, such a comparison is not straight forward. Furthermore, the regression models ignore the inherent heteroscedasticity of the data, and thus the largest trees have the most impact on the coefficient estimations. The resulting p-values are potentially overoptimistic, which in turn influences model selection. Overall, the tariff models shown in Figs. 12.7 and 12.8 show unbiased behaviour over all Switzerland and only minor biases within some production regions.

12.4 Volume Models for Large and Small Branches

The volumes of large branches (≥7 cm in diameter) and of twigs and small branches (<7 cm in diameter) are predicted as fractions of the stem volume (i.e. tariff volume (Sect. 12.3)).

$$branch\ volume_i = stem\ volume * pa_i \tag{12.11}$$

where i indicates large or small branches and pa_i is volume of branches as a fraction of stem volume.

The proportion pa_i is estimated with a logit model:

$$pa_i = exp(lga_i)/(1 + exp(lga_i)) \tag{12.12}$$

$$lga_i = b_{0i} + b_{1i}d_{1.3} + b_{2i}h_1 + b_{3i}h_2 \tag{12.13}$$

where b_{0i} to b_{3i} are species- and region-specific regression coefficients for large and small branches, $d_{1.3}$ is the measured diameter at breast height, and h_1 and h_2 are indicator variables (used for spruce and beech only) for elevation (m a.s.l.) in combination with production region, as indicated in Table 12.4.

The models were fitted to data collected on the permanent plot network of the Experimental Forest Management (EFM) trials conducted at WSL (N = 14,712).

Species-specific logit regression models were fitted to large and small branches for spruce, fir, larch, pine and oak. For beech, three different models were fitted based on region (Tables 12.5 and 12.6). For Norway spruce, the proportion of large

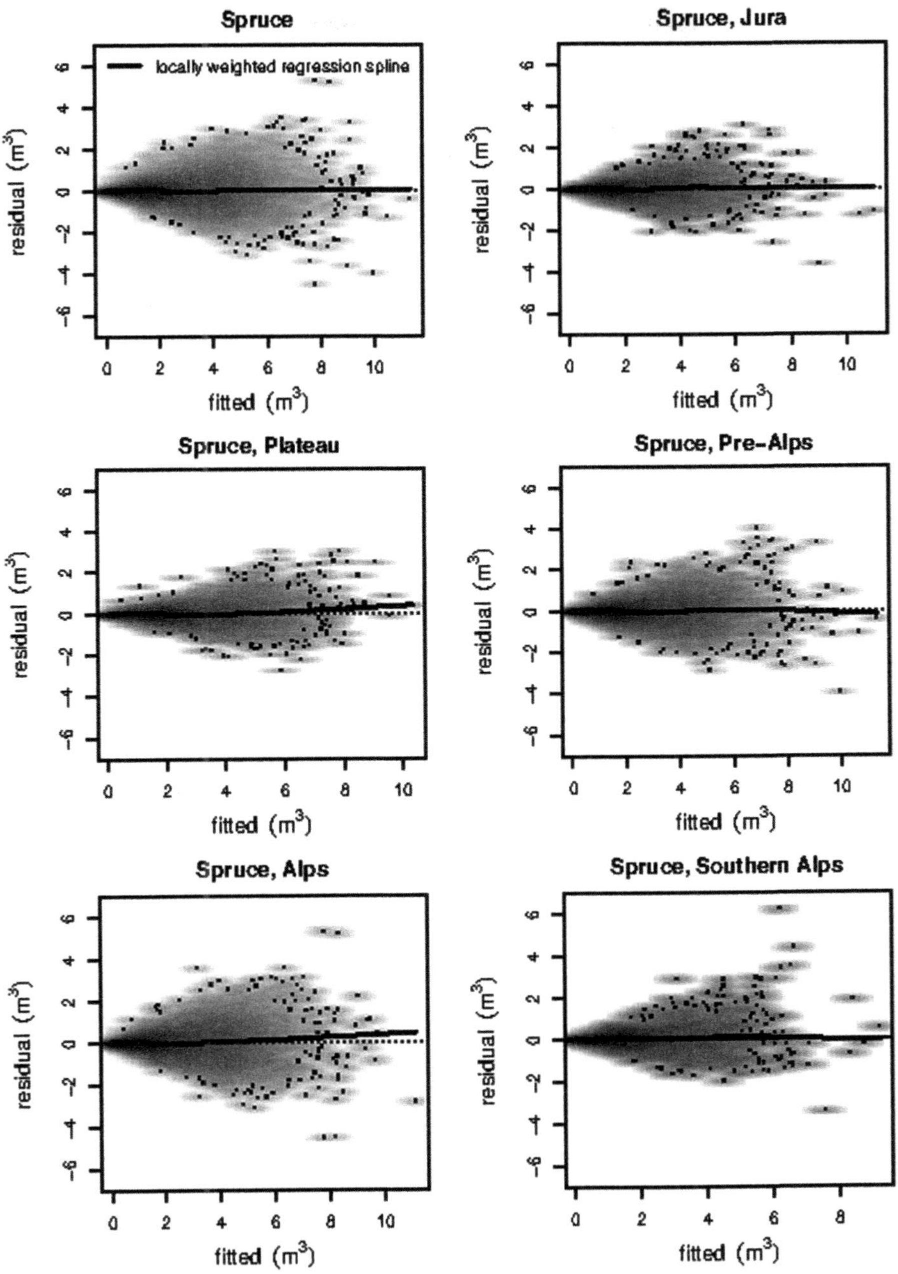

Fig. 12.7 Residual analysis of the tariff models for Norway spruce, shown for all regions together and for each region separately. The grey background is the two-dimensional density of observations (the darker the background the more observations) and the black line is a locally weighted regression spline

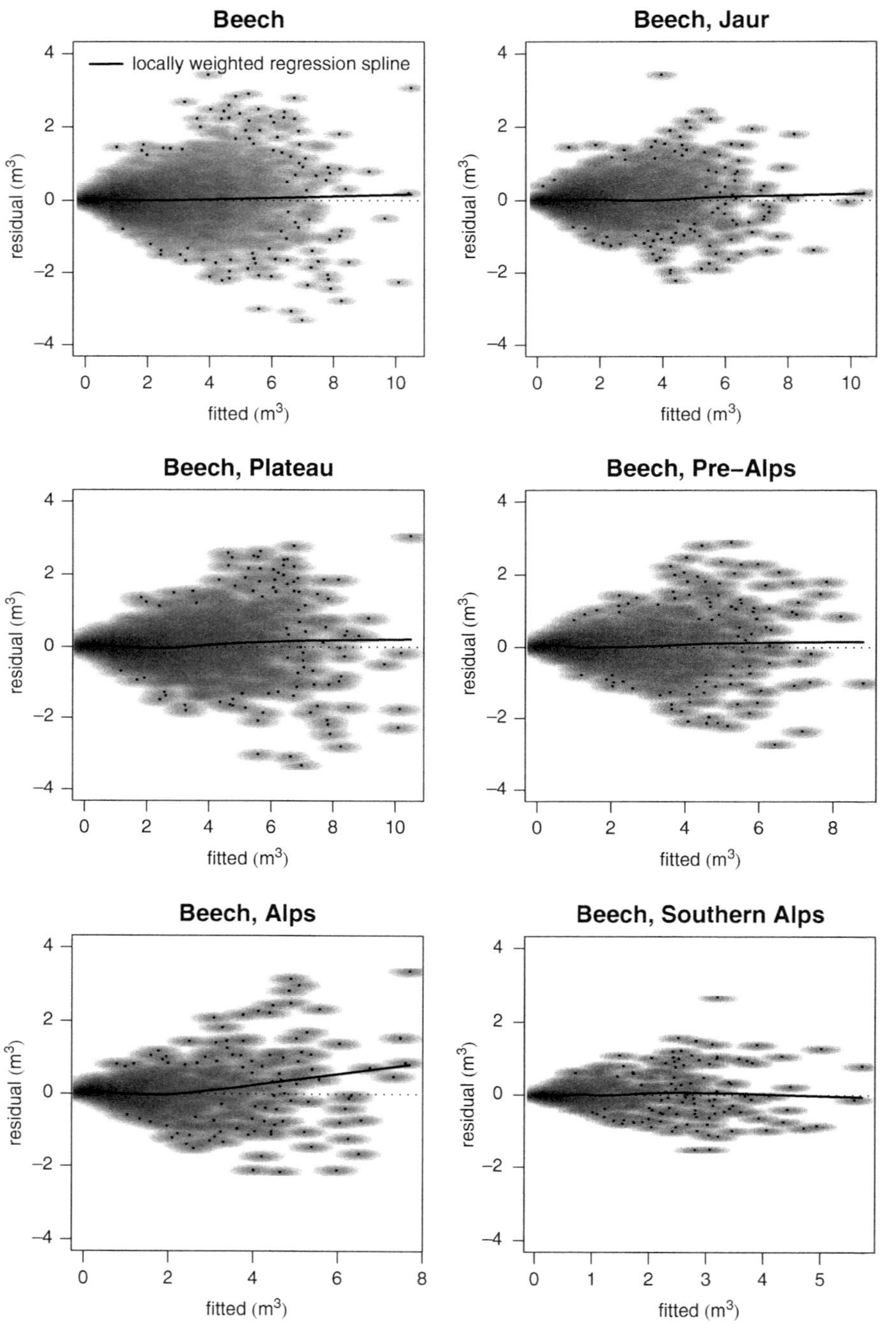

Fig. 12.8 Residual analysis of the tariff models for European beech, shown for all regions together and for each region separately. The grey background is the two-dimensional density of observations (the darker the background the more observations) and the black line is a locally weighted regression spline

Table 12.4 Indicator variables for production region and elevation (m a.s.l.), used in models to predict the volume of branches for Norway spruce and European beech

Indicator value	Production region	Elevation (m a.s.l.)
$h_1 = 1$	Alps	1000–1500
$h_1 = 0$	Alps	≤1000 and >1500
$h_1 = 1$	All other regions	600–1250
$h_1 = 0$	All other regions	≤600 and >1250
$h_2 = 1$	Alps	>1500
$h_2 = 0$	Alps	≤1500
$h_2 = 1$	All other regions	>1250
$h_2 = 0$	All other regions	≤1250

Table 12.5 Coefficients of the volume models for large branches

Tree species	Region	b_0	b_1	b_2	b_3
Silver fir	All regions	−8.7330758	0.059208154	–[a]	–[a]
Larch[b]	All regions	−5.8871184	0.010812163	–[a]	–[a]
Pine[c]	All regions	−7.7147742	0.072285665	–[a]	–[a]
European beech	Jura	−4.8322966	0.056314711	–[a]	–[a]
European beech	Plateau	−5.9903924	0.101889094	–[a]	–[a]
European beech	Pre-Alps, Alps and Southern Alps	−4.9853383	0.073941728	−0.7056977	–[a]
Other broadleaved	All regions	−4.9398872	0.061619224	–[a]	–[a]

[a]Variable not part of the model
[b]Larix decidua, L. kaempferi
[c]Pinus sylvestris, P. nigra, P. strobus, P. mugo arborea

Table 12.6 Coefficients of the volume models for small branches

Tree species	Region	b_0	b_1	b_2	b_3
Norway spruce and other conifers	All regions	−1.2064133	−0.01918645	–[a]	0.44297
Silver fir	All regions	−1.9411075	0.010967741	–[a]	–[a]
Larch[b]	All regions	−2.2772934	−0.00672607	–[a]	–[a]
Pine[c]	All regions	−1.7152468	−0.01391604	–[a]	–[a]
European beech	Jura	−0.84755833	−0.03342084	–[a]	–[a]
European beech	Plateau	−0.75961939	−0.03355523	–[a]	–[a]
European beech	Pre-Alps, Alps and Southern Alps	−2.2772572	−0.03117276	1.21051434	–[a]
Other broadleaved	All regions	−1.9339502	−0.01698668	–[a]	–[a]

[a]Variable not part of the model
[b]Larix decidua, L. kaempferi
[c]Pinus sylvestris, P. nigra, P. strobus, P. mugo arborea

branches is negligibly small, so trees of this species are assumed to have no large branches at all. Development of the volume models for large branches was documented by Kaufmann (2001). A similar procedure was used to develop volume models for small branches.

12.5 Changes in Forest Resources

As the NFI is based on permanent plots, changes such as growth, harvest and mortality are estimated by re-measuring single trees in these plots. Single-tree estimates of gains and losses per year can be directly estimated as the difference in tree volume between subsequent NFIs. Additionally, specific growth models for basal area increment (BAI) and – in the case of tariff trees – for stem volume increment (SVI) are applied to estimate the growth of trees that were cut, died, or reached the calliper threshold between two inventories.

12.5.1 *Model for Basal Area Increment*

For sample trees for which $d_{1.3}$ is only measured once in two consecutive NFIs, BAI is estimated based on a regression model fitted to data from NFI1 and NFI2. This model is used to predict BAI for trees that were cut or died between two inventories, using the $d_{1.3}$ value measured before this event took place. Analogously, the model is used to back-estimate BAI if $d_{1.3}$ was only measured at the end of the inventory interval, such as for trees that reached the calliper threshold of 36 cm in the outer circle (*nongrowth trees*; Sect. 2.9.2). No BAI is predicted for trees that reached the calliper threshold of 12 cm in the inner circle.

As done in the previous NFIs (Kaufmann 2001; pages 174–175), BAI was modelled as a function of several tree- and plot-specific variables separately for every tariff number (Table 12.3) according to Eq. 12.14 which is based on Teck and Hilt (1991) and Quicke et al. (1994):

$$\begin{aligned}\widehat{BAI} = {} & AS\exp(b_0 + \sum_{i=1}^{6} b_i B_i + b_7(AS+1)(1-\exp(b_8 d_{1.3})) \\ & + b_9(AS-1)(1-\exp(b_{10} d_{1.3})))\end{aligned} \quad (12.14)$$

where BAI is the basal area increment to be estimated (in m^2 per 10 years); *AS* indicates whether BAI is predicted forwards (+1, for trees that were cut or died) or backwards (–1, for nongrowth trees); the index *i* corresponds to the additional single tree and sample-plot attributes (1,. . ., 6), and $d_{1.3}$ is the measured diameter at breast height. The following additional attributes (B_1–B_6) are included:

B_1 stand basal area (m^2 ha^{-1}), calculated based on all standing trees in the plot that are living
B_2 basal area of larger trees (m^2 ha^{-1}), calculated based on all standing trees in the plot that are living and have a $d_{1.3}$ greater than the target tree
B_3 TMI: site quality expressed as the maximum of the total mean increment from stand establishment until the age of 50 years, in kg dry weight ha^{-1} $year^{-1}$ (Sect. 15.5)
B_4 elevation (m a.s.l.), taken from the digital elevation model with a 25 m grid
B_5 for even-aged forests: stand age (years), estimated according to a regression model based on tree-ring counts from stumps in the sample plots, as described in detail by Kaufmann (2001; pages 175–176)
for uneven-aged forests: dominant diameter d_{dom} (cm), calculated as the mean diameter of the 100 thickest trees per hectare (derived from the diameters of the sample trees in the plot)
B_6 stand layer to which the single tree belongs (0 = upper layer, 1 = understorey), based on field observations

The coefficients b_0 to b_{10} of the BAI model are presented in Table 12.8 for the case of even-aged forests and in Table 12.9 for the case of uneven-aged forests.

Based on the single measurement of $d_{1.3}$, either at the beginning (trees that died or were cut) or at the end (nongrowth trees) of the inventory interval, and on the predicted BAI, the $d_{1.3}$ that was not measured is estimated. Changes in individual-tree volumes are then calculated by applying the tariff models introduced in Sect. 12.3.

A similar BAI model is included in the forest development model MASSIMO (Chap. 17) but with some application-specific differences (e.g. growth boost). An alternative climate-sensitive BAI model was recently developed for MASSIMO (Rohner et al. 2017), but its potential future use in NAFIDAS for estimating changes in resources still needs to be evaluated.

12.5.2 *Model for Stem Volume Increment*

For the particular case of tariff trees (Sect. 2.3.4) with only one $d_{1.3}$ measurement in consecutive NFIs, a regression model for stem volume increment is applied in order to enable application of the two-stage volume prediction described in Sect. 12.2 and by Kaufmann (2001; page 176 ff.). Stem volume increment over bark is modelled according to the same nonlinear equation as presented for BAI (Eq. 12.14) including the same explanatory variables. The coefficients $b_0$to b_{10} of the stem volume increment model are presented in Table 12.10 for the case of even-aged forests and in Table 12.11 for the case of uneven-aged forests.

Appendix

Table 12.7 Coefficients of the tariff models for stem volume used in the NFI

Tariff number	b_0	b_1	b_2	b_3	b_4	b_5	b_6	b_7	b_8	b_9
201	−10.724153	3.256051	−0.0048393	1.8651E-05	0.0040682	−0.3614274	−0.0002185	–[a]	0.0223077	–[a]
202	−11.433229	3.5068775	−0.0061792	1.0366E-05	0.0052032	−0.2891386	−0.0001455	–[a]	0.0189578	–[a]
203	−11.557717	3.5567913	−0.0061486	2.6217E-05	0.0046213	−0.3471568	−0.0002382	−0.0681614	0.0176758	0.0103627
204	−11.128711	3.3859389	−0.0055547	3.8636E-05	0.0051585	−0.3029449	−0.0002215	−0.1157142	0.0218381	0.0214087
205	−11.216898	3.3166269	−0.0052951	7.3452E-05	0.004947	−0.2670809	−0.0001311	−0.1669029	0.0355638	–[a]
206	−12.645729	3.8521689	−0.0069649	1.996E-05	0.0036076	−0.0871577	−5.843E-05	–[a]	0.0213972	–[a]
207	−9.1920196	2.7776135	−0.0033026	1.1998E-05	0.0032396	−0.0954103	−9.42E-05	−0.0799185	–[a]	−0.0223729
208	−9.8212454	2.9689983	−0.0034168	2.2427E-05	0.0011562	−0.2396824	−6.907E-05	−0.0863353	–[a]	–[a]
209	−12.826018	3.9526722	−0.0066614	1.5048E-05	0.00504	−0.3413568	−0.000326	–[a]	–[a]	−0.0302187
210	−8.0864786	2.2899357	–[a]	2.2655E-05	0.0029153	−0.3086679	−0.0004536	–[a]	–[a]	–[a]
211	−10.94328	3.2717589	−0.0047869	2.6183E-05	0.0020279	–[a]	–[a]	–[a]	–[a]	0.0344802
212	−9.9217344	2.7309205	−0.0030583	6.5052E-05	0.0111059	−0.1667086	−4.188E-05	−0.135107	–[a]	–[a]
213	−11.820115	3.6012538	−0.0064718	3.8231E-05	0.003826	−0.1613262	−0.0001787	–[a]	–[a]	−0.0205343
214	−12.212199	3.7878699	−0.006457	–[a]	0.0017183	−0.578801	−0.0003175	–[a]	–[a]	–[a]
215	−11.669887	3.6109797	−0.0060115	–[a]	0.0052156	−0.1916867	−0.000475	–[a]	0.0556036	0.0540687

216	−9.9675636	2.9523813	−0.0035209	3.4071E-05	0.0046608	−0.2036949	−0.0001971	−0.0509308	0.0394829	−0.0187623
217	−11.08317	3.3529562	−0.0048976	2.3704E-05	0.0011135	−0.1103548	–[a]	−0.3674877	–[a]	–[a]
218	−11.901232	3.6164971	−0.0065386	2.4967E-05	0.0076413	−0.2283382	−0.0002516	−0.2855698	0.0352969	0.026345
219	−11.503753	3.4431726	−0.0058677	4.4736E-05	0.0095763	−0.1703535	−0.0003692	−0.4949966	–[a]	–[a]
220	−8.9986338	2.4789923	−0.0022845	–[a]	0.0082187	−0.2235004	–[a]	−0.1700668	0.0500578	0.13778
221	−11.240067	3.3752972	−0.0053652	–[a]	0.004923	−0.1366446	–[a]	–[a]	–[a]	–[a]
222	−9.8165892	2.8318847	−0.0036649	7.7658E-05	0.007115	−0.1737069	−0.0004692	−0.1952119	–[a]	–[a]
223	−7.8080883	2.1328846	–[a]	4.0574E-05	0.0067504	−0.1411788	−0.000194	−0.2800938	–[a]	–[a]
224	−10.532248	3.2857015	−0.0066407	–[a]	0.0054376	−0.2015521	−0.0003639	−0.0926404	0.0606351	0.0816757
225	−14.796222	4.7069939	−0.0110384	–[a]	–[a]	−0.1001387	–[a]	–[a]	–[a]	0.0509466
226	−8.9489701	2.5317642	−0.0019888	5.1401E-05	0.0072292	−0.2289778	−0.000182	−0.1746187	–[a]	–[a]
227	−6.4107181	1.7651895	–[a]	–[a]	–[a]	–[a]	−0.0004751	−0.2959489	0.1125232	0.0614846
228	−9.9483742	2.9543902	−0.00362	–[a]	0.0012139	−0.1706129	–[a]	−0.1849696	–[a]	0.0381569
229	−7.6634052	1.8681238	0.00181239	3.1734E-05	0.0118121	−0.1876773	–[a]	−0.2573701	–[a]	–[a]
230	−11.417793	3.3083886	−0.0058817	0.00012111	–[a]	−0.3370369	−9.689E-05	–[a]	0.0505664	0.0983005

[a]Variable not part of the model

Table 12.8 Coefficients of the model for basal area increment in even-aged forests

Tariff number	b_0	b_1	b_2	b_3	b_4	b_5	b_6	b_7	b_8	b_9	b_{10}
201	−5.9725569915	–[a]	−0.0141177298	0.0001238032	−0.0001432649	–[a]	−0.4419050465	1.2188805463	−0.0412665308	−1.5231641275	−0.0237900579
202	−5.3576716248	−0.0025350447	−0.0111731296	0.0001268912	−0.0001237766	−0.0080959830	–[a]	1.6531399040	−0.0309199696	−2.1006858679	−0.0193730741
203	−5.5999909162	−0.0043099417	−0.0031308013	0.0000464691	–[a]	−0.0061571625	−0.2207767935	2.0016539661	−0.0268785619	−2.5092794008	−0.0175886468
204	−5.9825581961	−0.0037218221	–[a]	0.0000380153	−0.0003578998	–[a]	−0.1401817972	1.7800081812	−0.0335025138	−2.1295719027	−0.0220781723
206	−6.8444118394	–[a]	−0.0093693450	0.0000595320	−0.0002954498	–[a]	–[a]	1.9166289572	−0.0524099113	−2.2333832935	−0.0318386482
207	−5.4476670286	−0.0059109694	–[a]	–[a]	0.0002315729	−0.0105309248	–[a]	2.2800097417	−0.0312465564	−3.0428097110	−0.0173548649
208	−6.4408829679	−0.0028410743	–[a]	–[a]	–[a]	−0.0060645499	–[a]	2.3633048416	−0.0390392191	−2.8302169872	−0.0244004409
209	−6.5055494813	–[a]	–[a]	–[a]	–[a]	−0.0101625094	–[a]	2.5696306474	−0.0389299081	−2.7872422289	−0.0309975118
210	−5.8575937635	–[a]	−0.0121997004	–[a]	−0.0009879144	0.0111109659	–[a]	0.7814336607	−0.0303123174	−10.0000000000	−0.0012457366
211	−6.9563871462	0.0070660957	−0.0320973525	0.0005112433	–[a]	0.0060150484	–[a]	−0.0197114881	−0.0985273172	0.0866565286	–[a]
212	−6.4261569817	−0.0074148414	–[a]	–[a]	–[a]	–[a]	−0.0001000000	1.6505499690	−0.0277870666	−2.2765817737	−0.0163653810
213	−6.7953609789	−0.0052489981	−0.0103129849	0.0001334874	–[a]	–[a]	–[a]	1.6772472515	−0.0509718006	−1.8539023598	−0.0353850006
215	−6.2673072979	–[a]	–[a]	–[a]	–[a]	–[a]	–[a]	1.2754456244	−0.0664971420	−3.1697298624	−0.0109958381
216	−6.0035754188	−0.0100298195	–[a]	0.0000592003	–[a]	−0.0060826314	–[a]	1.9892347814	−0.0361807091	−2.3349153659	−0.0252764159
217	−7.5435469469	–[a]	−0.0054871223	0.0001363784	−0.0003236908	–[a]	–[a]	2.1763432732	−0.0558397384	−2.3437735594	−0.0391380136
218	−6.9790127094	−0.0032962563	–[a]	0.0000421386	−0.0005080777	–[a]	–[a]	2.2032171585	−0.0487511502	−2.5636204245	−0.0305965922
219	−6.5431541989	–[a]	−0.0124751526	0.0000735144	−0.0007074670	0.0031207677	−0.5452301129	1.5517003191	−0.0753173816	−1.5899547937	−0.0556268702
221	−6.9921993719	0.0086011546	−0.0095388228	0.0001775024	–[a]	–[a]	–[a]	1.3732159062	−0.0471328782	−1.5589308056	−0.0304159108
222	−6.1045742673	–[a]	–[a]	–[a]	–[a]	–[a]	–[a]	3.6041963410	−0.0083661657	−7.0902276154	−0.0035660288
223	−6.5274474562	–[a]	−0.0172270729	0.0000907116	–[a]	–[a]	–[a]	1.3172166568	−0.0751507507	−1.4562757256	−0.0519940718
224	−5.8086525351	−0.0088521730	–[a]	0.0001087008	–[a]	–[a]	–[a]	3.1033997441	−0.0075478492	−10.0000000000	−0.0020843691
225	−5.7187214188	–[a]	–[a]	–[a]	–[a]	–[a]	–[a]	1.5401536430	−0.0363248815	−3.9809955692	−0.0077976708
226	−6.4276148516	–[a]	–[a]	0.0000520548	–[a]	−0.0080900070	–[a]	2.2956149040	−0.0351685160	−2.5650228462	−0.0268937547
227	−23.1177575591	–[a]	–[a]	–[a]	–[a]	–[a]	–[a]	9.4072550257	−0.1770963558	−10.0000000000	−0.0964787963
228	−6.3296482034	–[a]	–[a]	–[a]	–[a]	–[a]	–[a]	3.8433016946	−0.0116393351	−4.6639497862	−0.0082584971
229	−7.8887806029	−0.0189361901	–[a]	–[a]	−0.0004978637	–[a]	–[a]	2.8269973959	−0.0650354461	−3.0844060725	−0.0493176443

[a]Variable not part of the model

Table 12.9 Coefficients of the model for basal area increment in uneven-aged forests

Tariff number	b_0	b_1	b_2	b_3	b_4	b_5	b_6	b_7	b_8	b_9	b_{10}
201	−5.3463805127	–[a]	−0.0187785258	0.0000612228	–[a]	–[a]	0.0000000000	1.0287447250	−0.0405496467	−1.4472917416	−0.0179297677
202	−5.5534990481	−0.0094775685	–[a]	0.0001577019	0.0003495484	–[a]	−0.3370856275	1.9508277783	−0.0124437409	−3.0000000000	−0.0065142676
203	−5.0806115871	–[a]	−0.0095143248	0.0000690710	−0.0004799371	−0.0071544352	–[a]	1.3988472729	−0.0429638638	−1.7621742218	−0.0234704010
204	−5.4839564382	−0.0081505463	0.0049078212	0.0000292479	−0.0003921279	−0.0086835713	–[a]	1.7616296143	−0.0388817765	−2.4059857603	−0.0197084468
205	−4.5893146522	−0.0048346137	−0.0075398644	–[a]	−0.0005286772	0.0061500581	−0.5338777078	0.8454519540	−0.0519822681	−1.3600685185	−0.0174306781
206	−6.6103943593	−0.0098281726	0.0232563626	–[a]	–[a]	−0.0272062833	–[a]	2.8393884519	−0.0384469463	−3.2414553206	−0.0263771512
207	−6.9649489993	−0.0103624688	0.0109648412	–[a]	–[a]	–[a]	–[a]	2.3845980246	−0.0517128958	−2.6351473376	−0.0325415511
208	−5.9266305124	−0.0038519207	−0.0077257158	0.0001212667	–[a]	–[a]	−0.1737349481	1.5810843113	−0.0354557708	−1.8668678522	−0.0220897328
209	−6.9496764153	−0.0128876735	−0.0117087149	–[a]	−0.0003281241	0.0093460154	—[1]	2.0265882649	−0.0785205061	−2.1951196082	−0.0478887002
210	−4.7818446774	–[a]	–[a]	–[a]	–[a]	–[a]	–[a]	3.0000000000	−0.0011047318	−1.4187520167	−0.0015373376
211	−7.6402540921	–[a]	−0.0603585232	–[a]	–[a]	–[a]	–[a]	2.3109864961	−0.0487407286	−2.6139668701	−0.0310590012
212	−6.1058114909	–[a]	–[a]	−0.0001396340	−0.0005150931	–[a]	–[a]	1.5661264182	−0.0469866377	−1.6950020250	−0.0369951807
213	−5.5974573086	−0.0060007067	–[a]	0.0002170699	–[a]	–[a]	−0.2295110226	1.2305966730	−0.0191139809	−1.8379144009	−0.0103019105
214	−2.4512135510	−0.0076731853	−0.0174007601	−0.0003088548	−0.0009886755	–[a]	−0.3732842363	3.0000000000	−0.0039490260	−3.0000000000	−0.0040601338
215	−7.4889257646	–[a]	0.0203299522	–[a]	–[a]	–[a]	−0.6240826952	2.6758859496	−0.0219906826	−3.0951606559	−0.0161881377
216	−5.7690764286	–[a]	−0.0081976076	0.0000747968	0.0001565503	–[a]	–[a]	2.2241620640	−0.0120551377	−3.5527130438	−0.0062496692
217	−8.3686218200	–[a]	–[a]	0.0000948696	–[a]	–[a]	–[a]	2.5139879450	−0.0693457269	−2.7597035071	−0.0426371964
218	−6.3477822588	−0.0111179450	–[a]	–[a]	−0.0003807825	–[a]	–[a]	1.9096330293	−0.0552630020	−2.2694431841	−0.0339515038
219	−7.4164536583	–[a]	−0.0055648313	0.0000658307	–[a]	−0.0130392315	–[a]	2.3054479091	−0.0567858332	−2.5086384861	−0.0426036766
220	−7.1084697111	–[a]	−0.0001000000	–[a]	−0.0004985165	−0.0105495330	−0.2859337889	2.4993334571	−0.0524799097	−2.6250223977	−0.0431105050
221	−8.4022899987	0.0060541954	−0.0170831458	0.0001659352	–[a]	−0.0148269842	–[a]	2.4911454919	−0.0641389440	−2.6073997562	−0.0506235362
222	−7.8634955822	–[a]	–[a]	–[a]	–[a]	–[a]	–[a]	2.4026041833	−0.0483820806	−2.6279871204	−0.0369619186
223	−8.6963947922	–[a]	–[a]	0.0000708614	–[a]	−0.0120053170	–[a]	2.7697490346	−0.0672622887	−2.8371679713	−0.0583405991
224	−5.7467716000	–[a]	–[a]	–[a]	–[a]	–[a]	–[a]	1.8660330453	−0.0232199659	−3.0000000000	−0.0127560652
225	−5.2530765582	−0.0116087796	–[a]	–[a]	–[a]	–[a]	–[a]	3.3945739049	−0.0101681782	−2.8739321287	−0.0112585186
226	−7.2541434628	–[a]	–[a]	0.0000854853	–[a]	−0.0142416440	–[a]	2.3645807882	−0.0485700539	−2.5967763172	−0.0363075794
227	−10.5072740585	–[a]	–[a]	–[a]	–[a]	–[a]	–[a]	3.5396869228	−0.0964343943	−3.6228742135	−0.0755542458
228	−9.6558982180	−0.0218944051	–[a]	0.0001231541	–[a]	–[a]	–[a]	3.9363251304	−0.0447726900	−4.3240409439	−0.0321211317
229	−8.5346866377	−0.0160221363	–[a]	0.0000756011	−0.0013503759	−0.0071350973	–[a]	3.5559417470	−0.0720972566	−3.7935484613	−0.0528753369
230	−8.5012101613	−0.0125758616	0.0271552129	−0.0004131241	−0.0006962986	−0.0491419974	–[a]	4.2184520047	−0.0953190451	−4.7074136058	−0.0649524942

[a]Variable not part of the model

Table 12.10 Coefficients of the model for stem volume increment over bark in even-aged forests

Tariff number	b_0	b_1	b_2	b_3	b_4	b_5	b_6	b_7	b_8	b_9	b_{10}
201	−3.1862466089	0.0092009270	−0.0104988952	0.0001252450	–[a]	–[a]	–[a]	0.9683287000	−0.9044855329	−1.0481471252	−0.5427115359
202	−2.1200720881	–[a]	−0.0112944191	0.0001991331	0.0004150566	−0.0055370394	–[a]	0.8083285112	−0.5085677616	−1.1504208788	−0.2282178919
203	−2.4611717658	–[a]	–[a]	0.0001402713	0.0003199561	−0.0078657264	–[a]	1.2050726520	−0.6208337920	−1.5688057932	−0.3058353870
204	−2.6199517463	–[a]	–[a]	0.0000604492	–[a]	−0.0032551748	–[a]	1.1550063951	−0.9695188572	−1.3614255829	−0.5095657377
206	−2.5565347223	–[a]	–[a]	0.0000613960	–[a]	–[a]	–[a]	1.0539046744	−0.8374063743	−1.2890794184	−0.3905131862
207	−1.1180929299	–[a]	−0.0106346084	–[a]	–[a]	−0.0033177239	–[a]	0.7989323198	−0.7454273985	−1.1299642471	−0.2687610008
208	−1.9907008089	–[a]	–[a]	–[a]	–[a]	–[a]	–[a]	0.9828681521	−0.7973529551	−1.1202195523	−0.4123545657
209	−1.3565978491	–[a]	–[a]	−0.0002200785	0.0019210974	−0.0195913048	–[a]	3.3157999681	−0.0765474435	−2.4865301428	−0.1137584799
210	−1.2024720699	–[a]	−0.0335570979	–[a]	–[a]	–[a]	–[a]	–[a]	–[a]	–[a]	–[a]
211	−1.2024720699	–[a]	−0.0335570979	–[a]	–[a]	–[a]	–[a]	–[a]	–[a]	–[a]	–[a]
212	−1.2024720699	–[a]	−0.0335570979	–[a]	–[a]	–[a]	–[a]	–[a]	–[a]	–[a]	–[a]
213	−1.7514629101	–[a]	–[a]	0.0002349951	–[a]	–[a]	–[a]	–[a]	–[a]	–[a]	–[a]
215	−2.8517604971	–[a]	–[a]	0.0002846672	–[a]	–[a]	–[a]	1.0770804205	−0.6155649028	−1.1034269508	−0.4424810951
216	−1.3627608379	–[a]	−0.0087780685	–[a]	−0.0004140575	–[a]	–[a]	1.0163218373	−0.2975669315	−1.5137152541	−0.1410574199
217	−1.1511612691	0.0089813279	−0.0275305914	–[a]	–[a]	0.0071095059	–[a]	–[a]	–[a]	–[a]	–[a]
218	−1.0177571596	0.0106110357	−0.0376676477	0.0000984640	−0.0008842311	0.0050014278	–[a]	–[a]	–[a]	–[a]	–[a]
219	−1.0177571596	0.0106110357	−0.0376676477	0.0000984640	−0.0008842311	0.0050014278	–[a]	–[a]	–[a]	–[a]	–[a]
221	−4.1379287044	–[a]	–[a]	–[a]	0.0013509352	–[a]	–[a]	1.5250218847	−1.0000000000	−1.5827319160	−0.7451050074
222	−4.1379287044	–[a]	–[a]	–[a]	0.0013509352	–[a]	–[a]	1.5250218847	−1.0000000000	−1.5827319160	−0.7451050074
223	−0.5074755445	0.0273985148	−0.0625244058	–[a]	−0.0013493028	–[a]	–[a]	–[a]	–[a]	–[a]	–[a]
224	−0.5074755445	0.0273985148	−0.0625244058	–[a]	−0.0013493028	–[a]	–[a]	–[a]	–[a]	–[a]	–[a]
225	−1.3860688164	–[a]	–[a]	–[a]	–[a]	−0.0136376126	–[a]	1.3095281025	−0.8761970272	−1.5488068988	−0.5077292923
226	−1.3860688164	–[a]	–[a]	–[a]	–[a]	−0.0136376126	–[a]	1.3095281025	−0.8761970272	−1.5488068988	−0.5077292923
228	−1.1389097735	–[a]	–[a]	–[a]	–[a]	–[a]	–[a]	–[a]	–[a]	–[a]	–[a]
229	−1.1389097735	–[a]	–[a]	–[a]	–[a]	–[a]	–[a]	–[a]	–[a]	–[a]	–[a]

[a]Variable not part of the model

Table 12.11 Coefficients of the model for stem volume increment over bark in uneven-aged forests

Tariff number	b_0	b_1	b_2	b_3	b_4	b_5	b_6	b_7	b_8	b_9	b_{10}
201	−1.5402394206	–[a]	−0.0266740921	0.0000904223	−0.0004624973	0.0253903975	–[a]	–[a]	–[a]	–[a]	–[a]
202	−1.5402394206	–[a]	−0.0266740921	0.0000904223	−0.0004624973	0.0253903975	–[a]	–[a]	–[a]	–[a]	–[a]
203	−0.7116806330	−0.0042516658	−0.0286035731	–[a]	−0.0007156974	0.0277742594	–[a]	–[a]	–[a]	–[a]	–[a]
204	−1.0781841206	–[a]	−0.0348577102	0.0001031825	−0.0005051668	0.0185036077	–[a]	–[a]	–[a]	–[a]	–[a]
205	−3.7529887064	–[a]	–[a]	0.0002227484	–[a]	0.0168171038	–[a]	1.2796770363	−0.2456921716	−1.5582603792	−0.1536427575
206	−2.4402823250	–[a]	–[a]	0.0001466875	–[a]	–[a]	–[a]	0.9764040338	−0.7228907526	−1.0524134975	−0.4050528784
207	−2.4402823250	–[a]	–[a]	0.0001466875	–[a]	–[a]	–[a]	0.9764040338	−0.7228907526	−1.0524134975	−0.4050528784
208	−2.4402823250	–[a]	–[a]	0.0001466875	–[a]	–[a]	–[a]	0.9764040338	−0.7228907526	−1.0524134975	−0.4050528784
209	−2.4402823250	–[a]	–[a]	0.0001466875	–[a]	–[a]	–[a]	0.9764040338	−0.7228907526	−1.0524134975	−0.4050528784
210	−8.0535623547	−0.0256603607	−0.0886705485	0.0005046373	–[a]	0.1362866029	–[a]	–[a]	–[a]	–[a]	–[a]
211	−8.0535623547	−0.0256603607	−0.0886705485	0.0005046373	–[a]	0.1362866029	–[a]	–[a]	–[a]	–[a]	–[a]
212	−8.0535623547	−0.0256603607	−0.0886705485	0.0005046373	–[a]	0.1362866029	–[a]	–[a]	–[a]	–[a]	–[a]
213	−0.9493173384	–[a]	−0.0470198952	–[a]	–[a]	–[a]	–[a]	–[a]	–[a]	–[a]	–[a]
214	5.9199620429	–[a]	−0.0360266946	−0.0012222347	−0.0028575072	–[a]	–[a]	–[a]	–[a]	–[a]	–[a]
215	−8.0535623547	−0.0256603607	−0.0886705485	0.0005046373	–[a]	0.1362866029	–[a]	–[a]	–[a]	–[a]	–[a]
216	−1.7562023261	–[a]	–[a]	0.0002519269	–[a]	–[a]	–[a]	–[a]	–[a]	–[a]	–[a]
217	−1.7562023261	–[a]	–[a]	0.0002519269	–[a]	–[a]	–[a]	–[a]	–[a]	–[a]	–[a]
218	−2.1392803217	–[a]	–[a]	–[a]	–[a]	–[a]	–[a]	1.3675043273	−0.2973194495	−1.9407441561	−0.1705210706
219	−2.1392803217	–[a]	–[a]	–[a]	–[a]	–[a]	–[a]	1.3675043273	−0.2973194495	−1.9407441561	−0.1705210706
220	−1.5646451103	–[a]	–[a]	–[a]	–[a]	–[a]	–[a]	0.5013682402	−1.0000000000	−0.3578996563	–[a]
221	−2.7840727237	–[a]	–[a]	0.0004985905	−0.0030113676	–[a]	−0.9274380417	1.2519176398	−0.3458881151	−1.0326722733	−0.3929053333
222	−2.7840727237	–[a]	–[a]	0.0004985905	−0.0030113676	–[a]	−0.9274380417	1.2519176398	−0.3458881151	−1.0326722733	−0.3929053333
223	−2.0189165870	–[a]	–[a]	–[a]	–[a]	–[a]	–[a]	0.0168947458	–[a]	−10.0000000000	−0.0350066134
224	−2.0189165870	–[a]	–[a]	–[a]	–[a]	–[a]	–[a]	0.0168947458	–[a]	−10.0000000000	−0.0350066134
225	−1.6275582101	−0.0075711735	−0.0611526559	–[a]	–[a]	0.0389005799	–[a]	–[a]	–[a]	–[a]	–[a]
226	−1.6275582101	−0.0075711735	−0.0611526559	–[a]	–[a]	0.0389005799	–[a]	–[a]	–[a]	–[a]	–[a]
227	4.4583421722	–[a]	0.0091428195	−0.0012773259	−0.0030621450	–[a]	–[a]	1.3057925325	−0.3540028056	−1.2923499953	−0.3054081810
228	−2.5260783910	–[a]	–[a]	–[a]	–[a]	–[a]	–[a]	1.8153016379	−0.6458499180	−1.9470412312	−0.3820039239
229	−2.5260783910	–[a]	–[a]	–[a]	–[a]	–[a]	–[a]	1.8153016379	−0.6458499180	−1.9470412312	−0.3820039239
230	−0.6357439915	–[a]	–[a]	−0.0006260031	–[a]	–[a]	–[a]	2.7170453600	−0.2915851985	−1.5436936798	−0.5754856630

[a]Variable not part of the model

References

Efron B, Stein C (1981) The Jackknife estimate of variance. Ann Stat 9(3):586. (11 pp. https://doi.org/10.1214/aos/1176345462

Kaufmann E (1993) Einzelbaum-Voluminierung und Tarifprobebaum-Auswahl im LFI2. Internal report, unpublished. Swiss Federal Research Institute WSL, Birmensdorf, Switzerland

Kaufmann E (2001) Estimation of standing timber, growth and cut. In: Brassel P, Lischke H (eds) Swiss National Forest Inventory: methods and models of the second assessment. Swiss Federal Research Institute WSL, Birmensdorf, pp 162–196

Mandallaz D (1991) A unified approach to sampling theory for forest inventory based on infinite population and superpopulation models. Diss., Lehrstuhl Forstinventur und Planung, ETH, Zürich

Mandallaz D (1997) The anticipated variance: a tool for the optimization of forest inventories. Zürich: Chair of Forest Inventory and Planning, ETH, Zürich

Peck JL, Zenner EK, Brang P, Zingg A (2014) Tree size distribution and abundance explain structural complexity differentially within stands of even-aged and uneven-aged structure types. Eur J For Res 133:335–346

Quicke HE, Meldahl RS, Kush JS (1994) Basal area growth of individual trees: a model derived from a regional longleaf pine growth study. For Sci 40:528–542

Rohner B, Waldner P, Lischke H, Ferretti M, Thürig E (2017) Predicting individual-tree growth of central European tree species as a function of site, stand, management, nutrient, and climate effects. Eur J For Res 137:29–44

Teck RM, Hilt DE (1991) Individual-tree diameter growth model for the Northeastern United States. Radnor

Zell J (2018) Climate sensitive tree growth functions and the role of transformations. Forests 9 (7):382. (20 pp.). https://doi.org/10.3390/f9070382

Chapter 13
Assortments

Brigitte Rohner, Christoph Fischer, Anne Herold, and Erik Rösler

Abstract Besides reporting available resources, the Swiss National Forest Inventory (NFI) also reports potential merchantable assortments. For this purpose, the stem form of all recorded trees is estimated based on a combination of field measurements, regression models, splines and rotational integrals. Once the stem forms are described, the potential merchantable assortments are estimated by optimising cutting points along these hypothetical boles. This procedure is applied to NFI trees irrespective of their actual timber harvesting status, resulting in an estimate of the potential available assortments in the total Swiss growing stock.

13.1 Introduction

Besides reporting available resources, as described in Chap. 12, the Swiss NFI (NFI) also reports potential merchantable assortments. For this purpose, the stem form of all recorded trees is estimated based on a combination of field measurements, regression models, splines and rotational integrals. These predicted stem forms are split into different potential merchantable assortments. This procedure is applied to NFI trees irrespective of their actual timber harvesting status, resulting in an estimate of the potential available assortments in the total Swiss growing stock. The procedures for assorting trees were originally programmed in FORTRAN77, but during NFI4 they were reprogrammed in SAS© and SAS/IML© for their integration into the NFI Data Analysis System NAFIDAS (Chap. 20).

In Switzerland, guidelines for assorting merchantable timber are summarised and published as Swiss trading customs for raw timber (Waldwirtschaft Schweiz et al.

B. Rohner (✉) · A. Herold · E. Rösler
Resource Analysis, Swiss Federal Research Institute WSL, Birmensdorf, Switzerland
e-mail: brigitte.rohner@wsl.ch

C. Fischer
Scientific Service NFI, Swiss Federal Research Institute WSL, Birmensdorf, Switzerland

C. Fischer, B. Traub (eds.), *Swiss National Forest Inventory – Methods and Models of the Fourth Assessment*, Managing Forest Ecosystems 35,
https://doi.org/10.1007/978-3-030-19293-8_13

Table 13.1 Publications of Swiss trading customs for assorting roundwood, and their implementation in the NFI

Inventory	Name	Publication reference	Implemented in NAFIDAS
NFI1: 1983–1985	Schweizerische Holzhandelsgebräuche	Forstwirtschaftliche Zentralstelle der Schweiz (1976)	No
NFI2: 1993–1995	Schweizerische Holzhandelsgebräuche	Forstwirtschaftliche Zentralstelle der Schweiz 1976)	No
NFI3: 2004–2006	Schweizerische Handelsgebräuche für Rundholz	Waldwirtschaft Verband Schweiz et al. (2000)	Yes
NFI4: 2009–2017	Schweizer Handelsgebräuche für Rohholz	Waldwirtschaft Schweiz et al. (2010)	Yes

2010). The Swiss trading customs describe – separately for broadleaved and coniferous tree species – the dimensions (length and diameter) used in trading merchantable timber on the domestic market. For roundwood, assortments are categorised into types according to their length and into classes according to their diameter. For coniferous species, additional criteria regarding minimum diameters at the smaller end are applied.

The assortments reported for NFI3 (Brändli 2010) were based on the version of the Swiss trading customs published in 2000 (Waldwirtschaft Verband Schweiz et al. 2000). For NFI4, the procedures were adapted to correspond to the latest version published in 2010 (Waldwirtschaft Schweiz et al. 2010; Table 13.1). Assortments reported for NFI1 and NFI2, based on the Swiss trading customs published in 1976 (Forstwirtschaftliche Zentralstelle der Schweiz 1976), are not implemented in NAFIDAS (Table 13.1).

Table 13.2 summarises the definitions of the roundwood assortments implemented for NFI4. Assortment categories other than roundwood, for example wood chips used for energy production or pulpwood, were not considered in the NFI. Any timber product smaller than the assortments defined here is considered smallwood. Wood quality was not evaluated because reliable sampling methods are not currently available.

13.2 Approach for Estimating Assortments

In order to estimate merchantable assortments, stem forms of the trees recorded in the NFI have to be described. Once the stem forms are described, the assortments can be estimated by optimising cutting points along these hypothetical boles. In the NFI, this approach is implemented through the following steps: (1) estimate diameters at several locations along the bole, (2) fit cubic interpolation splines between these

Table 13.2 Summary of assortments considered in NFI4 (according to the Swiss trading customs; Waldwirtschaft Schweiz et al. 2010)

		Class 1	Class 2	Class 3	Class 4	Class 5	Class 6
Broadl. roundw.	Length	≥3 m	≥3 m	≥3 m	≥3 m	≥3 m	≥3 m
	Diameter under bark in the middle	10–19 cm	20–29 cm	30–39 cm	40–49 cm	50–59 cm	≥60 cm
	Diameter under bark at the smaller end	–	–	–	–	–	–
Conif. roundw.	Length type:						
	Very short	<3 m	<3 m	<3 m	<3 m	<3 m	<3 m
	Short	3–6 m	3–6 m	3–6 m	3–6 m	3–6 m	3–6 m
	Medium	6.5–14.5 m	6.5–14.5 m	6.5–14.5 m	6.5–14.5 m	6.5–14.5 m	6.5–14.5 m
	Long	15–22 m	15–22 m	15–22 m	15–22 m	15–22 m	15–22 m
	Diameter under bark in the middle	10–19 cm	20–29 cm	30–39 cm	40–49 cm	50–59 cm	≥60 cm
	Diameter under bark at the smaller end	7 cm[a]	18 cm	18 cm	22 cm	22 cm	22 cm

[a]This value differs from that in the Swiss trading customs (Waldwirtschaft Schweiz et al. 2010) because the NFI does not differentiate sub-classes.

diameters to achieve continuous stem profiles, (3) adjust the profiles at the stem base and top, (4) shift the profiles to better represent the measured diameters, (5) subtract bark estimates, (6) determine cutting points along the stem profiles to produce certain assortments, and (7) estimate the volume of each piece with rotational integrals. Only tree stems are considered, whereas all other tree compartments (branches, twigs, leaves, bark, stumps, roots) are not taken into account for assorting.

Step 1: Diameter Estimations Along the Stem Regression models are used to predict diameters at several locations along the stem (at 1 m height and at 5%, 10%, 20%, 30%, 50%, 70% and 80% of the total stem length). Explanatory variables included in these models are the measured diameter at breast height ($d_{1.3}$), a modelled diameter at 7 m height (D7MOD), and a modelled total tree height (HMOD). The regression models predicting the diameters at several locations were described in detail by Kaufmann (2001; pages 187–188). However, a description of the models providing the explanatory variables D7MOD and HMOD is presented below.

Both D7MOD and HMOD are predicted based on the same model formulation used for predicting tariff volumes (Sect. 12.3):

$$\hat{Y} = \exp\left(b_0 + b_1 \ln(d_{1.3}) + b_2 \ln^4(d_{1.3}) + \sum_{i=3}^{9} b_i B_i \right) \qquad (13.1)$$

where $\hat{Y}$ is either D7MOD (in cm) or HMOD (in m), b are the model coefficients, the index i indicates additional tree and sample-plot attributes (3, . . ., 9), and $d_{1.3}$ is the measured diameter at breast height. The following additional attributes ($B_3 - B_9$) are considered:

- B_3 TMI: site quality expressed as the maximum of the total mean increment from stand establishment until the age of 50 years, in kg dry weight ha^{-1} $year^{-1}$ (Sect. 15.5)
- B_4 d_{dom}: dominant diameter (in cm); mean diameter of the 100 thickest trees per hectare, derived from the diameters of the trees on the sample plot
- B_5 bifurcation of the stem (0: no bifurcation, 1: bifurcation) based on field observations
- B_6 elevation (in m a.s.l.) taken from the digital elevation model with a 25 m grid
- B_7 stand layer to which the single tree belongs (0 = upper layer, 1 = understorey) based on field observations
- B_8, B_9 inventory cycle in which the single tree was measured; for NFI3 and NFI4 $B_8 = 0$ and $B_9 = 1$ (Sect. 12.3)

The model coefficients b_0–b_9 are presented in Table 13.4 for D7MOD and in Table 13.5 for HMOD. As with the tariff models (Sect. 12.3) the models for D7MOD and HMOD were fitted to data from tariff trees measured in NFI1 to NFI3.

Step 2: Fitting of Cubic Interpolation Splines Continuous stem profiles are described by fitting cubic interpolation splines using the estimated diameters from step (1) as knots. This approach has a long tradition in forest science (Bruce and Max 1990; Kublin et al. 1984; Smaltschinski 1984; Sterba 1980). Details about the spline interpolation implemented for the assortment prediction in the NFI were described by Kaufmann (2001; pages 188–189).

Step 3: Adjustments at the Stem Base and Top At the stem base, the tree stump is assumed to have a fixed height of 0.3 m and is not described explicitly as part of the stem profile. From 0.3 to 1 m height a constant diameter is assumed, using the estimated diameter at 1 m height from step (1). From 1 m height upwards, the spline interpolation introduced in step (2) is used, with a starting curvature at 1 m height predicted for every tree according to a regression model described in detail by Kaufmann (2001; page 189). At the top of the stem, a curvature of −0.001 is assumed (Kaufmann 2001; page 189).

Step 4: Shifts to Measured Diameters As the estimated diameters from step (1) are used to describe the stem profiles, they do not exactly match the measured $d_{1.3}$ (and d_7 for tariff trees; measured diameter at 7 m height). Attempts to include $d_{1.3}$ and d_7 as additional knots in the spline interpolation (step 2) did not always result in realistic stem forms, because tree diameters did not decrease uniformly with increasing tree height in all cases (Kaufmann 2001). Therefore, the stem profiles derived in steps (1) to (3) are shifted so that they intersect with the measured diameters. This shift is described in detail by Kaufmann (2001; pages 189–190).

Step 5: Subtraction of Bark Estimates Based on several studies focusing on species-specific bark thickness (Altherr et al. 1974, 1975, 1976, 1978), the bark thickness r of the recoded NFI trees is estimated based on the height x along the stem according to Kaufmann (2001; page 193):

$$r(x) = b_{0k} + b_{1k}\, dm_k + b_{2k} dm_k{}^2 \tag{13.2}$$

where

$k = 1$ for $0 \leq x \leq 0.33\ H$
$k = 2$ for $0.33\ H < x \leq 0.66\ H$
$k = 3$ for $x > 0.66\ H$

and

dm_1 is the diameter over bark at $x = 0.25\ H$
dm_2 is the diameter over bark at $x = 0.5\ H$
dm_3 is the diameter over bark at $x = 0.75\ H$

where H is the total tree height. The coefficients b_{0k}, b_{1k}, and b_{2k} are given in Table 13.3. The estimated bark thickness is subtracted from the stem profiles obtained in step (4).

Table 13.3 Coefficients used to estimate bark thickness for three sections ***k*** located at different heights along the stem (according to Eq. 13.2)

Tree species	k	b_0	b_1	b_2
Picea abies, Pinus cembra, other conifers	1	1.5554	0.55475	−0.00225
Picea abies, Pinus cembra, other conifers	2	0.82652	0.59424	−0.00212
Picea abies, Pinus cembra, other conifers	3	0.1744	0.67905	−0.00247
Abies alba	1	0.82802	0.62504	0
Abies alba	2	1.67703	0.56074	0
Abies alba	3	0.67058	0.68492	0
Pinus sylvestris, P. nigra, P. mugo	1	5.43367	0.62571	0
Pinus sylvestris, P. nigra, P. mugo	2	0.05652	0.56149	0
Pinus sylvestris, P. nigra, P. mugo	3	4.17891	0.22292	0
Larix decidua, L. kaempferi	1	−6.46451	1.73845	−0.00943
Larix decidua, L. kaempferi	2	−6.45758	1.82516	−0.01176
Larix decidua, L. kaempferi	3	−9.74591	2.31981	−0.0225
Fagus sylvatica, Castanea sativa	1	1.97733	0.28119	0
Fagus sylvatica, Castanea sativa	2	2.25734	0.29724	0
Fagus sylvatica, Castanea sativa	3	2.69794	0.31096	0
Acer pseudoplatanus, A. platanoides	1	−0.60951	0.64014	−0.00329
Acer pseudoplatanus, A. platanoides	2	−3.53373	0.91611	−0.00707
Acer pseudoplatanus, A. platanoides	3	−4.573	1.06506	−0.00929
Fraxinus excelsior, F. ornus	1	−1.14181	0.96466	−0.00432
Fraxinus excelsior, F. ornus	2	−8.40201	1.41083	−0.00964
Fraxinus excelsior, F. ornus	3	−3.62803	1.21051	−0.00777
Quercus spp.	1	9.10974	0.66266	0
Quercus spp.	2	8.94454	0.71505	0
Quercus spp.	3	9.88377	0.75877	0

Step 6: Determination of Cutting Points To identify optimal cutting points along the modelled stem profiles, the intended assortment type according to Table 13.2 has to be determined. While for broadleaved trees a minimum length of 3 m is the only consideration, coniferous trees can be assorted into several defined length types (Table 13.2). For NFI1 to NFI3, the choice among the assortment types for conifers was made collectively for some defined regions based on a combination of region-specific assumptions and information collected in interviews with foresters (Duc et al. 2010; Kaufmann and Brassel 1999). All coniferous trees growing in the Plateau region were assorted as medium- or longwood (6.5–22 m length) and those in the Alps and Southern Alps were assorted as shortwood (4–6 m length, according to the previous version of the Swiss trading customs, Waldwirtschaft Verband Schweiz et al. 2000). For coniferous trees in the Jura and the Pre-Alps, the intended assortment type was defined separately for each forest district based on the interviews (for details see Duc et al. 2010; Kaufmann and Brassel 1999).

In NFI4, the generalised assumptions of the assortment type could be avoided by directly using plot-specific information gathered in the interviews with the forest service (Chap. 10). Therefore, the relevant question in the interview was adapted to correspond to the latest version of the Swiss trading customs (Waldwirtschaft Schweiz et al. 2010). The foresters are asked to state which assortment type, according to Table 13.2, they chose (or would choose) for each NFI plot. Based on this information, hypothetical cutting points along the stem profiles developed in steps (1) to (5) are defined in such a way that these plot-specific assortment types result. Between these hypothetically built sections a certain gap is included to reflect the loss caused by sawing (10 cm for coniferous assortments up to medium length, 30 cm for coniferous longwood, and 15 cm for broadleaved assortments). The hypothetically built sections are assigned to different assortment classes according to their diameters (Table 13.2).

Step 7: Rotational Integrals for Volume Prediction The volume of the individual assortments is estimated using rotational integrals (Kaufmann 2001; page 192). Specifically, the two-dimensional sections determined in the previous step (diameter over height) are rotated around the height axis to describe three-dimensional bole sections.

13.3 Further Development

The assortments considered in the NFI are currently limited to roundwood. Given the economic and ecological importance of pulpwood and energy wood assortments, however, it would certainly be desirable to expand the range of investigated assortments beyond roundwood. However, information about wood quality is important for assessing whether trees are potentially suitable for roundwood, pulpwood or energy wood assortments, and such knowledge is currently lacking. An alternative to conducting wood quality assessments in the field may be further expanding the questions related to assortments in the interviews with the forest service.

The assortment prediction methods reported by Kaufmann and Brassel (1999) and Duc et al. (2010) take into account the uncertainties involved in extrapolation from the plots to the national or regional scale, but they give no indication of the uncertainties inherent to the assortment prediction. A sensitivity analysis showing how, for example, uncertainties in the models predicting diameters at several heights (step (1), including models for D7MOD and HMOD) affect the assortment prediction would be beneficial.

Europe-wide comparisons of assortment quantities could give insights about (potentially) available wood products at a larger scale. However, a standardised European guideline for roundwood and other raw timber assortments was abrogated through the European parliament in 2007 (Guideline Nr. 68/89/EWG and Decision Nr. 714/2007/EG). Consequently, country-specific differences in rules for assorting make such comparison efforts rather challenging.

Appendix

Table 13.4 Coefficients of the regression model (Eq. 13.1) for the diameter at 7 m height (D7MOD)

Tariff number	b_0	b_1	b_2	b_3	b_4	b_5	b_6	b_7	b_8	b_9
201	−1.5208321	1.46710383	−0.0021327	2.43790E-06	0.00029299	−0.1843136	−2.37987E-05	0	0.01285451	0
202	−1.4962957	1.4739144	−0.0024017	0	0.0004031	−0.0896331	−1.69486E-05	0	0.00702941	0
203	−1.8183241	1.57335453	−0.0026872	3.04910E-06	0.00146176	−0.1991969	−6.00215E-05	−0.0744101	8.63E-05	−0.0131372
204	−1.7724182	1.55511792	−0.0025961	8.60720E-06	0.00163941	−0.1920661	−7.56833E-05	−0.124091	−0.002567	−0.009725
205	−2.5525958	1.74547685	−0.0034029	3.85857E-05	0.00265389	−0.1692707	−1.98827E-05	−0.1017157	0.0204808	0
206	−2.5465366	1.83509532	−0.003843	0	0.00021125	−0.0269428	0	0	0.01268219	0
207	−1.1956373	1.38180152	−0.0021367	0	0.00010433	−0.073014	0	−0.0336127	0	−0.0101419
208	−1.7220055	1.55290276	−0.0025664	0	0	−0.1073663	0	−0.0294029	0	0
209	−2.5185499	1.81357955	−0.0034741	8.04690E-06	0.00156987	−0.1775858	−8.97645E-05	0	0	−0.0266914
210	−0.6839117	1.11626538	0	1.01132E-05	0.00087663	−0.248238	−1.27605E-04	0	0	0
211	−1.3323806	1.34525915	−0.0013864	1.11941E-05	0.00079548	0	0	0	0	0.00061164
212	−2.6553861	1.74792226	−0.0035766	1.51527E-05	0.00592386	−0.0927801	−1.90530E-06	−0.0980758	0	0
213	−2.2652094	1.68513685	−0.0031421	1.23420E-05	0.00143661	−0.1382138	−1.65878E-05	0	0	−0.0191051
214	−2.7082263	1.84690799	−0.0035247	0	0.00139054	−0.3525086	−7.19749E-05	0	0	0
215	−3.0695687	2.0013031	−0.0043712	0	0.00192728	−0.1235632	−1.33916E-04	0	0.00487053	−0.0022147
216	−1.3883756	1.4178743	−0.0019011	7.61830E-06	0.00097847	−0.1225627	−4.69712E-05	−0.1083869	0.00803656	−0.0125473
217	−1.3163366	1.39760121	−0.0017356	0	0.00022697	−0.074693	0	−0.0855175	0	0
218	−2.1518767	1.70448512	−0.0035037	1.19020E-06	0.00279567	−0.1515962	−9.45879E-05	−0.1067611	0.01165515	0.00469282
219	−1.4952626	1.43940671	−0.0023369	1.39598E-05	0.00335424	−0.1299337	−8.88039E-05	−0.1954672	0	0
220	−1.9966223	1.55590876	−0.0032325	0	0.00548132	−0.1984397	0	−0.2086618	0.0280322	0.07896196
221	−1.3429282	1.38119791	−0.0017073	0	0.00168504	−0.0443392	0	0	0	0
222	−2.1637594	1.6570008	−0.0033708	2.48006E-05	0.00371692	−0.1630364	−1.60560E-04	−0.1162702	0	0
223	−0.4674011	1.03125615	0	1.00998E-05	0.00344886	−0.084829	−5.82144E-05	−0.151709	0	0
224	−2.7133685	1.94193834	−0.0055241	0	0.00410509	−0.1450716	−1.47186E-04	−0.0790774	0.01286658	0.01245871

225	−1.8463754	1.60825068	−0.0030088	0	0	−0.0832522	0	0	0	−0.0045452
226	−1.2107306	1.3424118	−0.0018456	4.75110E-06	0.0037936	−0.13599	−1.29318E-04	−0.0991702	0	0
227	0.36585142	0.73206156	0	0	0	0	−1.63450E-06	−0.2682128	0.17265618	0.14867898
228	−1.736514	1.50444869	−0.002493	0	0.00246717	−0.1446775	0	−0.1048929	0	−0.0042036
229	−2.1206585	1.53095801	−0.0026923	2.81247E-05	0.00660145	−0.1367753	0	−0.1857118	0	0
230	−2.5990139	1.83520052	−0.0043373	3.23911E-05	0	−0.2755522	−1.77049E-04	0	0.04162059	0.05192069

Table 13.5 Coefficients of the regression model (Eq. 13.1) for tree height (HMOD)

Tariff number	b_0	b_1	b_2	b_3	b_4	b_5	b_6	b_7	b_8	b_9
201	−0.106572	1.05329769	−0.0030819	2.17058E-05	0.00533756	−0.1590132	−2.53241E-04	0	0.01454928	0
202	−0.2608868	1.11552637	−0.0036256	1.22605E-05	0.00665318	−0.1334014	−1.79248E-04	0	0.00328735	0
203	−0.0483345	1.01652147	−0.00296	3.21761E-05	0.00578463	−0.0862507	−1.91396E-04	−0.1873082	0.01261734	0.00651411
204	0.05200309	0.96918987	−0.0026924	2.99887E-05	0.00485345	−0.1022876	−1.49212E-04	−0.1931868	0.02467849	0.02555364
205	−0.2388269	1.00916728	−0.0028753	3.38980E-05	0.00539691	−0.0992318	−1.06097E-04	−0.2031244	0.01133653	0
206	−0.5400951	1.16942054	−0.0033362	1.84546E-05	0.00541603	−0.0498618	−2.48248E-04	0	0.01462378	0
207	0.25337942	0.94056981	−0.0027857	1.30647E-05	0.00527609	−0.0890098	−1.72760E-04	−0.1679176	0	−0.0183461
208	0.33693956	0.89764267	−0.0022301	2.71699E-05	0.00426246	−0.0676475	−2.11919E-04	−0.1751395	0	0
209	−0.0996198	1.02459901	−0.0023255	1.86650E-05	0.00455536	−0.1388277	−2.76080E-04	0	0	0.01965426
210	1.54121072	0.337426	0	2.91197E-05	0.01191176	−0.192964	−4.60773E-04	0	0	0
211	−1.1304935	1.32390961	−0.0045133	3.69447E-05	0.00570385	0	0	0	0	0.03248706
212	0.01954302	0.7084251	−0.0023844	8.69360E-05	0.01310961	−0.0720778	−2.71110E-06	−0.2248376	0	0
213	−0.4846666	1.17282967	−0.0039301	3.14375E-05	0.00555166	−0.0807664	−1.88764E-04	0	0	0.01080404
214	−0.2501624	1.12581554	−0.0034994	0	0.00515648	−0.2270172	−2.70748E-04	0	0	0
215	−0.2753395	1.14579771	−0.0031633	0	0.0050698	−0.1317292	−4.01508E-04	0	0.05356718	0.04988256
216	1.02006902	0.63842661	−0.0017394	3.47158E-05	0.00584923	−0.0980331	−1.89906E-04	−0.1948662	0.02325846	0.00127808
217	0.97821338	0.65113455	−0.0016986	2.37977E-05	0.00452641	−0.0473186	0	−0.2193229	0	0
218	1.13698522	0.57912566	−0.0013513	2.70057E-05	0.00620111	−0.0668759	−1.40356E-04	−0.2314194	0.01965179	0.00535837
219	1.24818145	0.51361795	−0.0008803	2.78248E-05	0.00601154	−0.0973219	−1.58236E-04	−0.2180423	0	0
220	1.45947749	0.39570995	−0.0005272	0	0.00367604	−0.0701069	0	−0.1995287	0.08983298	0.10111693
221	0.637148	0.77439444	−0.0022645	0	0.00398495	−0.117503	0	0	0	0
222	0.82714226	0.64653187	−0.0017141	3.31596E-05	0.00692076	−0.1114008	−3.63343E-04	−0.1716821	0	0
223	1.7298286	0.32245275	0	2.58589E-05	0.00805293	−0.0713686	−1.55747E-04	−0.2033967	0	0
224	0.92866815	0.73375797	−0.0029542	0	0.00762556	−0.0654405	−3.09512E-04	−0.1425972	0.02834815	0.04453738
225	1.07117683	0.7089073	−0.0017513	0	0	−0.0310389	0	0	0	0.04428236
226	0.76946022	0.7308986	−0.002376	3.20524E-05	0.00675292	−0.104964	−2.01041E-04	−0.1110275	0	0
227	1.97240479	0.21846141	0	0	0	0	−4.69419E-05	−0.1642379	0.09827041	0.10230517
228	0.79038808	0.68154134	−0.0017332	0	0.00631477	−0.1221518	0	−0.1882053	0	0.01541504
229	0.73422941	0.50927732	−0.0007266	6.37755E-05	0.01033322	−0.0715221	0	−0.1956003	0	0
230	1.2107448	0.57978188	−0.0011117	−1.00000E-10	0	−0.1153318	−2.51920E-04	0	0.06960222	0.0727781

References

Altherr E, Unfried P, Hradetzky J, Hradetzky V (1974) Statistische Rindenbeziehungen als Hilfsmittel zur Ausformung und Aufmessung unentrindeten Stammholzes. Teil I: Kiefer, Buche, Hainbuche und Roterle. Mitteilungen der Forstlichen Versuchs- und Forschungsanstalt Baden-Württemberg 61

Altherr E, Unfried P, Hradetzky J, Hradetzky V (1975) Statistische Rindenbeziehungen als Hilfsmittel zur Ausformung und Aufmessung unentrindeten Stammholzes. Teil II: EuLä, JLä, Skie, StEi, TrEi, REi, Ahorn und Linde. Mitteilungen der Forstlichen Versuchs- und Forschungsanstalt Baden-Württemberg 68

Altherr E, Unfried P, Hradetzky J, Hradetzky V (1976) Statistische Rindenbeziehungen als Hilfsmittel zur Ausformung und Aufmessung unentrindeten Stammholzes. Teil III: Weykie, Robinie, Bergulme, Birke, *P. marilandica*, *P. robusta*. Mitteilungen der Forstlichen Versuchs- und Forschungsanstalt Baden-Württemberg 78

Altherr E, Unfried P, Hradetzky J, Hradetzky V (1978) Statistische Rindenbeziehungen als Hilfsmittel zur Ausformung und Aufmessung unentrindeten Stammholzes. Teil IV: Fichte, Tanne, Douglasie und Sitkafichte. Mitteilungen der Forstlichen Versuchs- und Forschungsanstalt Baden-Württemberg 90

Brändli U-B (ed) (2010) Schweizerisches Landesforstinventar. Ergebnisse der dritten Erhebung 2004–2006. Eidgenössische Forschungsanstalt für Wald, Schnee und Landschaft WSL, Birmensdorf; Bundesamt für Umwelt BAFU, Bern

Bruce D, Max TA (1990) Use of profile equations in tree volume estimation. Pacific Northwest Research Station. USDA Forest Service general technical report

Duc P et al (2010) Holzproduktion. In: Brändli U-B (ed) Schweizerisches Landesforstinventar. Ergebnisse der dritten Erhebung 2004–2006. Eidgenössische Forschungsanstalt für Wald, Schnee und Landschaft WSL, Birmensdorf; Bundesamt für Umwelt BAFU, Bern, pp 143–186

Forstwirtschaftliche Zentralstelle der Schweiz (1976) Schweizerische Holzhandelsgebräuche. Waldwirtschaft Verband Schweiz, Solothurn

Kaufmann E (2001) Estimation of standing timber, growth and cut. In: Brassel P, Lischke H (eds) Swiss National Forest Inventory: methods and models of the second assessment. Swiss Federal Research Institute WSL, Birmensdorf, pp 162–196

Kaufmann E, Brassel P (1999) Holzproduktion. In: Brassel P, Brändli U-B (eds) Schweizerisches Landesforstinventar. Ergebnisse der Zweitaufnahme 1993–1995. Eidgenössische Forschungsanstalt für Wald, Schnee und Landschaft WSL, Birmensdorf; Bundesamt für Umwelt, Wald und Landschaft BUWAL, Bern, Haupt, Bern

Kublin E, Hradetzky J, Hannak H (1984) Ein einheitliches Modell zur Beschreibung der Schaftform mit Spline-Funktionen. Mitteilung FVA Baden-Württemberg 108

Smaltschinski T (1984) Individuelle Baumschaftform und kubische Spline Interpolation. Allg Forst Jagdztg 155:193–197

Sterba H (1980) Stem curves – a review of the literature. For Abstr 41:141–145

Waldwirtschaft Schweiz, Holzindustrie Schweiz, Holzbau Schweiz, Lignum (2010) Schweizer Handelsgebräuche für Rohholz. Christoph Starck, Lignum, Holzwirtschaft Schweiz, Zürich

Waldwirtschaft Verband Schweiz, Schweizerischer Sägerei- und Holzindustrieverband, Verband Schweizer Rund- und Schnittholzhändler (2000) Schweizerische Handelsgebräuche für Rundholz. Waldwirtschaft Verband Schweiz

Chapter 14
Whole Tree Biomass and Carbon Stock

Markus Didion, Anne Herold, and Esther Thürig

Abstract Information on tree biomass and carbon (C) stock has become more relevant in the context of renewable energy production and reporting obligations, such as the Global Forest Resources Assessment of the FAO or the Greenhouse Gas Inventories under the UNFCCC and the Kyoto Protocol. Biomass and C stock cannot be measured easily in the field, and they are typically derived from existing estimates of volume or relationships to measured attributes such as tree diameter. In the Swiss NFI (NFI), whole tree biomass and its respective C content are derived separately for each individual tree element, including the aboveground elements stemwood, foliage, and large and small branches, and belowground coarse roots, based on volume estimates of stemwood and branchwood, and on foliage and root biomass correlations with tree diameter. Biomass and C stock are estimated at the individual tree level, thereby ensuring accuracy and consistency of the estimates.

14.1 Introduction

Information on tree biomass and carbon (C) stock has become more relevant in the context of renewable energy production and reporting obligations, such as the Global Forest Resources Assessment of the FAO or the Greenhouse Gas Inventories under the UNFCCC and the Kyoto Protocol. Biomass and C stock cannot be measured easily in the field, and they are typically derived from existing estimates of volume or relationships to measured attributes such as tree diameter. In the Swiss NFI (NFI), whole tree biomass and its respective C content are derived separately for each individual tree element, including the aboveground elements stemwood, foliage, and large and small branches, and belowground coarse roots (Fig. 14.1; Gschwantner et al. 2009), based on volume estimates of stemwood

M. Didion (✉) · A. Herold · E. Thürig
Resource Analysis, Swiss Federal Research Institute WSL, Birmensdorf, Switzerland
e-mail: markus.didion@wsl.ch

C. Fischer, B. Traub (eds.), *Swiss National Forest Inventory – Methods and Models of the Fourth Assessment*, Managing Forest Ecosystems 35,
https://doi.org/10.1007/978-3-030-19293-8_14

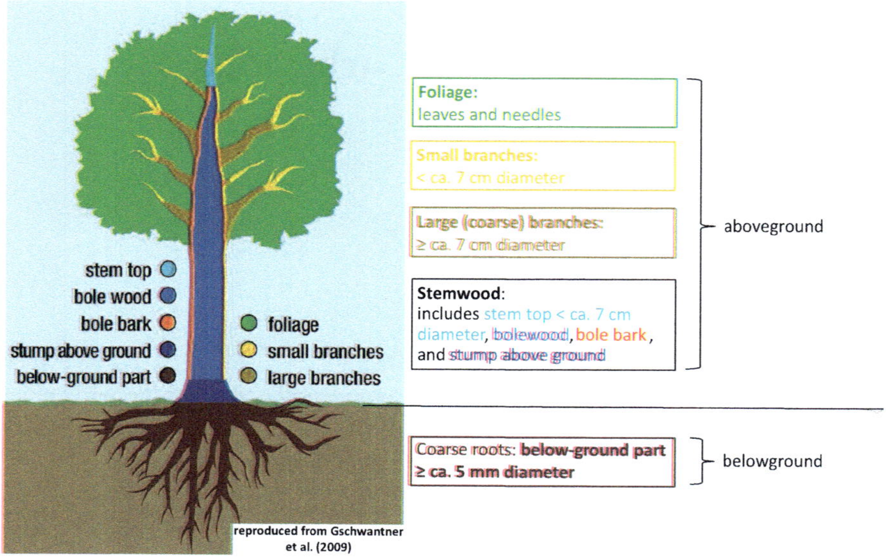

Fig. 14.1 Sketch of a whole tree, showing the separate elements used for volume and biomass estimates, reproduced from Gschwantner et al. (2009)

and branchwood, and on foliage and root biomass correlations with tree diameter. Biomass and C stock are estimated at the individual tree level, thereby ensuring accuracy and consistency of the estimates.

14.2 Conversion of Wood Volume to Wood Biomass

As described in Chap. 12 stemwood includes the tree stump, bolewood and top including bark, and branchwood consists of large (≥7 cm diameter) and small branches including bark. For living sample trees, the volume of each of these tree elements is converted to biomass based on tree-species-specific wood densities (Table 14.1) derived from a meta-analysis with a focus on data from Switzerland and Germany (Trendelenburg 1939). Dead sample trees are treated differently because the density of deadwood depends on its state of decay, which is recorded in the field (Table 14.2). Deadwood densities corresponding to the five decay classes were measured for *Picea abies* and *Fagus sylvatica*, the main coniferous and broadleaved tree species in Swiss forests (Dobbertin and Jüngling 2009, reproduced in Didion et al. 2014). The values for *P. abies* are used for all conifers and for cases where the tree species cannot be identified, whereas the values for *F. sylvatica* are used for all broadleaved species. The same approach is used for lying deadwood assessed using line-intersect sampling; converted biomass values are typically expressed using the standardised measurement of oven-dry weight.

Table 14.1 Species-specific basic wood densities for the 12 main tree species considered in the NFI. The densities shown are based on mean values estimated by Trendelenburg (1939) and partially reproduced by Assmann (1961, Table 28). Densities are assumed to also apply to bark

Species	Basic wood density [kg m^{-3}]
Picea abies	390
Abies alba	370
Pinus sylvestris	420
Larix decidua	470
Pinus cembra	420
Other coniferous species	410
Fagus sylvatica	560
Acer spp.	540
Fraxinus spp.	570
Quercus spp.	570
Castanea sativa	520
Other broadleaved species	520
Undetermined	450

Table 14.2 Basic wood densities of standing dead trees and lying deadwood, separated by decay classes, for coniferous and broadleaved species (Dobbertin and Jüngling 2009). Densities are assumed to also apply to bark

Decay class	Description	Species	Basic wood density [kg m^{-3}]
1 – Raw wood[a]	Trees that died recently, not yet dried out	Coniferous	390
		Broadleaved	560
2 – Solid deadwood	No sap, hard (knife blade penetrates wood but with difficulty), not decayed	Coniferous	394
		Broadleaved	521
3 – Rotten wood	No sap, wood is still solid (knife blade easily penetrates vertically but not horizontally), somewhat decayed	Coniferous	333
		Broadleaved	319
4 – Mould wood	Soft (knife blade penetrates easily in both directions), decayed	Coniferous	274
		Broadleaved	241
5 – Duff wood	Very soft, loose pieces, powdery	Coniferous	247
		Broadleaved	233

[a]Treated like living trees; see Table 14.1

14.3 Biomass Equations

Biomass estimates of belowground elements including coarse and fine roots and foliage are derived using the allometric relationship between biomass and tree diameter at breast height (dbh). The allometries are based on literature and are used to directly estimate biomass. For broadleaved species, the biomass of coarse roots (*roots*) larger than about 5 mm in diameter is calculated using Eq. 14.1 developed for European beech (*F. sylvatica*) (Wutzler et al. 2008). For conifers, the equation developed for Norway spruce (*P. abies*) by Wirth et al. (2004) and refitted by Zell and Thürig (2013) is used (Eq. 14.2). Both studies were carried out using data from several sources.

Broadleaved species:

$$roots = 0.0282\ dbh^{2.39} \tag{14.1}$$

Coniferous species:

$$roots = 0.04481998\ dbh^{2.28974783} \tag{14.2}$$

where dbh is the tree diameter at 1.30 m above the ground.

The biomass of fine roots is assumed to amount to 5% of the coarse root biomass (Perruchoud et al. 1999). Fine root biomass is not considered as part of the whole tree biomass but as part of the soil (following IPCC 2006; Gschwantner et al. 2009).

Foliage biomass (*foliage*), including leaves/needles but excluding reproductive organs (following Gschwantner et al. 2009), is estimated as a function of tree dbh. This equation (Eq. 14.3) was developed by Perruchoud et al. (1999) using Swiss tree data (Burger 1940, 1947, 1948, 1950, 1951, 1953; all cited in Perruchoud et al. 1999):

$$foliage = na(\text{nart}) + nb(\text{nart})\ dbh^2 + nc(\text{nart})\ dbh^4 \tag{14.3}$$

where dbh is the tree diameter at 1.30 m above the ground, and *na*, *nb* and *nc* are coefficients specific to individual species (nart; Table 14.3).

14.4 Whole Tree Biomass

For living trees, whole tree biomass is calculated as the sum of the biomass of the tree elements (stem, large and small branches, foliage, and coarse roots). It comprises more than the growing stock, which typically refers only to the volume of stemwood (Gschwantner et al. 2009). For *P. abies* individuals, it is assumed that the biomass of large branches is negligible (Mäkinen and Hein 2006).

Table 14.3 Coefficients for *na*, *nb* and *nc* for individual tree species (nart), used in Eq. 14.3; see also Table A1 in Perruchoud et al. 1999

nart	Species	na	nb	nc
1	Picea abies, Larix decidua, other coniferous and undetermined species	1.413701	0.024182	$-1.067*10^{-6}$
2	Abies alba	2.056995	0.024814	$-1.540*10^{-7}$
3	Pinus spp.	1.138795	0.007911	$8.639*10^{-7}$
4	All other broadleaved species	0.372238	0.006653	$9.780*10^{-7}$
5	Quercus spp., Castanea sativa	0.402990	0.007293	$2.800*10^{-7}$

In the case of standing dead trees, only the biomass of the stem, large branches, and coarse roots is considered. If stem damage is observed, the biomass of standing dead trees is the sum of stem biomass based on the measured dimensions of the remaining stem and coarse root biomass, and it is assumed that the remaining biomass is accounted for under lying deadwood.

14.5 Conversion of Whole Tree Biomass to Whole Tree Carbon Stock

The amount of C stored in trees depends on the C fraction in each tree element. For living trees, a mean C fraction of 0.5 in dry matter is assumed, based on table 4.3 in IPCC (2006), regardless of the tree species and element. The C content of living trees thus corresponds to half its biomass. The C concentration of dead *P. abies* and *F. sylvatica* trees in Swiss forests was analysed by Dobbertin and Jüngling (2009, reproduced in Didion et al. 2014). Based on this study, the C content of woody material in dead trees is calculated as a percentage of biomass regardless of the decay class: 49.3% for conifers and 47.6% for broadleaved species.

14.6 Objectives/Target Variables

The biomass and C stock of living and dead standing trees, as well as of deadwood, are required for several national and international reporting obligations, such as the Forest Resource Assessment (FAO 2015) and the State of Europe's Forests (FOREST EUROPE 2015). Data on living trees is also used for calculating the C balance reported in the Swiss greenhouse gas inventory (FOEN 2017). Based on the C stock in foliage and coarse roots of living trees, the annual production of non-woody litter (foliage and fine roots) is estimated using turnover rates. Litter produced by living trees and residues of dead trees serve as input for model simulations of the C balance in deadwood, litter and soil.

References

Assmann E (1961) Waldertragskunde: organische Produktion, Struktur, Zuwachs und Ertrag von Waldbestaenden. BLV Verlagsgesellschaft, München

Didion M, Frey B, Rogiers N, Thürig E (2014) Validating tree litter decomposition in the Yasso07 carbon model. Ecol Model 291:58–68

Dobbertin M, Jüngling E (2009) Totholzverwitterung und C-Gehalt. [Wood density and carbon content with changing degree of dead wood decay: first results]. Swiss Federal Research Institute for Forest, Snow and Landscape Research, Birmensdorf. avl. online at www.bafu.admin.ch/climatereporting

FAO (2015) Global Forest Resources Assessment 2015. FAO, Rome

FOEN (Federal Office for the Environment) (2017) Switzerland's greenhouse gas inventory 1990–2015. National Inventory Report including reporting elements under the Kyoto Protocol. Submission of April 2017 under the United Nations Framework Convention on Climate Change and under the Kyoto Protocol. Federal Office for the Environment, Bern. avl. online at www.bafu.admin.ch/climatereporting

FOREST EUROPE (2015) State of Europe's Forests 2015

Gschwantner T, Schadauer K, Vidal C, Lanz A, Tomppo E, di Cosmo L, Robert N, Englert Duursma D, Lawrence M (2009) Common tree definitions for national forest inventories in Europe

IPCC (2006) 2006 IPCC guidelines for national greenhouse gas inventories volume 4 agriculture, forestry and other land use. Available at: http://www.ipcc-nggip.iges.or.jp/public/2006gl/vol4.html

Mäkinen H, Hein S (2006) Effect of wide spacing on increment and branch properties of young Norway spruce. Eur J For Res 125:239–248

Perruchoud D, Kienast F, Kaufmann E, Bräker OU (1999) 20th century carbon budget of forest soils in the Alps. Ecosystems 2:320–337

Trendelenburg R (1939) Das Holz als Rohstoff: seine Entstehung, stoffliche Beschaffenheit und chemische Verwertung. J. F. Lehmanns Verlag, Münich

Wirth C, Schumacher J, Schulze E-D (2004) Generic biomass functions for Norway spruce in Central Europe–a meta-analysis approach toward prediction and uncertainty estimation. Tree Physiol 24:121–139

Wutzler T, Wirth C, Schumacher J (2008) Generic biomass functions for common beech (Fagus sylvatica) in Central Europe: predictions and components of uncertainty. Can J For Res 38:1661–1675

Zell J, Thürig E (2013) Root biomass functions for the GHG reporting under the UNFCCC and under the KP in Switzerland. Swiss Federal Institute for Forest, Snow and Landscape Research WSL

Chapter 15
Other Models

Christoph Fischer, Urs-Beat Brändli, and Markus O. Huber

Abstract Besides traditional models for tree volume and growth, other models have been developed in the Swiss National Forest Inventory (NFI). These models, described briefly in this section, cover four topics: (1) biodiversity indicators, (2) the protective function of the forest, (3) the potential recreational demand and (4) the site quality. These models were developed based on external user requirements and designed to meet actual reporting requirements at the national and international level.

15.1 Introduction

Besides traditional models for tree volume and growth, other models have been developed in the Swiss NFI (NFI). These models, described briefly in this section, cover four topics: (1) biodiversity indicators, (2) the protective function of the forest, (3) the potential recreational demand and (4) the site quality. These models were developed based on external user requirements and designed to meet actual reporting requirements at the national and international level. Three of these models were already used in NFI2 (1993–1995) and have been used, unchanged, since then within the NFI and by third parties. A detailed description of these models was given by Brassel and Lischke (2001).

15.2 Biodiversity

Since the world conference in Rio de Janeiro in 1992, biological diversity has been recognised as one of the most important issues of our time. In this context, forests are of great importance in Switzerland because, as relatively natural landscapes, they

C. Fischer (✉) · U.-B. Brändli · M. O. Huber
Scientific Service NFI, Swiss Federal Research Institute WSL, Birmensdorf, Switzerland
e-mail: christoph.fischer@wsl.ch

C. Fischer, B. Traub (eds.), *Swiss National Forest Inventory – Methods and Models of the Fourth Assessment*, Managing Forest Ecosystems 35,
https://doi.org/10.1007/978-3-030-19293-8_15

feature considerable plant and animal diversity. As a consequence, the data catalogue for the field survey of NFI2 (1993–1995) was extended to include a wide range of variables concerning the structure and composition of forest stands (biotopes). Additionally, methods for the world's first national inventory of forest edges were developed and implemented (Stierlin et al. 1994).

Assessing Ecotone and Biotope Values with NFI Data For evaluating the state and development of forest stands (biotopes) and forest edges (ecotones), a set of variables measured on NFI plots is used in the NFI. In addition, under the scope of NFI2, models for *biotope values* and *ecotone values* were developed and the set of resulting ecological indicators are included in NFI reports (Brändli et al. 2010). For example, the NFI Cockpit shows trends for 64 NFI indicators for sustainable forest management (Brändli et al. 2017). Further, this set of indicators is used to evaluate the sustainable development of Swiss forests (Altwegg et al. 2004) and is used in the Forest Report (Brändli and Brang 2015). The models have not changed since their first use in NFI2 and are described in detail by Brändli (2001). In this section, we thus only summarise its principles.

The biotope model comprises three sub-models related to the forest stand as a whole (1 to 3), whereas the ecotone model focuses specifically on aspects related to the forest edge (4 and 5):

1. The *closeness to nature* focuses on the proportion of coniferous tree species within stands in natural (mixed) broadleaved forests. For each forest community, threshold values are defined. The basal area proportion of sampled coniferous trees in a plot is compared with the corresponding threshold value and assigned a value between 1 and 4.
2. The *diversity of woody (ligneous) species within the stand* is calculated as a function of the number of sampled tree species in the plot and the occurrence of species with a special ecological value, such as oak species, sorbus and poplar species.
3. The *structural diversity of the stand* is based on a set of 13 variables: stage of development, crown structure, stand structure, proportion of trees with dbh >50 cm, degree of stand damage, presence of forest edge, presence of stand edge, type of forest gap, coverage of shrub layer, coverage of berry bushes, presence of stumps and lying dead trees, presence of standing dead trees, and presence of heaps of branches.
4. The *diversity of woody (ligneous) species* in the forest edge is calculated as a function of the number of tree and shrub species along an assessment line of 50 m (Sect. 9.7) and the proportions of briers and softwood/special species there.
5. The *structural diversity of the forest edge* is calculated based on a set of six variables: structure of forest edge, width of shelter belt, width of shrub belt, width of herb border, shape of forest edge, and density of forest edge.

The choice of the variables used, their weighting and the weighting value of their attributes are based on literature analyses. To develop and calibrate the sub-models and the full biotope and ecotone models, 351 forest stands and 79 forest edges were assessed by experts in the field to evaluate their individual ecological value.

15.3 Protection Forests

15.3.1 Definition of a Protection Forest

In a mountainous country like Switzerland, the protective effect of forests against gravitational hazards such as landslides, rockfall or avalanches is of high societal importance. According to the Swiss forest law of 1991, the cantons are responsible for determining which forests have protective functions and therefore need to be managed following specific guidelines. The Swiss federation grants subsidies for protection forest management, so it is crucial that the determination of protection forests is based on objective criteria. The Swiss Federal Office for the Environment (FOEN) worked with the individual cantons to develop such criteria in the course of the Project SilvaProtect-CH (Losey and Wehrli 2013). The resulting spatial determination of protection forests in Switzerland is based on cantonal forest maps, a spatially explicit simulation of gravitational hazards and a map of damage potential. A map of Swiss protection forests produced through this effort was used in the course of NFI4 to determine whether the primary forest function of a specific plot is protection against natural hazards.

15.3.2 Evaluation of Protection Forests with NFI Data

The protective function of a forest against natural hazards can be evaluated based on different indicators. These indicators, as well as their respective target values, were described by the FOEN in their guidelines for the sustainable management of forests with protective functions, NaiS (Frehner et al. 2005; Brang et al. 2006). The indicator values can be derived using NFI data at the reference unit level (e.g. canton). Applying these values does not lead to a spatially explicit evaluation of the protective effect of a specific forest stand with respect to a specific natural hazard process or damage potential. However, the state and development of the "mean protective function" of a certain forest within a given reference unit can be evaluated using the NFI data using the following indicators:

- Degree of crown coverage (derived from remote sensing data)
- Presence and size of gaps (derived from remote sensing data)
- Stand density (basal area per hectare)

In addition to the mean protective function of a forest, the sustainability of the provision of this service within the protection forest perimeter is assessed using the following set of indicators:

- Stand structure
- Crown closure
- Degree of species mixture
- Stand stability
- Degree of cover of regeneration
- Browsing intensity

15.4 Potential Forest Recreation Demand

Complementary to data on recreation from the interview survey (Chap. 10) a model on potential forest recreation (*ERHOLNA*) based on the national population census is used to quantify potential forest recreation demand. *ERHOLNA* was developed under the scope of NFI2 (Brändli and Ulmer 2001). Originally, the model applied data from the 1990 population census, but during the course of NFI3 data based on the 1990 census was replaced by data from the population census of 2000.

Model Description and Interpretation Based on the data from the population census of 1990, the number of households within a radius of 2 km around each NFI plot is calculated. Two types of households are recognised: permanently and temporarily inhabited. The potential recreational demand is highly dependent on the number of inhabitants per hectare. It was shown by Brändli and Ulmer (2001) that a close relationship exists between the number of inhabitants and the number of permanently occupied households, as well as between the number of beds in hotels and the number of temporarily occupied households. Thus, the model for the potential recreational demand is based solely on the number of households (Eq. 15.1).

$$WAE = wd2 + (5\ wt2) \tag{15.1}$$

ERHOLNA is the potential recreational demand in the forest, expressed in household equivalents (WAE), *wd2* is the number of permanently occupied households within a radius of 2 km, and *wt2* is the number of temporarily occupied or unoccupied households within a radius of 2 km.

Following Brändli and Ulmer (2001), the calculated WAE values can be interpreted as shown in Table 15.1.

Table 15.1 Household equivalents (WAE) and their corresponding potential recreational demand based on the *ERHOLNA* model (Brändli and Ulmer 2001)

Household equivalents (WAE)	Potential recreational demand
≤4	No recreational demand
5–1999	Low recreational demand
2000–4999	Moderate recreational demand
5000–11,999	High recreational demand
≥12,000	Very high recreational demand

15.5 Site Quality

Since NFI2, a model by Keller (1978, 1979) has been used as a proxy for site index in different modelling tasks, e.g. establishing tariff models (Sects. 6.2 and 6.5). The model has the following form:

$$TMI = k/(h - c) + m + p\,h \qquad (15.2)$$

where TMI is the site quality expressed as the maximum of the total mean increment from stand establishment until the age of 50 years in kilograms dry weight per hectare and year (Brassel and Lischke 2001), h is the elevation [m a.s.l.] and k, c, m and p are the model coefficients. The coefficients (Appendix Table 15.2) are estimated for different site types defined by a combination of factors such as region, acidity of the bedrock, geology, aspect and type of relief (Brassel and Lischke 2001). The database used for model parameterisation consists of yield-table-based estimates of TMI from permanent growth and yield plots, as well as from sample plots from experiments of plant sociology (Keller 1979).

Appendix

Table 15.2 Coefficients (*k*, *m*, *p*, *c*), boundaries of the independent variable (elevation) and maximum value of the dependent variable (TMI) for the site quality model (Keller 1978, 1979)

Site type	k	m	p	c	Lower elevation limit [m a.s.l.]	Upper elevation limit [m a.s.l.]	TMI maximum [kg ha^{-1} $year^{-1}$]
1	2,042,327.388	5074.689	1.4886	1873.9	0	1600	4376.94
2	817,233.6339	6859.791	−1.1286	1761.7	0	1600	6395.90
3	0.0672	6148.48	−3.8008	1600.001	0	1600	6148.48
4	39,685,578,736,360,704	24,050,405,171.849	14,590.3106	1,650,100.1	118	1600	4856.83
5	18,716,565.1295	9705.374	3.562	2815.001	0	1600	3402.26
6	744,875,972,826,732	1,241,022,014.9311	2075.2869	600,210.1	610	1600	848.88
7	1,119,175.4677	5733.256	−0.3396	2018.5	0	1800	5178.80
8	625,035,427.4645	73,368.735	6.561	9137.95	0	1800	4968.77
9	11,581,367.7085	5341.356	3.6909	2984.6	0	2100	3281.18
10	2420,947,686,491,376	2,250,872,009.3796	2096.7442	1,075,560	0	2100	2250.99
11	5,067,627,187,855,744	3,066,645,027.6	1857.367	1,652,500	0	2300	2826.21
12	13,026,340,700,812,160	8,273,839,084.3447	5264.7993	1,574,400.2	561	2300	2530.88
13	805,295.6	4830.524	−0.8411	2362.8	0	2100	4489.70
14	899,860.4	3291.198	0.1372	2351.4	0	2100	2908.51
15	6,199,444.6	9294.71	−0.1314	2484.4	0	1800	6799.36
16	516,638.2	5256.728	−0.0926	1901.5	0	1800	4985.03
17	315,005.1	4210.304	−0.4911	1894.7	0	1800	4044.05
18	30,106.3	3432.097	−0.4397	1811.4	0	1800	3415.48
19	19,521.1	2997.1	−0.6702	1810.9	0	1800	2986.32

References

Altwegg D, Roth I, Scheller A (2004) Monitoring sustainable development MONET. Final report-methods and results. Swiss Federal Statistical Office (SFSO), Swiss Agency for the Environment, Forests and Landscape (SAEFL), Swiss Federal Office for Spatial Development (ARE), Swiss Federal Statistical Office, Neuchâtel

Brändli U-B (2001) Nature protection function. In: Brassel P, Lischke H (eds) Swiss National Forest Inventory: methods and models of the second assessment. Swiss Federal Research Institute WSL, Birmensdorf, pp 265–282

Brändli U-B, Brang P (2015) Naturalness. In: Rigling A, Schaffer H-P (eds) Forest report 2015. Condition and use of Swiss forests. Federal Office for the Environment FOEN, Swiss Federal Research Institute WSL, pp 76–77

Brändli U-B, Ulmer U (2001) Recreational function. In: Brassel P, Lischke H (eds) Swiss National Forest Inventory: methods and models of the second assessment. Swiss Federal Research Institute WSL, Birmensdorf, pp 254–264

Brändli U-B, Abegg M, Duc P, Ginzler C (2010) Diversité biologique. In: Brändli U-B (ed) Inventaire forestier national suisse. Résultats du troisième inventaire 2004–2006. Birmensdorf, Institut fédéral de recherches sur la forêt, la neige et le paysage WSL, Federal Office for the Environment FOEN, Bern, pp 187–228

Brändli U-B, Brändli M, Brändli R, Speich S (2017) NFI cockpit – trends of forest indicators in Switzerland. http://www.lfi.ch/resultate/cockpit-en.php. Accessed 23 Jan 2018

Brang P, Schönenberger W, Frehner M et al (2006) Management of protection forests in the European Alps: an overview. In: For. Snow Landsc. Res. Citeseer, pp 23–4

Brassel P, Lischke H (eds) (2001) Swiss National Forest Inventory: methods and models of the second assessment. Swiss Federal Research Institute WSL, Birmensdorf

Frehner M, Wasser B, Schwitter R (2005) Nachhaltigkeit und Erfolgskontrolle im Schutzwald: Wegleitung für Pflegemassnahmen in Wäldern mit Schutzfunktion, Vollzug Umwelt. Bundesamt Für Umw Wald Landsch, Bern

Keller W (1978) Einfacher ertragskundlicher Bonitatsschlussel fur Waldbestande in der Schweiz. Mitteilungen Eidg Anst Für Forstl Vers 51

Keller W (1979) Una chiave di feracità auxometrica semplice per i soprassuoli forestali delle regioni al sud delle Alpi. Mitteilungen Eidg Anst Für Forstl Vers 55:179–232

Losey S, Wehrli A (2013) Schutzwald in der Schweiz: Vom Projekt SilvaProtect-CH zum harmonisierten Schutzwald. Bundesamt Für Umw BAFU Bern 29

Stierlin H-R, Brändli U-B, Herold A, Zinggler J (1994) Inventaire forestier national suisse Manuel d'instruction pour les relevés terrestres. Eidg. Forschungsanstalt WSL, Birmensdorf, pp 1993–1995

Chapter 16
Calculation of Potential Timber Harvesting Costs (HeProMo)

Christoph Fischer and Golo Stadelmann

Abstract As proceeds of timber selling are typically not surveyed or measured in national forest inventories, simulated timber harvesting costs are compared with mean timber prices to serve as a proxy for economic timber availability. Consequently, the profitability of timber harvest operations can be interpreted as an economic restriction when evaluating Forest Available for Wood Supply (FAWS). However, timber harvesting productivity and cost models are rare and therefore only a few European countries are able to calculate potential timber harvesting costs for their NFI sample plots.

16.1 Introduction

As proceeds of timber selling are typically not surveyed or measured in national forest inventories, simulated timber harvesting costs are compared with mean timber prices to serve as a proxy for economic timber availability. Consequently, the profitability of timber harvest operations can be interpreted as an economic restriction (Fischer et al. 2016; Alberdi et al. 2016) when evaluating Forest Available for Wood Supply (FAWS). However, timber harvesting productivity and cost models are rare and therefore only a few European countries are able to calculate potential timber harvesting costs for their NFI sample-plot network.

Within the Swiss NFI (NFI), the estimation of the potential timber harvesting costs is mostly based on the timber harvesting productivity model *HeProMo* (Erni 2003), which was originally developed to calculate timber harvesting interventions at the stand level. Therefore, HeProMo uses input parameters from the terrestrial inventory and from the interview survey with the forest service (Fischer et al. 2017;

C. Fischer (✉)
Scientific Service NFI, Swiss Federal Research Institute WSL, Birmensdorf, Switzerland
e-mail: christoph.fischer@wsl.ch

G. Stadelmann
Resource Analysis, Swiss Federal Research Institute WSL, Birmensdorf, Switzerland

C. Fischer, B. Traub (eds.), *Swiss National Forest Inventory – Methods and Models of the Fourth Assessment*, Managing Forest Ecosystems 35,
https://doi.org/10.1007/978-3-030-19293-8_16

Chap. 10). Based on the survey, the timber harvest system applied or potentially applied (in cases where no harvest has taken place yet) is known for each NFI plot.

HeProMo is a harvest productivity model that calculates the time needed for specific timber harvesting techniques and is based on time studies. Multiplying the total time needed for specific tasks by machine and personnel charges allows one to estimate timber harvesting costs in Swiss Francs per volume (CHF m^{-3}). HeProMo was first applied in NFI3 (Brändli 2010), while timber harvesting costs in NFI2 were calculated using a similar model (Ulmer 2001) that is limited to fewer harvest systems. HeProMo has been reprogrammed since its use in NFI3 but, for comparability reasons, the same version was used in NFI4. The modelling software is currently being extended and existing modules are being updated.

Besides calculating timber harvesting costs based on NFI results (Chap. 16) HeProMo is run with the forest development model MASSIMO (Chap. 17) which simulates forest management scenarios for the grid of the NFI. MASSIMO is intended to project sustainable timber harvesting potentials (Chap. 17) under a broad range of management scenarios at the national level (Hofer et al. 2011; Stadelmann et al. 2016; Fischer et al. 2017). MASSIMO has further been used to assess the effects of Swiss forests on carbon sequestration, considering the full timber chain (Werner et al. 2010) or regional management strategies (Stadelmann et al. 2015; Temperli et al. 2017a, b).

16.2 Calculation Framework

To calculate timber harvesting costs at the plot level, information is needed about the harvest system, which consists of a combination of cutting techniques (e.g. motor-manual felling with a chainsaw) and procedures for transporting timber to the closest road (e.g. skidding). Each element of the harvest system is defined as a module that is either included in HeProMo or directly programmed into the NFI storage and analysis system (Traub et al. 2017). The software HeProMo (Erni 2003) was originally programmed in .NET. However, in order to integrate the model into the NFI data analysis system NAFIDAS (Chap. 20), all model components were reprogrammed into SAS© and SQL, maintaining all model parameters and default settings. Further, four modules (Table 16.1) were not available through HeProMo, and instead they were programmed as internal modules in NAFIDAS.

The calculation framework is linked to a hierarchical database scheme comprising three levels to store the input parameters: the software (HeProMo or NFI-internal), the module (Table 16.1) and the process to which the parameters belong. The parametrisation is stored specifically for NFI4. To complete a harvest system, a sequence of modules is joined together (e.g. motor-manual felling followed by skidding with a forwarder), with each module consisting of a series of tasks

Table 16.1 Modules implemented in the NFI which are used for the calculation of potential timber harvesting costs

ID	Module source	Module name
1	HeProMo	Flying and storing assortments downhill with a helicopter
2	HeProMo	Mobile chipper
3	HeProMo	Harvester
4	HeProMo	Forwarder
5	HeProMo	Mobile cable crane
6	HeProMo	Skidding with a forwarder (as an additional skidding technique)
7	HeProMo	Flying and storing assortments uphill with a helicopter
8	HeProMo	Flying trees downhill with a helicopter
9	HeProMo	Flying trees uphill with a helicopter
10	HeProMo	Skidding (pre-skidding only)
11	HeProMo	Skidding with a skidder (as an additional skidding technique)
12	HeProMo	Skidding with a skidder (as the main skidding technique)
13	HeProMo	Conventional cable crane
14	HeProMo	Motor-manual felling and processing
15	HeProMo	Motor-manual felling
16	NFI	Lumber trailer (transport on small roads)
17	NFI	Skidding with other means
18	NFI	Hand skidding
19	NFI	Processor

(e.g. felling with a chainsaw, cutting the branches, cutting the bolewood into assortments). A total of 19 modules (Table 16.1) already available in NFI3, were used to calculate timber harvesting costs in NFI4. When run, each module outputs a productivity value in units of time. To calculate the potential timber harvesting costs, the productivity of each module is multiplied by the hourly rate of the related tasks, resulting in the total cost for a module. All module costs are then summed to obtain the total cost of a specific timber harvest system in a given NFI plot.

In order to calculate the potential timber harvesting cost for a given area, one needs to know which trees will be harvested. However, as this information is not available for NFI plots, the potential timber harvesting cost is defined as the cost arising if all standing timber in an NFI plot is harvested. The resulting costs are then presented in cost classes [CHF m^{-3}] assuming that all the growing stock in each field plot is completely harvested. This assumption clearly does not reflect the forest management practised in most cases, but the estimated costs are considered a reasonable approximation. In contrast, MASSIMO (Chap. 17) simulates thinning and shelterwood felling for each NFI plot, and costs are subsequently calculated based on the simulated harvest using HeProMo (Chap. 16).

16.3 Input Data

Besides tree, stand and site data, information on the selected harvest system is needed to calculate timber harvesting costs. During the interview survey (Chap. 4) local foresters are asked questions about the applied or potentially applied (in cases where no harvest has taken place yet) timber harvest system for each field plot. Once the system is identified, dependent input data is assessed through the interview (e.g. skidding routes and their lengths). In total, calculation of the potential timber harvesting costs in Switzerland requires up to 30 input variables Table 16.2.

16.4 Results and Discussion

According to NFI4, potential timber harvesting costs tend to be higher in the Southern Alps and Alps, where affordable harvest systems are less available than on the Plateau, in the Jura mountains and in the Pre-Alps (Table 16.3). Especially on the Plateau, nearly all timber (97.2%) can be harvested within the cost class of $\leq$80 CHF m^{-3}, while this percentage drops to only 13.4% in the Southern Alps.

Throughout Switzerland as a whole, the majority of NFI plots (63.6%) can be harvested at costs $\leq$80 CHF m^{-3}. Assuming a mean timber price of roughly 80 CHF m^{-3} is realistic, and we conclude that about 64% of the total growing stock in Switzerland can be harvested without financial loss. In contrast, about 59% of the growing stock in the Alps can only be harvested at costs between 81 and 150 CHF m^{-3}, and 13% of the growing stock in the Southern Alps at costs >150 CHF m^{-3}. However, part of the timber harvesting costs may be covered by subsidies to ensure appropriate forest management in forests with protective functions, as forest management may be required in spite of unfavourable economic circumstances. Further, the simulated timber harvesting costs may be underestimated, especially for sites where harvest systems are expensive and time consuming. This occurs in the cost estimation for timber harvesting a full NFI plot if the time needed for the establishment of machines (e.g. cable crane) is divided by an overestimated harvested volume. To conclude, there are large regional differences throughout Switzerland in timber harvesting costs, and harvests become more cost efficient with increasing infrastructure provisioning, as in the production regions Jura, Plateau and Pre-Alps.

Table 16.2 Input variables needed for the timber harvest modules in HeProMo

Input variable	Module ID
Population element: tree	4,18
Population element: living tree	4,18
Extrapolation factor for each tree	2,4,18
Diameter at breast height	2,4,18
Broadleaved or coniferous tree	2,4
Number of stems of living trees per hectare	3,4,14,15,19
Growing stock of living trees per hectare	1,2,3,4,5,6,7,8,9,10,11, 12,13,14,15,16,17,18,19
Dominant tree species with respect to basal area	4,14,15
Dominant tree species group (broadleaved/coniferous) with respect to basal area	1,7,8,9
Mean volume of assortment	5,10,11,12,13
Bole volume including bark	2,4
Development stage	18
Slope	4,14,15
Obstacles for timber harvest	10,11,12,14,15,18
Timber harvest system	1,2,3,4,5,6,7,8,9,10,11, 12,13,14,15,16,17,18,19
Location of the mobile cable crane	5
Length of cable crane line	5,13
Skidding distance on roads not suited for trucks	16
Skidding distance of the first skidding phase	1,2,4,6,7,8,9,10,11,12,18
Skidding distance of the second skidding phase	1,2,4,6,7,8,9,10,11,12
Skidding distance of the third skidding phase	1,2,4,6,7,8,9,10,11,12
Skidding technology of the first skidding phase	1,6,7,8,9,10,11,12
Skidding technology of the second skidding phase	1,6,7,8,9,11,12
Skidding technology of the third skidding phase	1,6,7,8,9,11,12
Skidding direction of the first skidding phase	4,13
Skidding direction of the second skidding phase	4,13
Skidding direction of the third skidding phase	4,13
Skidding route of the first skidding phase	2,4,6,10,11,12,13
Skidding route of the second skidding phase	2,4,6,10,11,12,13
Skidding route of the third skidding phase	2,4,6,10,11,12,13

The module IDs correspond to those shown in Table 16.1

Table 16.3 Percentage of growing stock for potential timber harvest cost classes in the five production regions in Switzerland (accessible forest without shrub forest, grid NFI4 2009–2013)

Production region	Jura		Plateau		Pre-alps		Alps		Southern Alps		Switzerland	
[CHF m^{-3}]	%	±	%	±	%	±	%	±	%	±	%	±
≤80	91.1	1.3	97.2	0.6	62.9	2.3	38.3	1.9	13.4	2.2	63.6	0.9
81–150	8.4	1.3	2.1	0.5	36.2	2.3	58.8	1.9	73.7	2.7	33.9	0.9
>150	0.5	0.4	0.5	0.3	0.8	0.3	2.9	0.5	12.9	1.8	2.4	0.2
Total	100	.	100	.	100	.	100	.	100	.	100	.

References

Alberdi I, Michalak R, Fischer C et al (2016) Towards harmonized assessment of European forest availability for wood supply in Europe. For Policy Econ 70:20–29. https://doi.org/10.1016/j.forpol.2016.05.014

Brändli U-B (ed) (2010) Schweizerisches Landesforstinventar. Ergebnisse der dritten Erhebung 2004–2006. Birmensdorf, Eidgenössische Forschungsanstalt für Wald, Schnee und Landschaft WSL. Bundesamt für Umwelt, BAFU, Bern

Erni V (2003) HeProMo – Produktivitätsmodelle für Holzerntearbeiten. Windows-Software. Version 1.01. Eidg. Forschungsanstalt WSL, Birmensdorf

Fischer C, Gasparini P, Nylander M et al (2016) Joining criteria for harmonizing European Forest available for wood supply estimates. Case studies from National Forest Inventories. Forests 7:104. https://doi.org/10.3390/f7050104

Fischer C, Camin P, Kaufmann E, Thürig E (2017) Switzerland. In: Barreiro S, Schelhaas M-J, McRoberts RE, Kändler G (eds) Forest inventory-based projection systems for wood and biomass availability. Springer, Cham, pp 303–313

Hofer P, Altwegg J, Schoop A et al (2011) Holznutzungspotentiale im Schweizer Wald. Auswertung von Nutzungsszenarien und Waldwachstumsentwicklung. Bundesamt für Umwelt, Bern

Stadelmann G, Temperli C, Conedera M et al (2015) Möglichkeiten zur Holzmobilisierung im Tessiner Kastaniengürtel. Schweiz Z Forstwes 166:291–298. https://doi.org/10.3188/szf.2015.0291

Stadelmann G, Herold A, Didion M et al (2016) Holzerntepotenzial im Schweizer Wald: simulation von Bewirtschaftungsszenarien. Schweiz Z Forstwes 167:152–161. https://doi.org/10.3188/szf.2016.0152

Temperli C, Stadelmann G, Thürig E, Brang P (2017a) Silvicultural strategies for increased timber harvesting in a Central European mountain landscape. Eur J For Res 136:493–509. https://doi.org/10.1007/s10342-017-1048-1

Temperli C, Stadelmann G, Thürig E, Brang P (2017b) Timber mobilization and habitat tree retention in low-elevation mixed forests in Switzerland: an inventory-based scenario analysis of opportunities and constraints. Eur J For Res 136:711–725. https://doi.org/10.1007/s10342-017-1067-y

Traub B, Meile R, Speich S, Rösler E (2017) The data storage and analysis system of the Swiss National Forest Inventory. Comput Electron Agric 132:97–107. https://doi.org/10.1016/j.compag.2016.11.016

Ulmer U (2001) Expenditure for timber felling and extraction. In: Brassel P, Lischke H (eds) Swiss National Forest Inventory: methods and models of the second assessment. Swiss Federal Research Institute WSL, Birmensdorf, pp 225–230

Werner F, Taverna R, Hofer P et al (2010) National and global greenhouse gas dynamics of different forest management and wood use scenarios: a model-based assessment. Environ Sci Pol 13:72–85. https://doi.org/10.1016/j.envsci.2009.10.004

Chapter 17
Forest Development Model MASSIMO

Golo Stadelmann, Christian Temperli, Brigitte Rohner, Markus Didion, Anne Herold, Erik Rösler, and Esther Thürig

Abstract Forest development models have been used to predict growth and yield for a long time. Yield tables, for example, have been optimised to predict the yield of pure even-aged stands on the basis of long-term observations from experimental forest plots. However, forests in Switzerland cover a broad ecological gradient and forest structures are diverse, with several even- and uneven-aged stand types that can include various species mixtures. Given such diverse forest ecosystems, we predict harvesting potentials and forest management reference levels using the Swiss National Forest Inventory-based individual-tree growth simulator MASSIMO.

17.1 Introduction

Forest development models have been used to predict growth and yield for a long time. Yield tables, for example, have been optimised to predict the yield of pure even-aged stands on the basis of long-term observations from experimental forest plots. However, forests in Switzerland cover a broad ecological gradient and forest structures are diverse, with several even- and uneven-aged stand types that can include various species mixtures. Given such diverse forest ecosystems, growth and yield are often more accurately predicted using Swiss National Forest Inventory-based individual-tree growth simulators (Barreiro et al. 2016).

To simulate growth and development of Swiss NFI (NFI) trees and plots, the empirical forest growth model MASSIMO (MAnagement Scenario SImulation MOdel) was developed (Stadelmann et al. 2019). MASSIMO is a distance-independent individual-tree growth model, and competition is thus respected in terms of distance-independent indices and stand-level characteristics (i.e. basal area of larger trees BAL, stand density index SDI, number of trees per ha, dominant diameter ddom). The model includes modules for growth, management, storm damage, individual-tree mortality and regeneration, and it is, to a large extent,

G. Stadelmann (✉) · C. Temperli · B. Rohner · M. Didion · A. Herold · E. Rösler · E. Thürig
Resource Analysis, Swiss Federal Research Institute WSL, Birmensdorf, Switzerland
e-mail: golo.stadelmann@wsl.ch

C. Fischer, B. Traub (eds.), *Swiss National Forest Inventory – Methods and Models of the Fourth Assessment*, Managing Forest Ecosystems 35,
https://doi.org/10.1007/978-3-030-19293-8_17

based on data and empirical models developed in the NFI (Kaufmann 2001b; Thürig et al. 2005a, b). The simulation of storm damage, single-tree mortality and harvest are based on empirical data and implemented using random draws (stochastic processes).

MASSIMO is intended to project timber harvesting under a broad range of feasible management scenarios at the national (Hofer et al. 2011; Stadelmann et al. 2016) and regional level (Stadelmann et al. 2015; Temperli et al. 2017a, b). Besides predicting timber harvesting potentials, MASSIMO has been used to assess the full timber wood products and energy chain (Werner et al. 2010). MASSIMO has additionally been applied to projects on timber harvesting potentials and forest management reference levels (Krug 2018) under the Kyoto framework (Sect. 19.3) proving the significance of this model for national forest policies. In the following sections we describe the MASSIMO framework.

17.2 Methods

17.2.1 Structure of the Model

MASSIMO simulates the growth of single trees in the NFI sample-plot grid based on individual-tree, stand and site data provided from the NFI Data Storage and Analysis System NAFIDAS (Traub et al. 2017; Chap. 20). The data needed to initialise a model run is stored as database tables and accessed by SQL scripts. MASSIMO simulations can be run for time-steps of 5 or 10 years. In addition to configuring the time-step, the simulation length (number of time-steps) and the number of replicates, the user can specify scenario settings such as timber harvesting restrictions (e.g. no timber harvesting in forest reserves, restricted timber harvesting in protection forests), timber harvesting targets (e.g. regulation of growing stock or timber harvesting amounts) and tending regimes (e.g. modification of species mixture in the regeneration, changing thinning intensities). Input tables are checked for missing values prior to model initialisation and, when necessary, average values specific to the sample-plot strata (e.g. production region, elevation belt, forest type) are imputed as default.

MASSIMO consists of five sub-models: (1) growth, (2) storm, (3) timber harvesting, (4) mortality, and (5) regeneration (Fig. 17.1). In order to vary simulated management (i.e. scenarios), the timber harvesting module is controlled by a routine that regulates the probability of a sample plot experiencing thinning and shelterwood felling, based on given timber harvesting or growing stock targets (not shown in Fig. 17.1) using an iterative approximation algorithm. Results are aggregated over simulation replicates and spatial strata (i.e. production regions or economic regions).

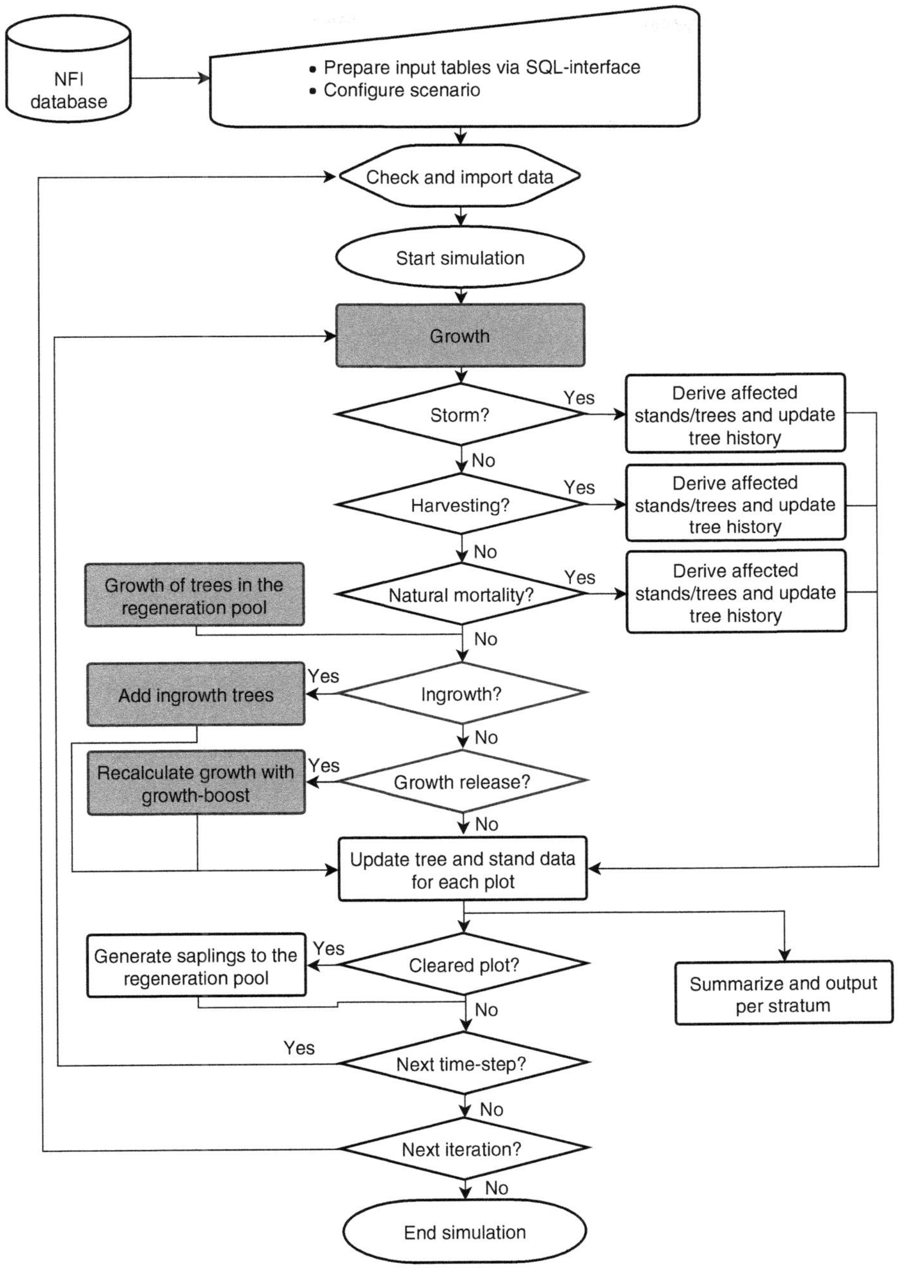

Fig. 17.1 Flowchart of the individual-tree simulator MASSIMO. Note: the grey shaded fields complement the growth sub-model; tree history is used as a state variable of each tree and is applied to deduce changes between two consecutive time-steps

17.2.2 Growth

The growth model of MASSIMO estimates the basal area increment (BAI) of (a) surviving trees, (b) young trees in the regeneration pool, i.e. trees with a dbh less than the calliper threshold of 12 cm, and (c) trees that show a growth reaction after cutting and thinning. Statistical models are fitted separately for the 12 main tree species (or species groups) in all five production regions of Switzerland, and for even-aged and uneven-aged forests. To obtain robust estimates, groups not containing enough observations are merged with similar groups, resulting in 47 specific BAI models for Switzerland (Kaufmann 2001a; Stadelmann et al. 2019). Some of these models have been evaluated by Thürig et al. (2005a).

Similar to the model presented in Sect. 12.5.1 the individual-tree BAI model is formulated as:

$$\widehat{BAI}_j = \exp\left(b_0 + \sum_{i=1}^{6} b_i B_i + b_7(1 - \exp(b_8 dbh))\right) + \varepsilon_j \tag{17.1}$$

where B_i are explanatory variables (B_1: stand basal area per ha (BAPH), B_2: basal area of trees larger than the subject tree (BAL), B_3: site quality expressed as the maximum of the total mean increment (TMI) (Sect. 15.5), B_4: elevation, B_5: stand age or dominant diameter (in cases of uneven-aged forests), B_6: growth boost in cases where cutting or thinning takes place (Boolean)), $b_0 - b_8$ are the model coefficients, and $\varepsilon_j \tilde{} N\left(0, \hat{\sigma}^2\right)$ is a normally distributed random variable per tree j accounting for stochastic effects with variance $\hat{\sigma}^2$ estimated from the residual distribution. This variance is incorporated into the simulations by multiplying it by a normally distributed random draw $N(0,1)$. As a consequence, tree growth in MASSIMO is modelled as a point estimate plus a stochastic variation.

The growth prediction is followed by the volume prediction of stemwood over bark using tariff models (Sect. 12.3). These models are fitted to the main tree species and production regions and, if necessary, aggregated to achieve robust estimations, resulting in 30 dbh-dependent tariff models for tree volume (Kaufmann 2001a).

Similar to the calculation presented in Sect. 12.3 the volume V is predicted as:

$$\widehat{V}_j = exp\left(b_0 + b_1 ln(dbh) + b_2 ln^4(dbh) + \sum_{i=3}^{7} b_i B_i\right) + \varepsilon_j \tag{17.2}$$

where B_i are the explanatory variables (B_3: site quality, total mean increment (TMI) (Sect. 15.5), B_4: dominant diameter, B_5: bifurcation (Boolean), B_6: elevation, B_7: overstorey (Boolean) and $b_0 - b_7$ are the model coefficients. Similar to the BAI prediction, $\varepsilon_j \tilde{} N\left(0, \hat{\sigma}^2\right)$ is a normally distributed random variable per tree j with variance $\hat{\sigma}^2$ estimated from the residual distribution. Consequently, the volume prediction also considers model variance.

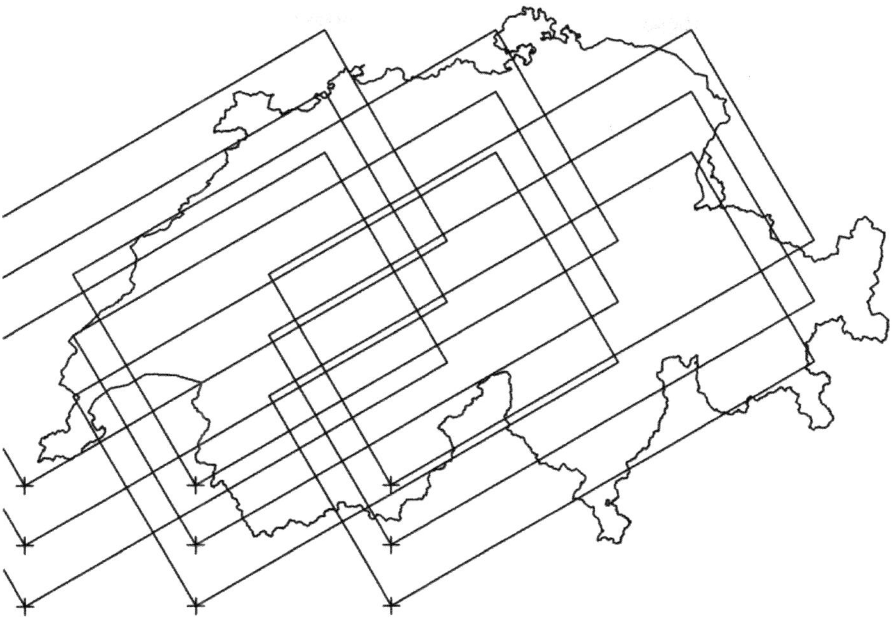

Fig. 17.2 Nine possible 200 km × 100 km storm tracks over Switzerland that are simulated in MASSIMO. The storm tracks are selected by randomly picking one out of nine lower-left (south-western) corners

17.2.3 Storm Damage

Storm damage is the only large-scale (affecting multiple sample plots in the same event) disturbance considered in MASSIMO, as this disturbance has proven to outrank other disturbance events in the past (Schelhaas et al. 2003). Storms are simulated as single events with a default return interval of 15 years, which corresponds to the estimated frequency of severe storms in Switzerland (Pfister 1999; Thürig et al. 2005b). Based on the observed severity of the three most severe past storms (1967, *Vivian* in 1990 and *Lothar* in 1999), MASSIMO recognises three storm severities with equal probabilities of occurrence. To simulate possible storm scenarios, e.g. under climate change, the return interval can be altered. The spatial extent of the simulated storm track is a 200 km × 100 km rectangle that is tilted by 30° and thus accounts for the prevailing west-south-western wind direction (Fig. 17.2). The location (easting and northing) of the storm tracks is determined by randomly selecting from nine possible south-western corners. This results in storms being most frequently simulated in the Bernese Pre-Alps, which roughly corresponds to storm hazard maps based on historic data (Dierer et al. 2014).

Table 17.1 Coefficients for storm probability models, estimated from observed storm damage on NFI plots

Forest structure	i_α	i_β	i_{dd}	Coniferous proportion	i_M
Even-aged	8.7756	8.9933	−1.7477	>90%	−0.6578
				≤90%	0
Uneven-aged	8.6399	9.1994	2.0584	>90%	0.6435
				≤90%	0

The probability that a sample plot within the storm track will suffer from partial or stand-replacing damage is modelled using logistic functions. The probability of partial damage P_{pd} is modelled based on storm severity (s, levels 0.1, 0.6 or 1); dominant diameter ddom; coniferous proportion; and forest structure (even-aged- vs. uneven-aged):

$$P_{pd} = s \, \frac{1 - e^{\alpha}}{1 + e^{\alpha}} \tag{17.3}$$

with:

$$\alpha = i_\alpha + i_{dd} \, \ln \left(\text{ddom}\right) + i_M \tag{17.4}$$

where ln(ddom) is the natural logarithm of plot-level dominant diameter and i_α, i_β, i_{dd} and i_M are coefficients specific to forest structure and coniferous proportion (Table 17.1). The probability of stand replacing damage P_{srd} is given by:

$$P_{srd} = P_{pd} \, \frac{e^{\beta} - e^{\alpha}}{1 + e^{\alpha}} \tag{17.5}$$

with:

$$\beta = i_\beta + i_{dd} \, ln(ddom) + i_M \tag{17.6}$$

Whether storm damage occurs on a sample plot within the simulated storm track is determined by comparing a uniform random number with P_{pd} and P_{srd}. If the selected number is $\leq P_{srd}$, all trees in the sample plot are harvested, and if the number is > P_{srd} but $\leq P_{pd}$, 20% of trees in the sample plot are removed.

17.2.4 Timber Harvesting

The timber harvesting module in MASSIMO represents thinning in even- and uneven-aged forests, shelterwood felling, timber harvesting in protection forests, unplanned thinning and felling interventions (e.g. sanitation felling, salvage logging, intervention for unknown reasons), and forest conversion from even-aged to uneven-aged forests.

Thinning in even-aged high forests is performed as soon as stand basal area is 10% larger than the basal area observed directly before the last thinning intervention. This surplus of the current basal area over the basal area before the last thinning intervention (delta thinning: Δ_{thin}) can be adapted to approximate scenario-specific timber harvesting targets using an iterative approximation routine (Sect. 17.2.7). By default, thinning removes 30% of the basal area of the sample plot. To maintain stable stand structures in uneven-aged forests, Δ_{thin} (5%) and thinning intensity (25% of basal area removed) are set to fixed values and cannot be varied by the iterative approximation routine.

The dbh-dependent selection probability that an individual tree is harvested during a thinning intervention is estimated from observed changes between NFI2 and NFI3 using Weibull-fits after stratifying by production region, ownership category (public or private), stand type and over- and understorey. These selection probabilities are estimated as the ratio of the dbh-distribution of the trees before thinning (initial stand) and the dbh-distribution of the logged trees (removals), multiplied by a scaling factor *f* that controls the thinning intensity (see p. 200 and 201 in Kaufmann 2001b). The scaling factor is estimated such that the summed basal area of the thinned trees matches the default 30% of the stand basal area in even-aged forests and 25% in uneven-aged forests. The trees to be thinned are then selected randomly according to the selection probabilities listed above.

Upon initialisation of MASSIMO, the basal area directly before the last intervention is unknown. Thus, the known basal area from an earlier state inventory (e.g. basal area measured in NFI2 for simulations initialised with NFI3 data) is used to approximate an initial value for the basal area directly before the last thinning intervention. This approximation assumes that stands are thinned on average every 10 years, which is an overestimate in many cases, especially in mountainous areas, but may be lower than the thinning frequency at very productive sites. As a consequence, the value of Δ_{thin}, which determines the probability that a thinning intervention occurs in a sample plot, has to be adjusted with the iterative approximation routine.

In the shelterwood system, stand regeneration is commonly initiated with an intensive thinning intervention (i.e. a larger portion of basal area is removed than during regular thinning) a few years before most of the overstorey trees are harvested. This last intervention is often referred to as a seed cutting. After timber harvesting is complete, the remaining shelter trees protect the natural regeneration for some years and are then removed in one or two interventions. In MASSIMO, however, thinning and shelterwood felling are not directly linked. While thinning is performed according to its basal-area-related rules, shelterwood cutting is controlled by strata-specific rotation periods that respect the tree species mixture and site index (Table 17.2). Note that the extraordinarily long rotation period in coppice forests is due to large-scale abandonment of coppicing since the 1950s (Stadelmann et al. 2015). To determine the sample plots where shelterwood felling should occur within

Table 17.2 Rotation periods of shelterwood cutting, organised by site index (TMI) and coniferous proportion

Site index (TMI)	Coniferous proportion (%)	Rotation period (years)
<1125	>50	180
	<50	180
1125–2250	>50	130
	<50	150
2250–4500	>50	110
	<50	130
>4500	>50	90
	<50	110
Coppice	0–100	60

a stratum, sample plots are ranked by stand age, with the oldest stands being given higher timber harvesting priorities. The number of stands to be harvested is calculated by:

$$n_{fell} = N \frac{rotper}{tsl} \Delta_{fell} \tag{17.7}$$

where N is the number of plots in a stratum, *rotper* is the rotation period (years) and *tsl* is the number of years in a time-step (5 or 10). The number of stands to be harvested (n_{fell}) can be adjusted to reflect timber harvesting or growing stock targets by adjusting Δ_{fell} using the iterative approximation routine. Shelterwood cutting removes 80% of the basal area (measured on the overstorey trees), and the remaining shelter is then removed in the first (colline – submontane elevation belt) or second (montane – subalpine) subsequent decade. Based on NFI observations, planned shelterwood cutting in uneven-aged forests is simulated in randomly selected sample plots with a probability of 0.0495.

Management in protection forests aims to maintain forest stability and natural regeneration in order to ensure a continuous forest cover in the long term. In MASSIMO, special management rules apply to sample plots within the protection forest perimeter SilvaProtect-CH (Losey and Wehrli 2013). Thinning is conducted as described above, but sample plots within the protection forest perimeter are not subjected to any shelterwood felling. This restriction allows these areas to maintain a continuous forest cover and thus preserve the protective function. In protection forests, stand regeneration is therefore simulated as a sequence of high-intensity thinning interventions (40% of basal area) with a fixed interval of 20 years that starts at a predefined, elevation-dependent stand age (colline and submontane: 120 years, montane: 140 years, and subalpine: 200 years). These thinning interventions are simulated in both even- and uneven-aged high forests and are repeated until the end of the simulation period.

In addition to simulating planned thinning interventions and shelterwood felling, unplanned thinning and felling are simulated. These interventions represent salvage logging operations and sanitation felling after abiotic (e.g. avalanche) and biotic (e.g. insect) disturbances. Note that interventions after storm damage are represented separately. Unplanned thinning (30% basal area removal) and felling (100% basal area removal) are simulated on randomly selected sample plots with observed probabilities of 0.01 and 0.001, respectively, in the production regions Jura and Plateau. The probabilities of thinning and felling are doubled in the Pre-Alps, Alps and Southern Alps. Unplanned thinning and felling are executed in the same way as the planned interventions.

In MASSIMO, it is possible to simulate the conversion of even-aged to uneven-aged stand structures. This conversion is applicable to sample plots without protective functions that have a dominant diameter 12 cm < ddom ≤ 50 cm and where there are at least 30 years left to realise this process (i.e. stand age < (length of rotation period – 30 years)). The proportion of sample plots to be converted can be specified in the scenario settings. The sample plots to be converted are randomly selected. Conversion is initiated in the first simulation decade and is performed through thinning. To this end, the starting point for conversion thinning (Δ_{thin}) is set to 1.08 for the first decade and to 0.9 for the following decades. Stands under conversion are thinned with an intensity of 25% basal area removal. The removal probabilities of individual trees are derived similarly as for other thinning interventions, but with specific fits of the Weibull distribution. Neither unplanned thinning nor shelterwood felling is simulated in conversion forests.

17.2.5 Individual-Tree Mortality

In addition to storm-induced mortality, which affects trees throughout the storm track, two processes of mortality are simulated at the level of individual trees. First, background mortality due to unknown reasons is estimated from the NFI and occurs with a probability $P_{bm} = 0.03$. Second, density-dependent mortality is formulated as a mortality factor *fm* with four levels (1, 1.5, 2, 3) depending on stand age, basal area per hectare, coniferous proportion, stand type and production region. The tree-level probability of small-scale mortality P_m is defined as:

$$P_m = P_{bm}(P_{bm} + fm\ (1 - P_{bm})) \tag{17.8}$$

resulting in values from 0.03 ($fm = 1$) to 0.0882 ($fm = 3$). Consequently, the probability of density-related mortality, $P_{drm} = P_m - 0.03$, can have values from zero to 0.0582.

Table 17.3 Decadal survival rates and replication numbers of new young trees separated by production region

Production region in Switzerland	Time-step survival probability	Survival probability at calliper threshold	Number of new young growth trees upon ingrowth
Jura	0.15 (0.50)	0.6 (0.8)	7 (15)
Plateau	0.15 (0.45)	0.4 (0.8)	5 (15)
Pre-Alps	0.20 (0.50)	0.5 (0.8)	7 (15)
Alps	0.30 (0.60)	0.7 (0.8)	8 (18)
Southern Alps	0.30 (0.60)	0.7 (0.8)	10 (21)

Numbers in parentheses show survival rates and replication numbers if the time-step is set to 5 years

17.2.6 Regeneration

In MASSIMO, regeneration is simulated to represent two processes: (1) understorey regeneration of trees below the (closed) canopy growing under limited light conditions and (2) advance regeneration of trees from the seedling bank that profit from improved light conditions following shelterwood cutting and stand-replacing (i.e. all trees damaged) storm damage. New trees are added to the pool of regenerating trees (<12 cm dbh, young growth) with a dbh of 1 cm, and the same increment model is used for young growth as for the trees with dbh >12 cm (i.e. the calliper threshold). For consistency with the NFI procedures, increments and volumes of young growth are not included when calculating MASSIMO results. Every decade, and whenever saplings grow beyond the calliper threshold, the survival of young growth is determined using a uniform random selection. The survival probability is specific to each production region and length of time-step (Table 17.3). If a tree grows beyond the calliper threshold and survives the random selection, it becomes a 'regular' tree in MASSIMO. At this point, a given number of new young growth trees of the same species (with dbh = 1 cm) are added to the regeneration pool to represent the process of understorey regeneration.

In contrast to understorey regeneration, the process of regeneration due to improved light conditions following shelterwood cutting or stand-replacing storm damage is initiated using empirical data from NFI regeneration subplots in the young growth and thicket stages. For a plot to be regenerated, MASSIMO randomly selects a regeneration data set from the same production region and elevation belt and adds those trees to the regeneration pool. The new young growth trees are assigned a dbh of 1 cm, irrespective of the diameter that was recorded in the NFI regeneration subplots. This procedure leads to a tree species composition in the regeneration pool that reflects data from the last inventory. However, the observed proportion of conifers in the regeneration pool can be modified for each sample plot using the scenario settings. This functionality has been implemented to simulate the effects of NaiS recommendations on plant-community-specific coniferous proportions (Frehner et al. 2005, 2007). The Swiss-specific NaiS recommendations include

Table 17.4 Proportions of conifers in the regeneration pool

Community	Spruce	Fir	Pine	Larch	Swiss stone pine
1–17	0.8	0.2	0	0	0
18–20	0.2	0.8	0	0	0
21–44	0.6	0.133	0.133	0.133	0
45	0	0	1	0	0
46–52	0.5	0.5	0	0	0
53–57	1	0	0	0	0
58	0.5	0	0	0.5	0
59	0	0	0	0.5	0.5
60	1	0	0	0	0
61–71	0.15	0	0.85	0	0

When altering species composition towards more conifers, the proportions of conifers are specific to forest community unit

three levels of coniferous proportions (natural, recommended, maximum tolerable), and the user can specify which of these levels is used in the scenario settings.

If species composition is modified in the scenario settings, the original tree species proportions within the NFI sample plots are changed. As long as the observed coniferous proportion is lower than the user-specified recommendation, broadleaved trees are replaced by coniferous trees. The coniferous species identity (Norway spruce, silver fir, Scots pine, European larch or Swiss stone pine) is randomly determined based on plant-community-specific probabilities (Table 17.4). If the observed coniferous proportion is higher than the NaiS recommendation, conifers are replaced by beech trees.

17.2.7 Scenario Configuration and Iterative Scenario Approximation Routine

The setup of a MASSIMO simulation consists of determining the general configuration of the simulation, assigning scenario-specific settings (Table 17.5) and predefining growing stock or timber harvesting targets by economic region and time-step. These standard options allow simulation of a relatively broad range of scenarios, as in the Project on timber harvesting potentials (Chap. 19). Additionally, the program structure allows users to define more detailed scenarios by introducing scenario-specific intervention rules into the program code of the management module. These highly specific options will not be described here. Typically, MASSIMO is run with time-steps of 10 years, but for the Forest Management Reference Level FMRL (Chap. 19) a time-step of 5 years can be simulated as well. As simulation errors increase over time, MASSIMO is intended to simulate a maximum of 10 time-steps. When running national scenarios with a time-step of 10 years, the simulation is run with 20 replicates (i.e. simulations using identical start conditions). Results of

Table 17.5 Configuration and scenario-specific settings in MASSIMO

Variable	Description	Allowed values
tsl	Length of a time-step	5/10 years
numTS	Number of time-steps	Up to 10
numIter	Number of replicates	Typically 20, any value allowed
mort	Simulated mortality rates	0/10/15/20%
res	Simulation of natural forest reserves	1/0
aktMix	Species mixture in regeneration (coniferous proportion levels)	Observed/natural/recommended/maximum
stSim	Simulation of storm damage	1/0
stPer	Periodicity of storm damage	10/15 years
stSan	Salvage logging of storm damage	1/0

the replicates are averaged and standard errors of the mean are calculated as a proxy for variation. Simulations with a time-step of 5 years or a smaller sample-plot network may need more replicates.

Besides general configuration and scenario settings, growing stock or, alternatively, timber harvesting targets can be specified for each economic region and each time-step. When MASSIMO is run in the iterative scenario adaption mode, the parameters delta thinning Δ_{thin} and delta felling Δ_{fell} are adjusted to reach the predefined values. A decrease in Δ_{thin} leads to an increase in the number of sample plots that experience thinning, while a decrease in Δ_{fell} reduces the number of felled plots. Thus, to balance between thinning and felling interventions when adapting to predefined targets, values of Δ_{thin} and Δ_{fell} are adjusted simultaneously in opposite directions. The approximation routine may alter the simulated rotation period and the return interval of thinning, while thinning intensities, i.e. basal area thinning proportions, are maintained. To keep management interventions within a silviculturally plausible range, Δ_{thin} cannot be adjusted to values below 0.9. This restriction sets an upper limit to the thinning frequency, in that stand basal area in the focal decade must be at least 90% of the basal area prior to the previous intervention.

The approximation routine is limited to even-aged forests outside the forest reserve perimeter. For protection forests, only the thinning parameter Δ_{thin} can be adjusted because shelterwood felling is not allowed. This restriction is therefore needed to simulate plausible management in protection forests. In uneven-aged forests, the degree to which both Δ_{thin} and Δ_{fell} can be adjusted is limited while there is even no timber harvesting in forest reserves; this limitation is especially relevant to strata with a high proportion of these stand types. Thus, in extreme scenarios with a considerable increase or decrease in timber harvesting, predefined targets may not be reached because the adaption only influences the even-aged sample plots in an economic region. Typically, the adaption mode is run with five iterations, i.e. Δ_{thin} and Δ_{fell} are adjusted five times for each time-step.

In MASSIMO, it is possible to simulate growth-, mortality- and harvest-related derivations from individual-tree data, and stand characteristics similar to those measured in the NFI can then be deduced. Thus, the model allows users to project future inventory data.

17.3 Considerations and Further Development

Most European countries use NFI-oriented growth simulators to predict and report woody biomass. Like MASSIMO, most of these models have empirically based individual-tree growth modules that are distance independent (Barreiro et al. 2016). Unlike similar models, MASSIMO not only accounts for NFI sampling errors but also respects random errors and uncertainties in the basal area increment and volume estimations. MASSIMO is therefore highly effective for representatively projecting individual-tree and forest growth, timber harvest and growing stock at the landscape scale and for Switzerland as a whole. Furthermore, the BAI growth models have been validated using independent data and sensitivity analysis has been performed (Thürig et al. 2005a). However, these models do not account for climate effects, which may limit the validity of long-term projections with MASSIMO. To resolve this issue, new climate-sensitive increment models have been developed (Rohner et al. 2018) and are planned to be incorporated into the next version of MASSIMO. Further, species mixture may affect individual-tree growth (Mina et al. 2018), and plans are therefore in place to evaluate such effects in long-term projections.

The simulation of ingrowth is a crucial part of forest growth models, as stand growth reacts very sensitively to the number of simulated ingrowth trees, especially in long-term projections. However, empiric growth models fitted from trees in the seedling to thicket stage are not available because there are no repeated measurements of these small trees. As a consequence, the increment of small trees is simulated in MASSIMO with the same growth equations as for trees exceeding the calliper threshold. To this end, seedlings in regeneration subplots are assigned a dbh of 1 cm because the growth models depend on dbh. While this starting diameter is an overestimation, increment growth may still be underestimated because of non-linear relationships of the growth model. To solve this issue, survival rates are adjusted to reach ingrowth rates observed in the NFI. When simulating small landscapes, these adjustments may need to be modified over time to stay within a plausible range of ingrowth (Temperli et al. 2017b). This issue implies that direct simulation of ingrowth might be more appropriate. To this end, a new ingrowth module is currently under evaluation that will make it possible to account for climatic effects, thereby establishing plausible ingrowth rates and tree characteristics (i.e. dbh, species) in long-term projections.

When modelling tree growth from empiric starting conditions, there are two challenges on the simulated time axis. First, the initialisation of sample data into model data may require some assumptions to maintain the plausibility of projected forest development. This initialisation process typically affects the simulation of the first time-step, which is, however, often assumed to be the most precise. While the single processes of growth, ingrowth and mortality have been validated, validation of the initialisation process is challenging because scenario assumptions influence the simulated forest development already in the first time-step. However, starting simulations from known NFI2 or NFI3 state inventory data enable users to evaluate and improve the initialisation process by comparing the first simulated time-step with NFI4 results. Second, owing to model uncertainties and stochastic processes, the precision of projections decreases over time. This issue can partly be resolved by increasing the simulation replicates. However, the highly consuming post processing in the current version of MASSIMO means that the number of iterations is limited by computational power. Consequently, the projection timespan is restricted to 10 simulated time-steps.

In summary, while MASSIMO is accurate for representatively projecting tree growth in Swiss forests, its projection timespan is limited by content-related (e.g. climate sensitivity is not yet considered) and software-related constraints. MASSIMO has been programmed in the SAS© environment and is thus capable of handling large data sets in a straight-forward manner. However, the implementation of numerous enhancements has led to an inflexible structure, making further developments increasingly difficult. As the planned improvements to the processes in MASSIMO require a complete reconsideration of the software architecture, we are currently developing the next MASSIMO version under a completely different environment. This shift will make it possible to flexibly complete further developments to MASSIMO as any new requirements arise.

References

Barreiro S et al (2016) Overview of methods and tools for evaluating future woody biomass availability in European countries. Ann For Sci 1–15. https://doi.org/10.1007/s13595-016-0564-3

Dierer S, Müller S, Stucki P, Brönnimann S, Romppainen-Martius O (2014) Karten der Sturmgefährdung in der Schweiz. Meteotest and Geographisches Institut der Universität Bern Bundesamt für Umwelt (BAFU), Bern

Frehner M, Wasser B, Schwitter R (2005) Nachhaltigkeit und Erfolgskontrolle im Schutzwald. Wegleitung für Pflegemassnahmen in Wäldern mit Schutzfunktion. Bundesamt für Umwelt, Wald und Landschaft, Bern

Frehner M, Wasser B, Schwitter R (2007) Sustainability and success monitoring in protection forests. Guidelines for silvicultural interventions in forests with protective functions. Environmental Studies no. 27/07. Federal Office for the Environment (FOEN), Bern

Hofer P et al (2011) Holznutzungspotentiale im Schweizer Wald. Auswertung von Nutzungsszenarien und Waldwachstumsentwicklung, vol 1116. Bundesamt für Umwelt, Bern

Kaufmann E (2001a) Estimation of standing timber, growth and cut. In: Brassel P, Lischke H (eds) Swiss National forest inventory: methods and models of the second assessment. Swiss Federal Research Institute WSL, Birmensdorf, pp 162–196

Kaufmann E (2001b) Prognosis and management scenarios. In: Brassel P, Lischke H (eds) Swiss National forest inventory: methods and models of the second assessment. Swiss Federal Research Institute WSL, Birmensdorf, pp 197–206

Krug JHA (2018) Accounting of GHG emissions and removals from forest management: a long road from Kyoto to Paris. Carbon Balance Manag 13:1

Losey S, Wehrli M (2013) Schutzwald in der Schweiz. Vom Projekt SilvaProtect-CH zum harmonisierten Schutzwald. Bundesamt für Umwelt, Bern

Mina M, Huber MO, Forrester DI et al (2018) Multiple factors modulate tree growth complementarity in Central European mixed forests. J Ecol 106:1106–1119

Pfister C (1999) Wetternachhersage. 500 Jahre Klimavariationen und Naturkatastrophen (1496–1995). Verlag P. Haupt, Bern

Rohner B, Waldner P, Lischke H et al (2018) Predicting individual-tree growth of central European tree species as a function of site, stand, management, nutrient, and climate effects. Eur J For Res 137:29–44. https://doi.org/10.1007/s10342-017-1087-7

Schelhaas MJ, Nabuurs GJ, Schuck A (2003) Natural disturbances in the European forests in the 19th and 20th centuries. Glob Chang Biol 9:1620–1633. https://doi.org/10.1046/j.1529-8817.2003.00684.x

Stadelmann G, Temperli C, Conedera M, Gómez A, Brang P (2015) Timber mobilization possibilities for the chestnut belt of Ticino | [Möglichkeiten zur Holzmobilisierung im Tessiner Kastaniengürtel]. Schweiz Z Forstwes 166:291–298. https://doi.org/10.3188/szf.2015.0291

Stadelmann G, Herold A, Didion M, Vidondo B, Gómez A, Thürig E (2016) Timber harvesting potential of Swiss forests: simulation of management scenarios | [Holzerntepotenzial im Schweizer Wald: Simulation von Bewirtschaftungsszenarien]. Schweiz Z Forstwes 167:152–161. https://doi.org/10.3188/szf.2016.0152

Stadelmann G, Didion M, Herold A, Rohner B, Rösler E, Temperli C, Thürig E (2019) Presenting MASSIMO: a management scenario simulation model to project growth, harvests and carbon dynamics of Swiss forests. Forests 10:94. https://doi.org/10.3390/f10020094

Temperli C, Stadelmann G, Thürig E, Brang P (2017a) Silvicultural strategies for increased timber harvesting in a Central European mountain landscape. Eur J For Res 136:493–509. https://doi.org/10.1007/s10342-017-1048-1

Temperli C, Stadelmann G, Thürig E, Brang P (2017b) Timber mobilization and habitat tree retention in low-elevation mixed forests in Switzerland: an inventory-based scenario analysis of opportunities and constraints. Eur J For Res 136:711–725. https://doi.org/10.1007/s10342-017-1067-y

Thürig E, Kaufmann E, Frisullo R, Bugmann H (2005a) Evaluation of the growth function of an empirical forest scenario model. For Ecol Manag 204:53–68. https://doi.org/10.1016/j.foreco.2004.07.070

Thürig E, Palosuo T, Bucher J, Kaufmann E (2005b) The impact of windthrow on carbon sequestration in Switzerland: a model-based assessment. For Ecol Manag 210:337–350. https://doi.org/10.1016/j.foreco.2005.02.030

Traub B, Meile R, Speich S, Rösler E (2017) The data storage and analysis system of the Swiss National Forest Inventory. Comput Electron Agric 132:97–107. https://doi.org/10.1016/j.compag.2016.11.016

Werner F, Taverna R, Hofer P, Thürig E, Kaufmann E (2010) National and global greenhouse gas dynamics of different forest management and wood use scenarios: a model-based assessment. Environ Sci Pol 13:72–85. https://doi.org/10.1016/j.envsci.2009.10.004

Chapter 18
Model of Carbon Cycling in Dead Organic Matter and Soil (Yasso07)

Markus Didion and Jürgen Zell

Abstract While the Swiss NFI (NFI) delivers detailed information on the state of forest resources at the time of the field visit, data on the annual carbon (C) balance in dead organic matter (DOM) and soil are beyond the scope of the NFI. The annual C balance of DOM and soil on NFI sample plots is thus estimated with the C cycling model YASSO07.

18.1 Introduction

While the Swiss NFI (NFI) delivers detailed information on the state of forest resources at the time of the field visit, data on the annual carbon (C) balance in dead organic matter (DOM) and soil are beyond the scope of the NFI. The annual C balance of DOM and soil on NFI sample plots is thus estimated with the C cycling model YASSO07 (Tuomi et al. 2009, 2011; Didion et al. 2014).

18.2 YASSO07

YASSO07 (Tuomi et al. 2009, 2011) is a model of C cycling in mineral soil, litter and deadwood. For estimating stocks of organic C in mineral soil down to a depth of about 100 cm and temporal dynamics of the C stocks in litter and deadwood, YASSO07 requires information on C inputs from DOM components (i.e. non-woody inputs, including foliage and fine roots, and woody inputs, including standing and lying deadwood and dead roots) and climate (temperature, temperature amplitude and precipitation). The Swiss NFI is the source of input data on C stored in individual elements of sample trees (deadwood ≥12 cm in diameter and litter

M. Didion (✉) · J. Zell
Resource Analysis, Swiss Federal Research Institute WSL, Birmensdorf, Switzerland
e-mail: markus.didion@wsl.ch

C. Fischer, B. Traub (eds.), *Swiss National Forest Inventory – Methods and Models of the Fourth Assessment*, Managing Forest Ecosystems 35,
https://doi.org/10.1007/978-3-030-19293-8_18

including woody matter <12 cm and non-woody matter; Chap. 14) and on the state change of each sample tree between two consecutive inventories (i.e. survivor, cut, mortality, ingrowth and nongrowth trees; Sect. 2.9).

Decomposition of the different DOM components of C inputs is modelled based on their chemical composition, the size of woody parts and climate (Tuomi et al. 2009, 2011). Decomposition rates of C that is either insoluble (N) or soluble in ethanol (E), in water (W) or in acid (A), as well as flow rates of C between these four chemical compound groups and to a more stable humus compartment (H) and out of the soil, are derived from a global data set. Coefficients defining C decomposition rates and mass flows between the four chemical compound groups and to the humus are obtained probabilistically using Markov Chain Monte Carlo sampling and are fitted based on data from litterbag studies (Tuomi et al. 2009) and woody litter decomposition experiments (Tuomi et al. 2011). This approach makes it possible to calculate the maximum posterior density and credible intervals for the 21 coefficients including their variance-covariance matrix. Presently, three independent sets of these coefficients have been published (Tuomi et al. 2009, 2011; Rantakari et al. 2012).

Currently, the YASSO07 release 1.0.1 is used with the coefficients presented in Rantakari et al. (2012). The implementation of the model for generating data on the C balance of DOM and soil in Swiss forests is described in detail in FOEN (2017) and by Didion and Thürig (2017); see also Chap. 19 for more information on the model and developers is available online at http://en.ilmatieteenlaitos.fi/yasso. The YASSO07 Fortran source code (available at https://github.com/JariLiski/yasso07ui) was compiled for the Windows 7 operating system. The statistical software R (R Core Team 2016) version 3.0.2 (64 bit) is used to run YASSO07 simulations.

The applicability of the YASSO07 model to Swiss forests was examined by Didion et al. (2014), who analysed the accuracy of YASSO07 for reproducing observed C decomposition in litter and deadwood in Swiss forests. Using the set of coefficients presented in Rantakari et al. (2012), they found no statistically significant differences between simulated and measured carbon mass loss in foliage and fine root litter based on data from a 10-year litterbag experiment (Didion et al. 2014). The same study also showed a close agreement between simulated and observed C mass loss in individual standing and lying dead trees over periods of 14–21 years.

Estimates of C inputs for the simulations with YASSO07 are derived from NFI data separately for coniferous and broadleaved tree species based on the state of individual sample trees in two consecutive inventories. The tree state determines the type and quantity of DOM that is produced (i.e. net primary production). As trees grow, foliage and fine root litter biomass is produced and estimated as described in Chap. 14. Turnover rates reflecting the longevity of leaves, seeds and fine roots are used to determine the fraction of the total foliage and fine root biomass that is produced annually. It is assumed that surviving trees contribute to biomass production during the whole period between two inventories, whereas ingrowth trees and

Table 18.1 Coefficients for Eq. 18.1 used to estimate the annual production of seeds and fruits. Values were determined based on expert judgement

			Production region				
bagr	gw	Co-efficient [kg ha^{-1}]	Jura	Plateau	Pre-Alps	Alps	Southern Alps
Pinus spp.	1162	1	460	330	480	580	500
	1162	2	1100	1700	1200	800	600
All other conifers	1034	1	390	390	410	500	380
	1034	2	1200	1400	1400	900	1000
All broadleaved trees	602	1	370	370	370	270	210
	602	2	1100	1100	1100	2100	2500

those that died, including harvested trees, are assumed to contribute only for half the period. Stemwood, branchwood (large and small branches) and coarse roots are assumed to accrue only as the result of mortality. Depending on the cause of mortality, i.e. natural or timber harvesting, either the total mass of these tree elements or only the non-merchantable fraction is considered in the DOM pool.

The foliage pool includes the biomass of reproductive organs (fruits) derived based on data from Rohmeder (1972):

$$fruits = gw/fru_{bagr,1,prodreg}\ dbh^2/fru_{bagr,2,prodreg} \tag{18.1}$$

where *fru* is a table with two coefficients (1, 2) for 15 combinations of three tree species aggregations (*bagr*) and five production regions (*prodreg*), *gw* is the mass of fruits in kg ha^{-1} for each species group, and *dbh* is tree dbh (Table 18.1).

Observed annual climate data (i.e. mean annual temperature, temperature amplitude, and mean annual precipitation sum) is obtained from spatially gridded data prepared by the Federal Office of Meteorology and Climatology, MeteoSwiss. Using measured data from its network of weather stations, MeteoSwiss interpolates a nationwide data set of several climate variables with a spatial resolution of approximately 2.2 km available starting in 1961 (MeteoSwiss 2016).

18.3 Objectives/Model Outputs

The primary purpose of applying YASSO07 is to provide estimates of annual C fluxes resulting from DOM decomposition and soil respiration (Sect. 19.3). Every year, data on the C stocks and the C stock changes in deadwood, litter and mineral soil down to a depth of 100 cm are prepared for permanent forest land in Switzerland and reported in the annual Swiss National Inventory Report (FOEN 2017).

References

Didion M, Thürig E (2017) Data on soil carbon stock change, carbon stock and stock change in surface litter and in coarse deadwood prepared for the Swiss GHGI 2018 (1990–2016). Swiss Federal Institute for Forest, Snow and Landscape Research WSL, Birmensdorf. Available online at: www.bafu.admin.ch/climatereporting

Didion M, Frey B, Rogiers N, Thürig E (2014) Validating tree litter decomposition in the Yasso07 carbon model. Ecol Model 291:58–68. https://doi.org/10.1016/j.ecolmodel.2014.07.028

FOEN (Federal Office for the Environment) (2017) Switzerland's greenhouse gas inventory 1990–2015. National inventory report including reporting elements under the Kyoto protocol. Submission of April 2017 under the United Nations framework convention on climate change and under the Kyoto protocol. Federal Office for the Environment, Bern. Available online at: www.bafu.admin.ch/climatereporting

MeteoSwiss (Federal Office of Meteorology and Climatology) (2016) Produkte [Products] (Available in German, French and Italian only). http://www.meteoschweiz.admin.ch/home/service-und-publikationen/produkte.html. Accessed 29 Sept 2016

R Core Team (2016) R: a language and environment for statistical computing. R Foundation for Statistical Computing, Vienna. https://www.R-project.org/

Rantakari M et al (2012) The Yasso07 soil carbon model – testing against repeated soil carbon inventory. For Ecol Manag 286:137–147. https://doi.org/10.1016/j.foreco.2012.08.041

Rohmeder E (1972) Das Fruchten der Waldbäume. In: Das Saatgut in der Forstwirtschaft. Parey, Hamburg, pp 35–58

Tuomi M et al (2009) Leaf litter decomposition—estimates of global variability based on Yasso07 model. Ecol Model 220:3362–3371. https://doi.org/10.1016/j.ecolmodel.2009.05.016

Tuomi M, Laiho R, Repo A, Liski J (2011) Wood decomposition model for boreal forests. Ecol Model 222:709–718. https://doi.org/10.1016/j.ecolmodel.2010.10.025

Chapter 19
Scenario Simulations

Golo Stadelmann, Markus Didion, and Esther Thürig

Abstract National forest inventory based growth simulators are an important tool to assess long-term consequences of forest management in many European countries. MASSIMO is the empirically-based growth simulator used in Switzerland. This individual-tree model has been developed to simulate the growth of trees using the spatial grid of the Swiss National Forest Inventory (NFI) and its repeated measurements. MASSIMO has been used at the national scale to predict timber harvesting potentials, to assess the CO2 effects of Swiss forests and their potential for carbon sequestration, especially regarding the full timber chain, and for simulating the forest management reference level (FMRL) under the Kyoto protocol. Further, MASSIMO has been used to evaluate different timber mobilisation scenarios in a mountainous landscape and to evaluate timber-mobilisation strategies and habitat-tree retention in low-elevation Swiss forests.

19.1 Introduction

The growing stock in Swiss forests has been increasing since the first Swiss NFI (NFI) (1983–1985), although with large regional variation. While this increase has been considerable in alpine regions, the growing stock on the Plateau showed a clear decrease between NFI3 and NFI4 (Abegg et al. 2014). Short-term over- and under-harvesting may not influence timber harvesting potentials in the long term, but they may decrease forest productivity and thus indirectly impact timber harvesting potentials, which may influence carbon (C) sequestration.

G. Stadelmann (✉) · M. Didion · E. Thürig
Resource Analysis, Swiss Federal Research Institute WSL, Birmensdorf, Switzerland
e-mail: golo.stadelmann@wsl.ch

C. Fischer, B. Traub (eds.), *Swiss National Forest Inventory – Methods and Models of the Fourth Assessment*, Managing Forest Ecosystems 35,
https://doi.org/10.1007/978-3-030-19293-8_19

NFI-based growth simulators are an important tool to assess long-term consequences of forest management in many European countries (Barreiro et al. 2016). MASSIMO is the empirically-based growth simulator used in Switzerland (Stadelmann et al. 2019). This individual-tree model has been developed to simulate the growth of trees using the spatial grid of the Swiss NFI (Chap. 17) and its repeated measurements. MASSIMO has been used at the national scale to predict timber harvesting potentials (Stadelmann et al. 2016), to assess the CO_2 effects of Swiss forests and their potential for carbon sequestration, especially regarding the full timber chain (Thürig and Kaufmann 2008; Werner et al. 2010), and for simulating the forest management reference level (FMRL) under the Kyoto protocol. Further, MASSIMO has been used to evaluate different timber mobilisation scenarios in a mountainous landscape (Temperli et al. 2017a) and to evaluate timber-mobilisation strategies and habitat-tree retention in low-elevation Swiss forests (Temperli et al. 2017b).

While MASSIMO simulates the development of forests and their growing stock, changes in the annual C balance of dead organic matter (deadwood and litter) and soil are simulated with the C cycling model YASSO07 (Chap. 18). Combining the two models enables estimations of total C budgets above- and belowground.

In this section, two applications of the model MASSIMO and its scenarios are described: MASSIMO was used to estimate timber harvesting potentials of Swiss forests (Sect. 19.2) and it was combined with YASSO07 to estimate the carbon budget of the Swiss forests for the Swiss Greenhouse Gas Inventory and the Kyoto Protocol (Sect. 19.3).

19.2 Timber Harvesting Potentials of Swiss Forests

On the Swiss Plateau, growing stocks have decreased since NFI2 (Abegg et al. 2014). While the decrease between NFI2 and NFI3 was driven by damage caused by the severe windstorm named Lothar, later decreases resulted from forest management. The overall timber harvesting potential in Switzerland may increase despite losses in the Plateau region, but timber availability might suffer from regional shifts into alpine regions that lead to higher timber harvesting costs and, in turn, may prevent forest managers from timber harvesting. NFI-based tree growth simulators are promising tools to predict timber harvesting potentials over the long term (Barreiro et al. 2016), especially when combined with timber harvesting productivity models, as applying this model combination not only makes it possible to estimate timber harvesting potentials but also hints at economic consequences.

In Switzerland, the management scenario simulation model MASSIMO (Chap. 17) is applied to predict forest growth and yield under given timber harvesting scenarios, and timber harvesting costs of produced assortments are calculated using the productivity model HeProMo (Chap. 16). MASSIMO is an empirical model, in that the basic processes of growth, storm damage, timber harvesting, mortality and regeneration are statistically fitted to NFI data. The selection of trees for timber harvesting, small-scale mortality and storm damage involves random draws with given selection probabilities, meaning that MASSIMO incorporates stochastic and dynamic components. MASSIMO can simulate management scenarios at the national and regional scale, but it cannot be used to interpret forest development at the local (i.e. sample plot or stand) scale.

The Swiss Timber Harvesting Potential (THP) Project aims to ensure long-term timber availability from Swiss forests at the national scale. A first study was completed based on the results of the second and third NFIs (Hofer et al. 2011), and changing circumstances and new NFI results have prompted a continuation of the Project. Based on the results of NFI4 (2009–2013), a new set of management scenarios has been defined, in collaboration with the Swiss Federal Office for the Environment (FOEN) and with contributions from invited stakeholders from forest research and practice, as well as from the timber industries. Hence, the intention is to define relevant scenarios that cover current forest policy and needs of the timber industries. Specifically, the following questions have been posed: (a) how will growing stock and timber harvesting potential develop under different management scenarios? (b) does a short-term increase in timber harvesting compensate for decreasing timber harvesting potentials, and could it even accelerate growth rates and therefore lead to a higher timber harvesting potential in the long term? (c) How do timber harvesting costs develop under different management scenarios? Answering these questions will provide the basis for discussions on national timber harvesting targets. However, in order to propose conclusive implications of management scenarios, social and ecological factors should also be considered.

19.2.1 Model Settings and Scenario Definition

MASSIMO consists of five sub-models: (1) growth, (2) storm, (3) timber harvesting and management scenario, (4) mortality and (5) regeneration (Chap. 17). First, individual-tree growth is simulated as basal area increment using specific models for the main tree species and regions of Switzerland (Thürig et al. 2005a). These models are applied for the growth of surviving trees at the beginning of a time-step, as well as for the growth of small trees in the regeneration pool, and for the simulation of growth releases after timber harvesting or mortality.

Second, while large-scale storm damage and subsequent timber harvesting can be switched on or off in the scenario settings, simulation of small-scale mortality is compulsory in order to represent mortality due to unknown reasons and mortality in relation to stand density. As observed in the NFIs, total losses between two consecutive inventories equal approximately 3% of the growing stock in the first inventory and mortality corresponds to 15% of the harvested volume (Chap. 17). The configuration of scenarios makes it possible to define three fixed levels of mortality, which correspond to 10, 15 and 20% of the harvested volume. If switched on, storm damage can be configured to match the observed return interval of 15 years or a shorter return interval of 10 years if storm damage is assumed to increase with climate change (Thürig et al. 2005b).

Third, timber harvesting is implemented to represent thinning (in even- and uneven-aged forests) and shelterwood felling at the end of a rotation period. Further, forests with protective functions against natural hazards are regenerated by means of heavy thinning in order to meet guidelines for silvicultural interventions (Frehner et al. 2007). Overall timber harvesting or growing stock targets are defined for each economic region, and then thinning and shelterwood felling are adjusted to meet the growing stock targets using an iterative adaption algorithm (Chap. 17).

Fourth, at the end of each time-step young trees are added to the regeneration pool after storm damage or shelterwood cutting. To do so, a regeneration subplot from the same production region and vegetation zone of the NFI is randomly selected from the database. Further, for each tree species growing beyond the calliper threshold of ingrowth (12 cm), a number of saplings of the same species are added to the regeneration pool to allow ingrowth to be taken into account, also in later development stages of the forest. The proportion of coniferous and broadleaved trees on newly regenerated sample plots can be defined in the scenario settings according to four levels (as observed in NFI3; natural, recommended or maximum tolerable proportion of conifers) based on the guidelines presented by Frehner et al. (2007). Thus, if the coniferous proportion is less or more than the defined level, trees type is switched accordingly (Temperli et al. 2017a).

For the Timber Harvesting Potential Project, MASSIMO was run for a period of 100 years starting from NFI3, with a time-step of 10 years. Thus, the first time-step 2007–2016 was configured to represent the growing stocks observed in NFI4b (2009–2013). For the period after 2016, growing stock and/or timber harvesting targets differed among scenarios (Stadelmann et al. 2016; Table 19.1). However, all other scenario-specific key factors (i.e. settings) were applied from the beginning of the simulations. As MASSIMO contains stochastic and dynamic model components, such as individual-tree selection in the modules storm, timber harvesting and mortality, simulations were run with 20 replicates to balance possible extreme outcomes of a single simulation. Results are presented for both mean and variance of the 20 replicates, as well as the extrapolation errors from sample plots to the forest areas classified as accessible, which were determined using a double sampling algorithm (Köhl 2001).

Scenarios were defined in collaboration with an expert group coordinated by FOEN to guarantee their credibility. This iterative procedure resulted in five scenarios that comprise a broad range of feasible forest developments and therefore are expected to support decision making and forest policy (Table 19.1). In the base scenario, growing stocks were maintained based on the level observed in NFI4b, and thus the total loss is equal to the gross growth. The business as usual (BAU) scenario aimed to keep timber harvesting stable for each region. While timber harvesting was already above the volume increment in the lowlands of Switzerland, the overall growing stock increased in Alpine regions. Further scenarios aimed to decrease growing stock rapidly in order to improve forest growth over the long term.

Table 19.1 Definition of management scenarios considered in the Project about timber harvesting potential in Swiss forests

Scenario	Growing stock and timber harvesting targets after 2016	Other specifications
A: Base scenario	Stable growing stock, as observed in NFI4b	Mortality: 15% of growing stock
		Regeneration: recommended coniferous proportion
		Storm damage: 15-year periodicity
		Reserves: no
B: Business as usual	Stable timber harvesting, as observed between NFI3 and NFI4b	Mortality: 15% of growing stock
		Regeneration: recommended coniferous proportion
		Storm damage: 15-year periodicity
		Reserves: no
C: Constantly high increment	Decreasing growing stock to 300 m^3 ha^{-1} until 2046, then maintaining	Mortality: 15% of growing stock
		Regeneration: recommended coniferous share
		Storm damage: 15 years periodicity
		Reserves: no
D: High demand on coniferous roundwood	Decreasing growing stock to 250 m^3 ha^{-1} until 2046, then increasing slowly to 300–330 m^3 ha^{-1}	Mortality: 10% of growing stock
		Regeneration: maximum tolerable coniferous share
		Storm damage: 15 years periodicity
		Reserves: no
E: High demand on timber for energy and chemical uses	Decreasing growing stock to 250 m^3 ha^{-1} until 2046, with regional differences: Plateau: 200 m^3 ha^{-1}; Jura, Pre-Alps, Valais, Southern Alps: 250 m^3 ha^{-1}; Alps without Valais: 300 m^3 ha^{-1}, then maintaining growing stock in all regions	Mortality: 10% growing stock
		Regeneration: recommended coniferous proportion
		Storm damage: 15-year periodicity
		Reserves: yes; rare species communities or no management for a long time (Jura >55 years, Plateau >25 years, Pre-Alps >110 years, Alps >120 years, Southern Alps >150 years)

19.2.2 Results

The proportions of broadleaved and coniferous trees remained quite stable under all scenarios, even E, which allowed higher coniferous proportions in the regeneration. Thus, results in graphs combining broadleaved and coniferous trees were more pronounced than those in graphs showing only one tree type. Under all scenarios, timber harvesting and growing stock targets were met for all individual production

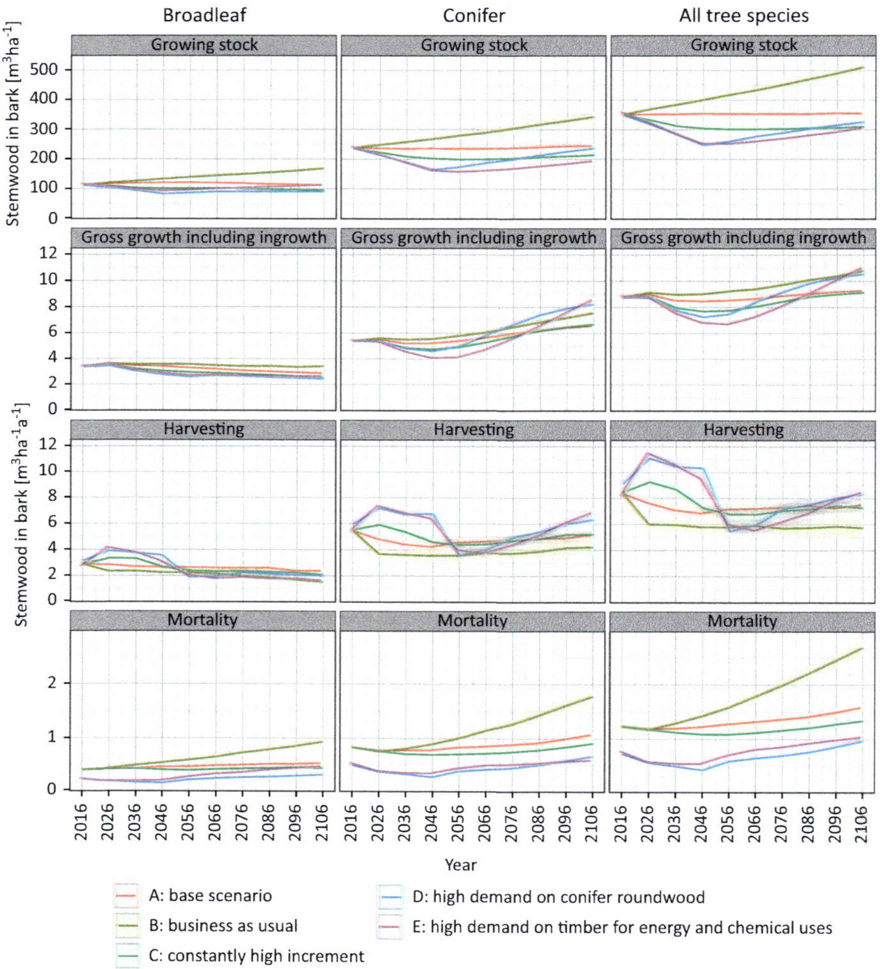

Fig. 19.1 Development of growing stock, gross growth including ingrowth, timber harvesting and mortality under five different management scenarios. Error bands indicate variation between sample plots and simulation replicates, calculated as standard errors of means

regions (not shown here) and for Switzerland as a whole. Except for scenario E, where predefined (stable) growing stocks could not be maintained after 2046 (Fig. 19.1) increasing growing stocks in forest reserves and in protection forests with timber harvesting restrictions were overcompensated by timber harvesting in the remaining forest area. In scenario B, overall growing stock increased strikingly, reaching 510 m^3 ha^{-1} by the end of the century. On the Plateau, however, growing stocks decreased to 240 m^3 ha^{-1}, continuing the trends observed after NFI2. The strong growing stock reductions observed for scenarios C, D and E until 2046 were followed by a phase of reduced timber harvesting. Nevertheless, because timber

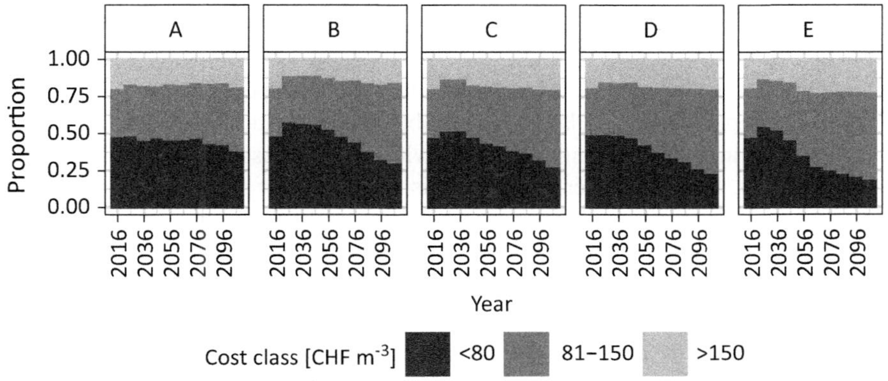

Fig. 19.2 Proportion of timber that was harvested in different cost classes depending on scenario. (**a**) base scenario; (**b**) business as usual; (**c**) constantly high growing stock; (**d**) high demand on coniferous roundwood; (**e**) high demand on timber for energy and chemical uses

harvesting was tremendous in the beginning of the simulation and increasing growth rates allowed more frequent timber harvesting until the end of the century, the total merchantable timber harvested under these scenarios (C: 7.8, D: 8.5, E: 8.1 million m^3 $year^{-1}$) increased in comparison to the base scenario A, which involved constant timber harvesting of 7.7 million m^3 $year^{-1}$. In contrast, the business as usual (BAU) scenario B only allowed an overall timber harvesting of 6.3 million m^3 $year^{-1}$.

For all scenarios, mortality increased with increasing growing stock. In contrast, increasing timber harvesting led to a decrease in mortality but the effect was less pronounced. In scenario B, the very small timber harvesting amounts in comparison to the gross growth led to notably high mortality rates. Owing to predefined lower mortality, mortality rates in scenarios D and E showed a downwards shift in comparison to the other scenarios, while the mortality trends followed growing stock and timber harvesting development.

Timber harvesting was simulated according to predefined targets that did not consider economic criteria (i.e. timber harvesting costs). However, achieving the simulated timber harvesting values may be prevented by forest managers because of increasing timber harvesting costs, as interventions are not cost effective when proceeds are lower than costs. Timber harvesting costs on the Plateau increased over time in all scenarios, although this effect was less substantial in scenario A. Hence, both timber harvesting and timber harvesting costs remained more-or-less stable in scenario A (Fig. 19.2) while a shift towards more wood harvested in the middle cost class instead of the cheapest class was observed in the remaining scenarios. This shift was more pronounced under increased timber harvesting scenarios (C–E) than under the BAU scenario. However, as BAU was associated with the lowest total timber harvesting amounts, absolute timber harvesting costs remained lower than in the constant growing stock scenario A. Nevertheless, owing to lower timber harvesting amounts, possible proceeds, i.e. income from timber selling, were lowest as well.

19.2.3 Discussion

In scenario A, growing stock is stable over the simulation period for each production region. Except for on the Plateau, this stable growing stock is associated with an increase in timber harvesting compared with the business as usual scenario (B). Nevertheless, overall timber harvesting is higher under scenario A and timber harvesting costs remain more-or-less stable over the simulation period. Even greater increases in timber harvesting are simulated under scenarios C–E but fluctuations are also observed: timber harvesting increases until mid-century and then declines to below the current level. For scenarios D and E, increasing the increment allows timber harvesting to increase again by the end of the century. The growing stock reductions observed until mid-century lead to increasing timber harvesting costs per harvested m^3, but they result in accelerated growth rates in the long term. However, strong fluctuations in timber harvesting amounts require similar fluctuations in lumberjacks and timber harvesting machines, which in turn necessitates further investments in timber industries. Thus, extreme timber harvesting scenarios do not guarantee social guidelines, nor safe amortisations of necessary investments. However, these implications are limited by the fact that we did not specifically consider social, ecological or economic indicators. Further, increased timber harvesting scenarios might only be feasible with strong subsidies if substantial amounts of timber harvesting shift from the cheapest to the intermediate cost class.

In conclusion, the constant growing stock scenario A is the most balanced scenario regarding the development of timber harvesting costs and timber amounts, while scenario B is the cheapest scenario. However, reduced timber harvesting in mountainous regions may lead to decreased stability of protection forests, and overall decreasing timber harvesting amounts may lead to job losses. Thus, there is not a best scenario for all of Switzerland, and a combination of the given management scenarios is needed instead. To this end, the scenarios show a wide scope of action that may help to define suitable regional and national strategies for growing stock targets and their development.

19.3 Greenhouse Gas Inventory

As a signature state to the United Nations Framework Convention on Climate Change (UNFCCC), Switzerland is required to maintain a comprehensive Greenhouse Gas Inventory (GHGI), including emissions and removals of CO_2 from Land Use, Land-Use Change and Forestry (LULUCF). Under the UNFCCC, the reporting of emissions and removals on forest land in Switzerland is based on measured data for living biomass and on use of the YASSO07 model for deadwood, litter and mineral soil (Chaps. 14 and 18). In addition to the reporting obligations under the UNFCCC, Switzerland has committed to the Kyoto Protocol, which is an international agreement linked to the UNFCCC and for which supplementary information has to be reported. This includes data on the effect of forest management activities (Article 3.4 of the Kyoto Protocol) on the carbon sequestration of forests including, for example, regeneration, fertilisation, and pest and fire control but excluding afforestation, reforestation and deforestation, as defined in Article 3.3 of the Kyoto Protocol.

In the second commitment period of the Kyoto Protocol from 2013 to 2020, a reference level approach was adopted for forest management. The *Forest Management Reference Level* (FMRL) should reflect average annual net emissions and removals from forest management. In this second commitment period, only emissions and removals exceeding this reference level can be counted as sources or sinks. Switzerland opted for a model-based approach to estimate its FMRL and, therefore, was required to estimate net emissions and removals for the second commitment period prior to the start of the period. Following the guidance on how to construct the FMRL (Appendix II of the Decision 2/CMP.6[1]), Switzerland's FMRL took into account forest characteristics and actual forest policy implemented until the end of 2009, which were to represent a so-called business as usual (BAU) timber harvesting scenario.

The BAU scenario in Switzerland's submission of the FMRL is described in FOEN (2011) and in Switzerland's National Inventory Report (FOEN 2015). This scenario involved an increase in harvested timber volume by about 30% for the period 2013–2020 compared to 1990–2007. The methodology used by Switzerland for preparing its FMRL was based on an integrated modelling approach that combines the models MASSIMO (Chap. 17) and YASSO07 (Chap. 18) described by Didion et al. (2014). For harmonising the YASSO07 simulation with the MASSIMO simulation, a one-way link between the two models was implemented (Brandmeyer and Karimi 2000). The link was achieved via an interface that compiles the MASSIMO data to estimate the annual production of woody and non-woody litter for each simulated NFI plot. The two models were spatially linked based on NFI plots to ensure consistency between the results.

The FMRL estimate is based on changes in C stocks in the five pools, including above- and belowground living biomass, deadwood, litter and soil. Carbon stock changes in living biomass were modelled with the stochastic, empirical individual-tree forest management scenario model MASSIMO (Kaufmann 2001a, b, 2011; Thürig and Kaufmann 2010; Stadelmann et al. 2019). The model version applied in this study was based on MASSIMO (Kaufmann 2011; Fischer et al. 2017) using a 5-year time-step. The model was initialised with NFI3 data (2004–2006) and run for four time-steps (2007–2011, 2012–2016, 2017–2021 and 2022–2026). The increase in harvest was implemented by shortening the rotation periods and by intensifying the frequency of thinning. In protection forests and uneven-aged forests, the following model assumptions limited the change in rotation periods and thinning frequency: to preserve the vertical and horizontal structure of these forests and to ensure their protective function in the future, the model allows only small changes in their management. The scenario also incorporates a constant natural mortality rate

[1]https://unfccc.int/resource/docs/2010/cmp6/eng/12a01.pdf

of 14–15% of the total drain, as described by Kaufmann (2011), which includes damage caused by regular storm events but not by extreme events. For the FMRL, the calculations were restricted to accessible forest in Switzerland, not including 'shrub forest'. The model was run 20 times to consider the stochastic model components related to storm intensity, frequency and location, and estimates were calculated as average values. Estimates of C inputs for the YASSO07 model (Chap. 18) were derived from the simulated forest development.

YASSO07 was used to obtain estimates of C stock changes in deadwood, litter and soil. The model was initialised with C pool information for 2006, as estimated for the GHGI submission under the UNFCCC (Chap. 18). Starting in 2006, the simulations were continued with projected C inputs obtained from the MASSIMO simulation of the BAU scenario. For the simulation, the C balance of deadwood, litter and soil, based on C inputs for the Swiss FMRL until 2026, and observed climate data until the end of 2013 were used. To estimate annual climate data for the years 2014–2026, means of randomly selected contiguous 3-year slices from the reference norm period 1981–2010 were used.

19.4 General Discussion and Conclusions

NFI data can be used for much more than the analysis of the current forest state and its short-term changes. The NFI data forms the basis for empirical simulations of mid- to long-term forest development. As forests are slow growing systems, the impact of forest management cannot easily be investigated by classical experiments. Forest simulation models such as MASSIMO and YASSO07 offer possibilities to explore different silvicultural interventions and assess their consequences on forests and their services.

The study on timber harvesting potentials in Swiss forests revealed not just one but a suite of potential management scenarios for the different regions of Switzerland that reflect different interventions depending on timber harvesting costs, protective functions and growth potential.

By coupling the empirical model MASSIMO with the litter decomposition model YASSO07, a carefully tested and documented methodology was developed to analyse the development of carbon estimates for Swiss forests and for the FMRL. Future developments aim towards the flexible implementation of natural mortality and natural disturbances.

References

Abegg M, Brändli U-B, Cioldi F, Fischer C, Herold-Bonardi A, Huber M, Keller M, Meile R, Rösler E, Speich S, Traub B, Vidondo B (2014) Fourth national forest inventory – result tables and maps on the internet for the NFI 2009–2013 (NFI4b). Eidg. Forschungsanstalt WSL, Birmensdorf. http://www.lfi.ch/resultate/

Barreiro S, Schelhaas M-J, Kändler G, Antón-Fernández C, Colin A, Bontemps J-D, Alberdi I, Condés S, Dumitru M, Ferezliev A, Fischer C, Gasparini P, Gschwantner T, Kindermann G, Kjartansson B, Kovácsevics P, Kucera M, Lundström A, Marin G, Mozgeris G, Nord-Larsen T, Packalen T, Redmond J, Sacchelli S, Sims A, Snorrason A, Stoyanov N, Thürig E, Wikberg P-E (2016) Overview of methods and tools for evaluating future woody biomass availability in European countries. Ann For Sci 73:823–837

Brandmeyer JE, Karimi HA (2000) Coupling methodologies for environmental models. Environ Model Softw 15:479–488

Didion M, Kaufmann E, Thürig E (2014) Estimating carbon stock changes in living and dead biomass and soil for Switzerland's forest management reference level – modeling methodology and results. Swiss Federal Institute for Forest, Snow and Landscape Research WSL, Birmensdorf

Fischer C, Camin P, Kaufmann E, Thürig E (2017) Switzerland. In: Barreiro S, Schelhaas M-J, McRoberts RE, Kändler G (eds) Forest inventory-based projection systems for wood and biomass availability. Springer International Publishing, Cham, pp 303–313

FOEN (2011) Switzerland's submission on reference levels as an accounting approach for forest management under the Kyoto protocol. Federal Office for the Environment, Bern

FOEN (2015) Switzerland's greenhouse gas inventory 1990–2013. National inventory report 2015 including reporting elements under the Kyoto protocol submission of 15 April 2015 under the United Nations framework convention on climate change and under the Kyoto protocol. Federal Office for the Environment, Bern

Frehner M, Wasser B, Schwitter R (2007) Sustainability and success monitoring in protection forests. Guidelines for silvicultural interventions in forests with protective functions, Bern

Hofer P, Altwegg J, Schoop A, Hässig J, Rüegg R, Kaufmann E, Frutig F, Ulmer U, Rosset C, Camin P (2011) Holznutzungspotentiale im Schweizer Wald. Auswertung von Nutzungsszenarien und Waldwachstumsentwicklung. Bundesamt für Umwelt, Bern, p 80S

Kaufmann E (2001a) Estimation of standing timber, growth and cut. In: Brassel P, Lischke H (eds) Swiss National Forest Inventory: methods and models of the second assessment. Swiss Federal Research Institute WSL, Birmensdorf, pp 162–196

Kaufmann E (2001b) Prognosis and management scenarios. In: Brassel P, Lischke H (eds) Swiss National Forest Inventory: methods and models of the second assessment. Swiss Federal Research Institute WSL, Birmensdorf, pp 197–206

Kaufmann E (2011) Nachhaltiges Holzproduktionspotenzial im Schweizer Wald. Schweiz Z Forstwes 162:300–311

Köhl M (2001) Inventory concept NFI2. In: Brassel P, Lischke H (eds) Swiss National Forest Inventory: methods and models of the second assessment. Swiss Federal Research Institute WSL, Birmensdorf, pp 19–44

Stadelmann G, Herold A, Didion M, Vidondo B, Gomez A, Thürig E (2016) Holzerntepotenzial im Schweizer Wald: Simulation von Bewirtschaftungsszenarien. Schweiz Z Forstwes 167:152–161

Stadelmann G, Didion M, Herold A, Rohner B, Rösler E, Temperli C, Thürig E (2019) Presenting MASSIMO: a management scenario simulation model to project growth, harvests and carbon dynamics of Swiss forests. Forests 10:94. https://doi.org/10.3390/f10020094

Temperli C, Stadelmann G, Thürig E, Brang P (2017a) Silvicultural strategies for increased timber harvesting in a Central European mountain landscape. Eur J For Res 136:493–509

Temperli C, Stadelmann G, Thürig E, Brang P (2017b) Timber mobilization and habitat tree retention in low-elevation mixed forests in Switzerland: an inventory-based scenario analysis of opportunities and constraints. Eur J For Res 136:711–725

Thürig E, Kaufmann E (2008) Waldbewirtschaftung zur Senkenerhöhung? Mögliche Konfliktfelder und Synergien. Schweiz Z Forstwes 159:281–287

Thürig E, Kaufmann E (2010) Increasing carbon sinks through forest management: a model-based comparison for Switzerland with its Eastern Plateau and Eastern Alps. Eur J For Res 129:563–572

Thürig E, Kaufmann E, Frisullo R, Bugmann H (2005a) Evaluation of the growth function of an empirical forest scenario model. For Ecol Manag 204:53–68
Thürig E, Palosuo T, Bucher J, Kaufmann E (2005b) The impact of windthrow on carbon sequestration in Switzerland: a model-based assessment. For Ecol Manag 210:337–350
Werner F, Taverna R, Hofer P, Thürig E, Kaufmann E (2010) National and global greenhouse gas dynamics of different forest management and wood use scenarios: a model-based assessment. Environ Sci Pol 13:72–85

Part VI
The Swiss National Forest Inventory Data Analysis System

Chapter 20
The Swiss National Forest Inventory Data Analysis System

Berthold Traub, Rolf Meile, Simon Speich, and Erik Rösler

Abstract The NAtional Forest Inventory Data Analysis System (NAFIDAS) is a multitier application to store and analyse Swiss National Forest Inventory (NFI) data. It provides a basis to publish findings about the current state and development of the Swiss forest on a regional and national level. The system supports the complete chain of data processing of a sample-based forest inventory, starting with the upload of field data and ending with the publication of tables and maps on the Internet. It also provides the necessary tools for administering and maintaining the system and offers access to lists of variables, derivations and dependencies, as well as to documentation of code and production processes. NAFIDAS creates concise result tables and maps with reproducible statistics on NFI data including standard errors and allows the publication of multilingual results based on the most recent NFI data. The efficient long-term storage facilities of raw and derived inventory data form the backbone of the system.

B. Traub (✉) · S. Speich
Scientific Service NFI, Swiss Federal Research Institute WSL, Birmensdorf, Switzerland
e-mail: berthold.traub@wsl.ch

R. Meile
GIS, Swiss Federal Research Institute WSL, Birmensdorf, Switzerland

E. Rösler
Resource Analysis, Swiss Federal Research Institute WSL, Birmensdorf, Switzerland

C. Fischer, B. Traub (eds.), *Swiss National Forest Inventory – Methods and Models of the Fourth Assessment*, Managing Forest Ecosystems 35,
https://doi.org/10.1007/978-3-030-19293-8_20

20.1 Introduction

The NAtional Forest Inventory Data Analysis System (NAFIDAS) is a multitier application to store and analyse Swiss NFI (NFI) data, providing a basis to publish findings about the current state and development of the Swiss forest on a regional and national level. The system supports the complete chain of data processing of a sample-based forest inventory, starting with the upload of field data and ending with the publication of tables and maps on the Internet. It also provides the necessary tools for administering and maintaining the system. The following requirements are covered by NAFIDAS:

- Documentation: lists of variables, derivations and dependencies, as well as code documentation and production processes
- Analysis: creation of concise result tables with reproducible statistics on NFI data including standard errors
- Reporting: publication of multilingual results on www.lfi.ch based on the most recent NFI data in the form of result tables and interactive maps
- Storage: efficient long-term storage and organisation of raw and derived inventory data
- Management: maintenance of the system and of derived inventory data

Technically, the NAFIDAS architecture consists of three main components (tiers): (1) web applications for management, documentation and reporting, (2) databases for storage, and (3) the data analysis application (DAA) for analysis. A general overview of the system and its components is given in Fig. 20.1 and in Traub et al. (2017). The article, however, uses data warehouse nomenclature as its target audience is more IT-oriented. In this chapter we adopt a perspective that is focused more on inventories, and use terminology from the NFI. Table 20.1 contains the most important terms and concepts used in the subsequent sections.

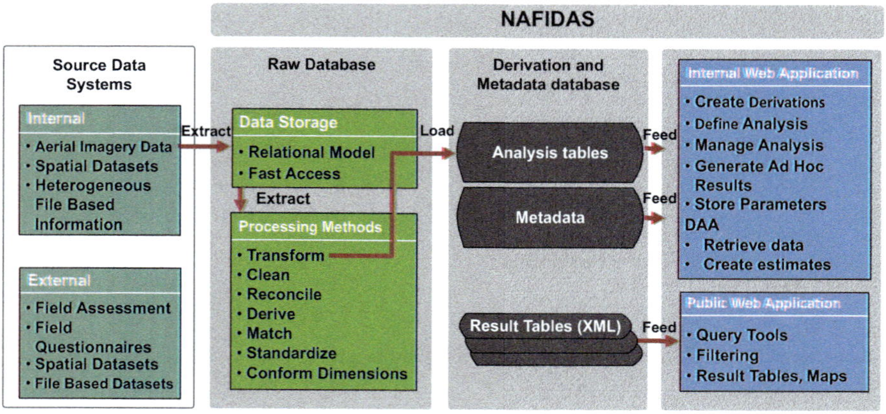

Fig. 20.1 Architecture of NAFIDAS. The system includes the components: raw database, derived database and web applications including the DAA. (Adapted from Traub et al. 2017)

Table 20.1 Most relevant NAFIDAS terms

NAFIDAS term	Content/concept	Parameter	Relation to derivations	Example	German term
Variable	Definition of structure and semantic specification	No	Mandatory base of all derivations	Tree diameter at breast height (dbh)	Variable
Derivation	Calculation instruction (script) for a variable by inventory	No	–	dbh for inventory NFI4	Ableitung
Inventory	Timeframe and method	Yes	Inventory-dependent implementations of a variable by derivations	NFI4 (2009 to 2017)	Inventur
Target variable	Algorithm linking multiple derivations	Yes	Algorithm consisting of one or more derivations or a dummy	Tree basal area	Zielgrösse
Analysis table	Structure containing derivation data	No	Data container for calculated values	Inventory tree table	Ableitungstabelle
Reference unit	Polygonal geographical extent	Yes	References a derivation	Production region	Aussageeinheit
Reference domain	Forest (sub) domain	Yes	References a derivation	Accessible forest without shrub forest	Auswertungseinheit
Sampling grid	Spatially distributed sample points	Yes	References a derivation	1.4 km × 1.4 km	Netz
Classification unit	Attributes of plots and trees assessed in sampling	Yes	Each classification unit is a derivation and vice versa	List of tree species	Befundeinheit

The data model used in NAFIDAS is predominantly based on the concept of an inventory. From the point of view of data-analysis processing, the term inventory defines a timeframe and determines a set of appropriate raw data to be used for processing derivations and calculating result tables. Typically, this contains field survey data collected on a certain sampling grid during a certain period of time that is accumulated in the inventory's data set. In an inventory, homogenised methods for deriving data and stable statistical algorithms are applied. All result tables finally generated are specific to one of three different kinds of inventories:

1. A *state* inventory, which usually represents a single timeframe for the data collecting labelled, for example as 'NFI1'.
2. A *change* inventory, which calculates the gross change (balance) in any attribute between two state inventories, i.e. the difference between the overall values for the attribute, labelled for example as 'NFI2–NFI1'. The sampling grid and forest population definition do not have to be the same in both inventories.
3. An inventory of *change components,* which directly compares individual trees, each visited in a different state inventory. It applies only to special types of target variables, such as increment, cut and mortality. In this type of inventory, the most detailed tree information for statistical analysis is collected. As a prerequisite, the sampling grid and forest population definition of both timeframes must be identical. It is labelled for example as 'NFI12'.

Conceptually, type (1) and (3) are stored in the same database structure (Sect. 20.3.2). Type (2) is calculated during runtime by the DAA (Sect. 20.4.3).

We first introduce NAFIDAS from two different perspectives: (a) following the course of the variable diameter at breast height (dbh) through the system to the target variable basal area until it is finally published as a result table and map on the Internet; (b) listing all features of NAFIDAS from the perspective of a user and the functions required by the system to provide them. In Sects. 20.2, 20.3 and 20.4, the databases, the web applications and the DAA are described in detail.

The data used in NAFIDAS comes from different sources. Most of them, such as dbh, is collected by teams in the field (Chap. 9), but some comes from the aerial-photo interpretations (Chap. 6) or the forest road survey (Sect. 10.3). In a few cases, input data is generated from GIS applications (Chap. 7) or otherwise modelled externally (Part V).

All input data is stored in the raw database (Sect. 20.3.1). From there, it is copied over into the derivation database for each inventory along with metadata. Where necessary, the data is also transformed. Further (new) data is created using the raw and derived data, and saved as derived data. Additionally, dependencies on other data is registered in the system.

Upon request, the data from the derivation database is pulled into the DAA to estimate the basal area and corresponding standard errors. Subsequently, the output of the DAA is processed by the internal web application (Sect. 20.2.1) to display or export the estimated data as a table or a map in real-time. The table can also be stored again in the database, where it is used for cached display of tables or for generating interactive maps on the public website (Sect. 20.2.2). Figure 20.2 shows the

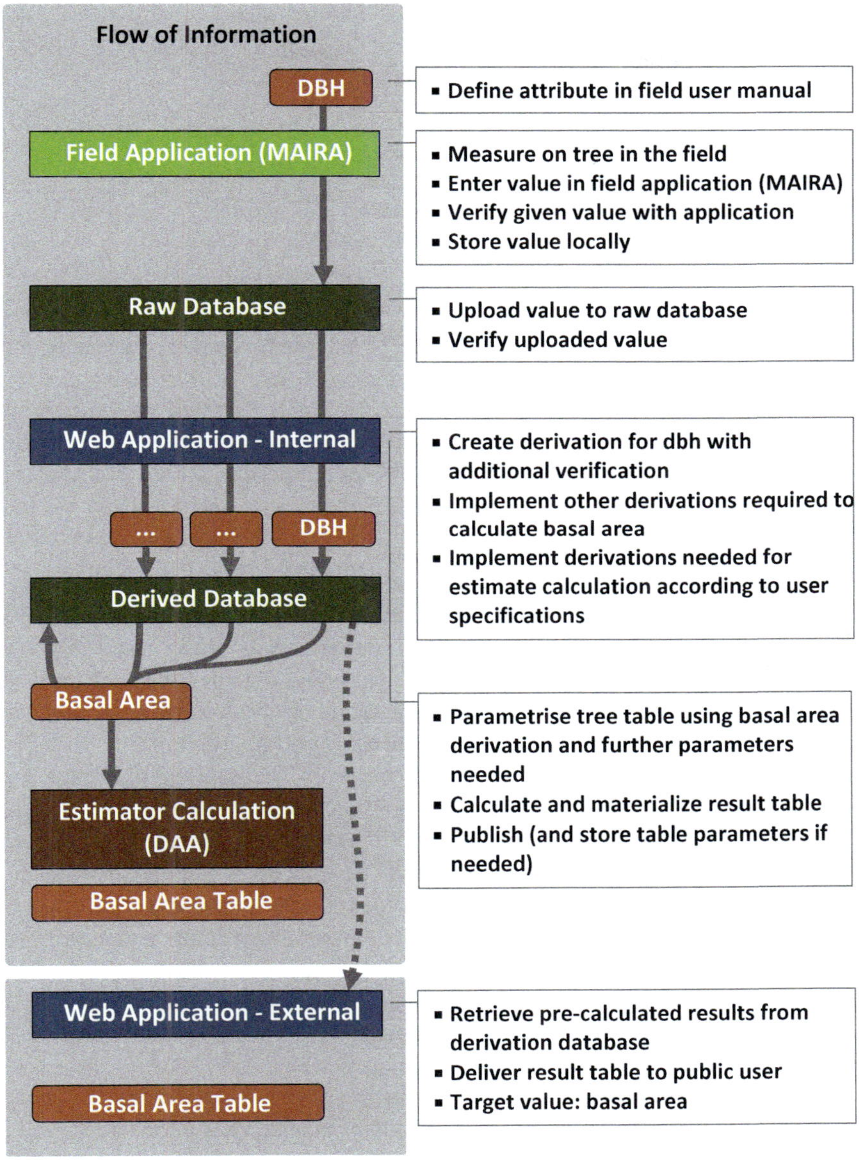

Fig. 20.2 Example: from the dbh to the result table of basal area, with the full flow of information from data collected in the field to the final result tables depicted, including all software and data storage components

complete flow of data with dbh and basal area as examples. Most features of NAFIDAS are accessible through graphical user interfaces. As illustrated in Fig. 20.2, access to NAFIDAS is provided by web applications hosted on two different websites, either by the public web application on the website www.lfi.ch

or by the internal web application over WSL's intranet. These are described in Sect. 20.2 with the relevant features and parameters listed in Table 20.2. First, however, the main terms used in NAFIDAS are explained in Table 20.1.

20.2 Web Applications

20.2.1 Internal Web Application

The internal web application enables the user to analyse NFI data and generate tables and maps, provides tools to manage the system, visualise the NFI data and display documentation. This application is only available on the institute's intranet or over VPN for NFI experts. Access is controlled by user accounts and group memberships. Depending on the user's credentials, the intranet application allows access to both real-time NFI data analyses and the full set of analysis and management tools. Table 20.2 lists the features available to internal users only and the functionality required behind the scenes to provide these features.

Analysing NFI Data and Producing Result Tables and Maps The heart of the internal web application is the collection of parameters to create result tables and maps in a process called parameterisation. The system guides the user through a series of pages, where they choose between a number of different parameters, such as *inventory period*, *target variable* or *sampling grid* (Fig. 20.3).

Logical correctness of the parameter combinations is ensured in two ways. First, the user interface is populated with selectable values only successively, based on database queries filtered by previously selected values. Depending on the parameter in question, each query checks whether all dependencies are satisfied (such as meeting a hierarchical constraint or complying with the same lookup values in two inventories). Second, before proceeding to the next page, the parameter set is validated for completeness with an XML Schema. Once all parameters, both required and optional, have been selected, they are posted to the DAA, where the estimates, including standard errors, are calculated. The result is then returned to the web server and displayed in the browser as a table (Fig. 20.4).

The user may save this set of parameters permanently in the database, but not the calculated result itself. Saved parameter sets can be loaded back into memory and the result table is recalculated immediately. This guarantees that each result is always based on the most current data. Furthermore, it facilitates semantical validation of the table values and it allows quick updating of the results in case a certain set of derivations was re-calculated.

A description of the most important parameters is given in Table 20.3, and Table 20.7 in section "Equations for Point Estimators and Standard Errors" in Appendix. To create a (time) snapshot of the actual results, the tables can be stored permanently and compiled into packages called products. Result tables can also be exported to a spreadsheet or rendered as a map (Fig. 20.5). The user interface of the

Table 20.2 Internal web application – Features available to the user and functionality required by the system to provide them

Feature	Functions required
General aspects	
Process of analysis parameterisation	Check completeness and conformity to derived data model
	Provide hierarchy between parameters
	Prevent invalid combinations of parameters
	Guide and assist the user
	Multiply existing parameter sets
Analysis and result tables	Evaluate state and change of national and regional inventory data available in NAFIDAS according to NFI methods.
	Calculate estimates based on pre-checked combination of parameters and create result tables
	User can launch the analysis from the internal web interface or run batch jobs
	After termination through internal web application the tables are passed back to the web application
View system information and metadata	Display among other things metadata and status of derivations, inventories and target variables
	Display metadata about the system
	Provide links between system components
Manage parameter sets	Save, load and delete from database
	Assign to products or projects
	Share individual parameter sets between users
	Assign to criteria and indicators
	Maintain backwards compatibility
	Export data to spreadsheet
Manage products and collections	Define result table sets
	Archive former products in collections (decouple results from metadata)
	Batch process parameter sets to create persistent result tables
	Export complete sets of result tables as zip files
	Provide DOI service for timber harvesting result tables
Manage users and permissions	Define availability of tools
	Create user roles concerning access to different results
	Track ownership of derivations
Documentation and help system	Maintain code documentation
	Use version control for code
	Record system concepts/components
	Document format of parameter set including exchange interface
Tools	
Labelling of tables	Translate labels of variables and target variables
	Keep track of translation status
Collect and execute derivations	Collect user input to write metadata for derivations
	Handling status and ownership of derivations

(continued)

Table 20.2 (continued)

Feature	Functions required
	Generate database structure from metadata
	Save values to database calculated from derivation algorithm
	Validate user input, database structure and written data
Register and visualise dependencies	Parse SQL code to auto-detect possible dependencies
	Register dependencies for each derivation
	Create list of graph nodes and adjacency lists
	Visualise as graph
Visualise raw data	Provide queries to aggregate data
	Produce visualisations
Order new derivations	Order and track derivations to be implemented
	Check validity of user input
Remove unused/outdated derivations	Detect possible dependencies on derivations still used
	Detect occurrence in valid parameter sets
	Remove possible corresponding lookup tables and language entries
	Use Data Definition Language (DDL) for complete removal and ensure consistency

internal web application is in German, but result tables and maps can also be generated in French, Italian and English.

20.2.2 *Public Web Application*

The main purpose of the publicly accessible website is to provide general information about the NFI. A user login is not required. The website allows pre-calculated NFI results to be queried and provides access to the results produced with the internal application through an extensive query system. It also includes a data catalogue of the derivations available in the query system, as well as general information on the NFI. For the cantonal authorities an additional access-restricted area is available to check their own results. The whole NFI website is available in German, French and Italian, and parts of it also in English. Table 20.4 lists the features the public may use and the functions required by the system.

The web application to query and display results follows the principles of an online shop. It uses filtering to limit the possible result set, paging to display them and a temporary shopping cart to accumulate potential or interesting results.

To ensure finding the right result is as easy as possible, the user is given several entry points for querying the results. The choices can be according to theme (e.g. basal area), region (e.g. a canton). inventory (e.g. for the period 2009/13), MCPFE criteria or indicators (e.g. Maintenance of forest ecosystem health and

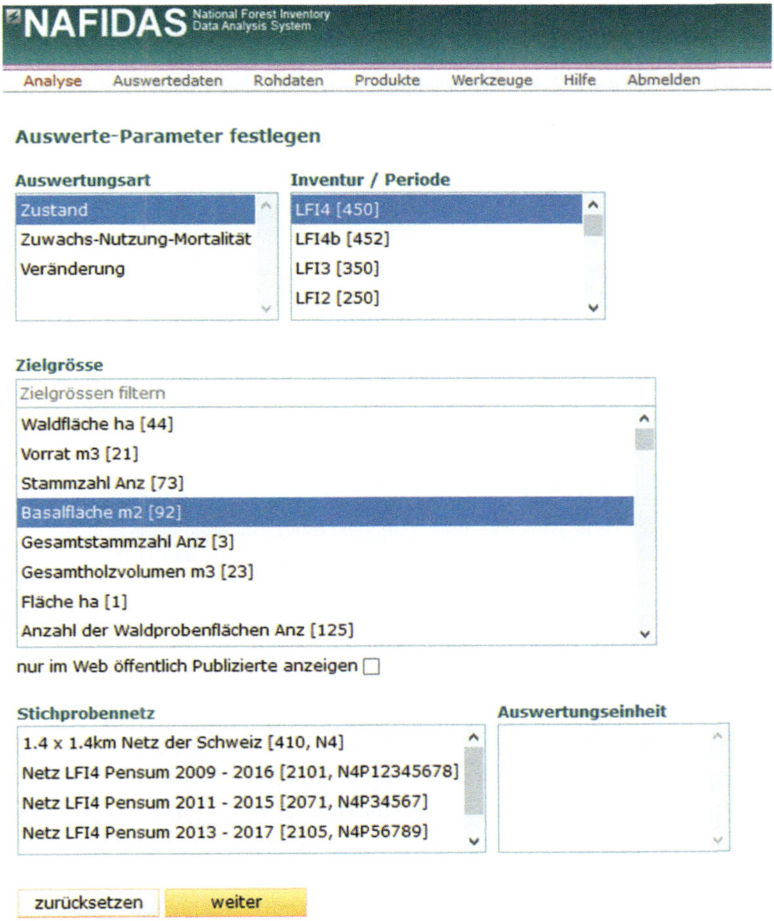

Fig. 20.3 Screenshot showing the first page of parameterisation. The last selection field *Auswertungseinheit* (reference domain) is not yet filled with data because nothing has been selected in the field *Stichprobennetz* (sampling grid)

vitality), or based on a full-text search with real-time suggestions. The search also supports the use of synonyms.

These entry points act as a filter to already reduce the number of possible results. Additional filters can be set on the following page to further limit the result set (Fig. 20.6). Each individual filter setting comes with its own description to facilitate choosing the right one. The filter and the accompanying description are generated directly from the derivation metadata in the database. At the end of the process, a list of results is displayed. Each item in the list leads either to a table with additional meta-information or an interactive map, or it can be added to the shopping cart. The tables and maps can be exported to various formats or printed out.

NFI4

basal area
altitudinal vegetation zone

unit of reference: production region
unit: m²/ha
unit of evaluation: accessible forest without shrub forest
grid: 1.4 x 1.4 km grid
state 2009/17

	production region											
	Jura		Plateau		Pre-Alps		Alps		Southern Alps		Switzerland	
altitudinal vegetation zone	m²/ha	± %	m²/ha	± %	m²/ha	± %	m²/ha	± %	m²/ha	± %	m²/ha	± %
alpine/nival	-	-	-	-	31.3	20	25.0	30	12.7	50	24.8	19
upper subalpine	-	-	-	-	-	-	27.3	5	22.6	8	26.0	4
lower subalpine	27.0	6	29.2	13	31.2	5	31.6	2	27.7	6	30.7	2
upper montane	35.6	3	34.5	13	36.7	3	31.7	3	28.7	4	33.4	2
lower montane	33.8	3	31.9	3	38.9	3	31.2	4	29.4	7	33.6	2
colline/submontane	28.7	3	28.1	2	37.4	5	25.9	5	29.4	5	28.8	2
total	31.8	2	29.4	2	35.9	2	30.5	1	28.0	2	31.2	1

© WSL, Swiss National Forest Inventory, 02.08.2018

Fig. 20.4 Basal area table available on the public website. The table header, footer and column header contain basic information about the type of derivation parameters used

Result tables available on the public website www.lfi.ch are not calculated in real-time by the DAA. They are served from the database, where they were stored directly as HTML files together with a reference to the parameter set used to create the result table. Storing pre-calculated results has two main advantages: first, loading results directly from the database is much faster than producing results on demand. Calculating can take up to 20 s or even longer, depending on the type of analysis and selected parameters, whereas simply displaying takes less than half a second. We decided that waiting times of more than 1 or 2 s are not acceptable for our public web users.

Second, underlying data, models and methods may be updated at any time. Thus real-time analysis is not guaranteed to produce exactly the same results when repeated. Storing the calculated result instead of the parameter set allows us to create snapshots of results and control their lifetime on the public website.

Each result table is stored, as mentioned above, along with a reference to the original parameter set, which in turn references the underlying derivations. Thus, each published result can be fully traced back to the raw data, and all the metadata stored alongside can be displayed on the website. Previous (legacy) results which are no longer used can, however, be decoupled from the original parameter set and moved to so-called 'collections' for archiving purposes to prevent having to keep referenced derivations and other parameter data permanently. They can then still be accessed on the public website via a simple search function.

Currently, more than 57,000 tables, each in four languages, are publicly available on the NFI website. A further 3600 results are available for registered users only. The result tables were created with one batch process per language (Sect. 20.2.3).

Table 20.3 Main NAFIDAS parameters

Parameter	Description
Type of analysis	Defines type of analysis (state, change or change components) and controls computational processing steps
Inventory	Defines the inventory cycle(s) to be evaluated
Target variable(s)	The target variable is defined as a combination of derivations and inventory specific algorithms. Information about target variables is stored in the respective concept in the database. The DAA reads this metadata during runtime. The target variable consists of one or several independent derivations (e.g. basal area)
Reference domain	The reference domain restricts the analysis to a certain type of forest domain. The restrictions are threefold and refer to:
	1. Accessibility
	2. The domain (of interest), such as forest, shrub forest, young growth to be considered or excluded for analysis
	3. Commonality, e.g. restriction to common forest, i.e. to plots which were forest in several NFIs
	Example: common accessible forest without shrub forest in NFI4 and NFI3
	The area of a reference domain unlike a reference unit is unknown and must be estimated by statistical methods
Sampling grid	The sampling grid restricts the analysis to a regular (rectangular) geometrically determined set of sample points in a horizontal projection within a region, identified by a unique pair of abstract coordinates. Sampling grids may differ in density and geometrical layout. Their definition has no dependency on an inventory!
	Defining regions and sampling grids as immutable is strongly recommended since fixed region perimeters and common sampling grids are necessary in change analysis for statistical reasons. Example: The 1.4 km × 1.4 km grid
Classification unit (CU)	Up to three classification units may be selected which determine the cells of the result table and hence the granularity of statistical results. Metric units are classified into categories according to user or automatically by the DAA
Reference unit	Syn. area domain of interest. The reference unit is a geographically defined extent with clearly demarcated borders and known area, such as the whole of Switzerland, a canton or any administrative district. It is not necessary for the reference unit to be continuous as long as the allocation of sample points is unambiguous and the area of each sub-unit is exactly known. Example: Switzerland maybe divided into disjunctive sub-units such as production regions

20.2.3 System Architecture

General The internal web application architecture is designed as a client-server system executed on two tiers: one part of the application runs on the web server, the other in the web browser on the client. Users do not need to install any software apart from an up-to-date Firefox web browser[1]. The program logic on the server is mainly

[1]To limit the amount of programming and testing for using the internal web application, only the Firefox Browser on the desktop is supported.

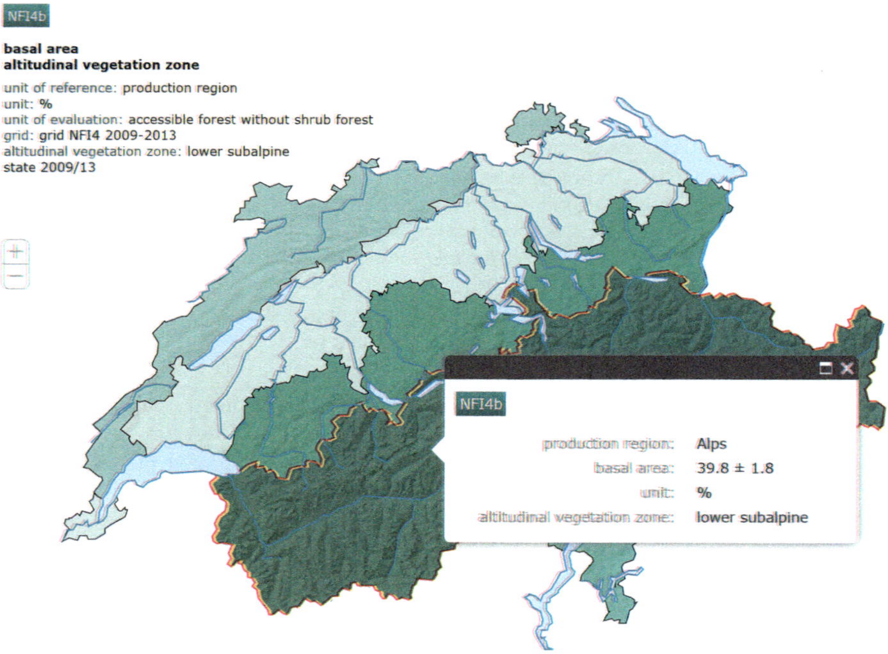

Fig. 20.5 Results displayed as an interactive map

used to access the database to read, store and update data, as well as to handle user permits and communication with the DAA. The web browser on the client creates the user interface and handles user input. The design principle *separation of concerns* (Laplante 2007) is strictly followed. Only the database is used for permanent storage and only the DAA to calculate results. The source code is managed with the version control system Apache Subversion (SVN).

The architecture of the web application mostly follows REST principles (Fielding 2000), but not very strictly. For example, it makes use of cookies.

Client-Side Programming Languages used are JavaScript, HTML, SVG and CSS relying on web standards only, JavaScript files are managed with the AMD format (AMD, https://github.com/amdjs/amdjs-api/wiki/AMD, James Burke 2014). To minimise dependencies, the only JavaScript libraries used are: prismjs (prismjs.com) for code highlighting, CodeMirror (https://codemirror.net/) for code editing, and ArcGIS JavaScript API for the mapping application. In addition, the dojotoolkit (https://dojotoolkit.org/) mainly serves as an AMD loader when code is shared with the public web application to ensure compatibility with older web browsers.

Server-Side Programming An Apache 2 web server is used together with PHP 5.3. The code follows the PHP Standards Recommendations (PSR) and fully complies with the auto-loading Standard (PSR-4) and the Coding Style Guide (PSR-2). The Basic Coding Standard (PSR-1) is loosely followed. The external libraries used are: the Composer Autoloader, Firebase's PHP implementation of the JSON Web Token Standard (JWT) https://tools.ietf.org/html/rfc7519 and the portable PHP password hashing framework phpass. (http://www.php-fig.org/ https://getcomposer.org/, http://www.openwall.com/phpass/).

Performance In NAFIDAS, some resources are requested from the client (web browser) by a user simply following links as well as by sending forms. However, to provide the user interface with a more seamless desktop-like quality, many resources are loaded or updated in the background by making direct calls to the server using the JavaScript Fetch or EventSource API. This improves responsiveness by only updating a part of the user interface without reloading the whole page and thereby losing the application status. Data between the frontend and backend is generally passed on in JSON format, and sometimes directly in HTML format as well. Parameter sets are stored in XML format. To save and to visualise the spatial context of the several thousand result tables stored, a spatial extension is used on the database side. The saved spatial data sets are published as web services through a geometry server. The web map application showing the geographical context of the results consumes the prepared services using the http protocol and REST interfaces.

Batch Processing of Result Tables The batch process is implemented by using the EventSource API for communication between the browser and server, while a socket connection is maintained between PHP and the DAA for reading and writing data using an event loop. This allows for real-time updating of the calculation status and concurrent, non-blocking exchanges of data between PHP and the DAA. To prevent PHP session timeouts during a process, that might run for days. JSON web tokens are used instead of PHP session variables for user access control and to maintain the application status.

The parameters selected during parameterisation are stored temporarily in a PHP session variable as serialised XML. They are de-serialised into the DOM for processing. The set of collected parameters is validated with an XML schema, which is broken down into three subparts to validate each page of the parameterisation. Before posting the parameters, the XML document is serialised again and sent as a simple HTTP POST with key value pairs to the DAA for further processing. The XML Schema is also used to automatically create the initial XML to store the parameter data in.

DAA The DAA is an additional analysis tier embedded in the classic three-tier architecture of NAFIDAS. It is embedded in the internal web application of NAFIDAS (Sect. 20.2.1 and Fig. 20.7).

The DAA can be directly launched from the parameter interface of the internal web application (after finalisation of the parameter selection). After DAA invocation, the parameters are transferred to the DAA via the Apache web server. The DAA processes the data analysis and sends the result tables in HTML format back to the web application. The program runs and terminates autonomously, no user-interaction is possible once it is started.

Special Aspects of the Public Web Application This application uses the same architecture and programming languages as the internal application. However, security issues are of much more concern, especially those compromising the database by SQL injection. It runs in a less controlled environment where the application has to support different browsers from different generations and with

Table 20.4 Public web application – Features available to the user and functionality required by the system to provide them

Feature	Functions required
Result tables	Maintain permalinks and DOIs
	Provide meta-information about parameters used
	Convert to different formats for export
	Shopping basket to display a set of result tables
Map visualisations of result tables	Provide geodata and map services
	Provide token service to access map services
	Expand with thematic data
	Convert to SVG or CSV
	Create interactive graphical representation dynamically
Query results (filter and search)	Materialise results in database (calculated) for fast querying
	Provide data model for filtering according to product, region, inventory and theme
	Register results according to criteria and indicator
	Maintain indices for suggesting search- and keywords
	Provide semantic search service (synonyms)
	Maintain hierarchy (data structure between parameters)
Complete multilingual content	Provide labelling (internationalisation) system
	Provide language resolver and redirect
Restricted user area (Cantons)	User administration and permissions
	Access control
	Register public and non-public (private) results
Data catalogue (with history)	Query database for variables used in results only
	Compare changes in variables between inventories
	Provide links to field manual (pdf) and historical catalogue
Collections of legacy result tables	Decouple legacy data from live data

different operating systems. To mitigate the cross-browser development effort, the dojotoolkit is used for most JavaScript programming.

The whole query and filtering system is based on a materialised view,[2] where the XML of the parameter data is normalised into table columns for fast querying. This materialised view is updated by the database once a day. The user interface is implemented in such a way that the search engine crawlers can simply follow links to index all result tables. In contrast to the result tables, maps are constructed on-the-fly from different REST endpoints. Spatial services are provided by geometry data servers, whereas thematic data is served by parsing the HTML of the result tables stored in the database and returning it as JSON.

Digital Object Identifiers, DOIs, for each result table are exposed over a REST endpoint that fully implements the Open Archives Initiative Protocol for Metadata

[2](Wikipedia 2018).

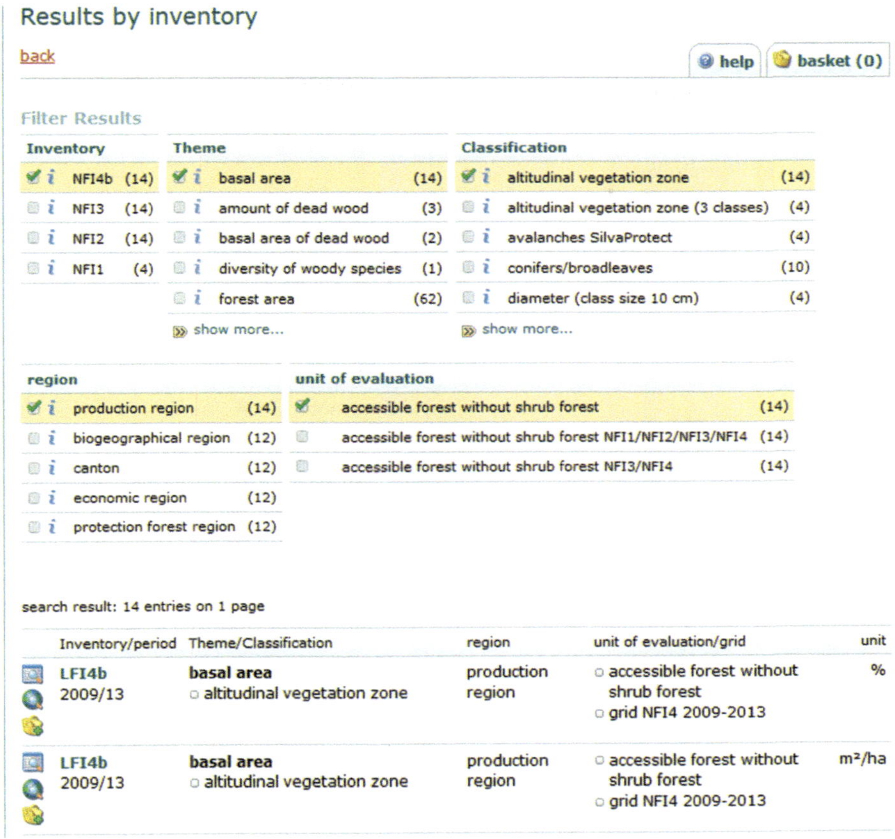

Fig. 20.6 Screenshot of the query system on the public website www.lfi.ch

Harvesting (OAI-PMH). Dublin Core Simple is used as the metadata container format. The abstract PHP library Picturae\Oai-Pmh was used to create the repository endpoint (http://www.dublincore.org/, http://www.openarchive.org/, https://github.com/picturae/OaiPmh).

20.3 Database

20.3.1 Raw and Derivation Database

Raw Database Data is collected in the field with the field application MAIRA, an application specifically developed and designed for the NFI (Sect. 23.2). The acquisition of over 280 attributes in the field is controlled by a highly formalised protocol described in Chap. 9. After recording a full plot, the data is uploaded to the

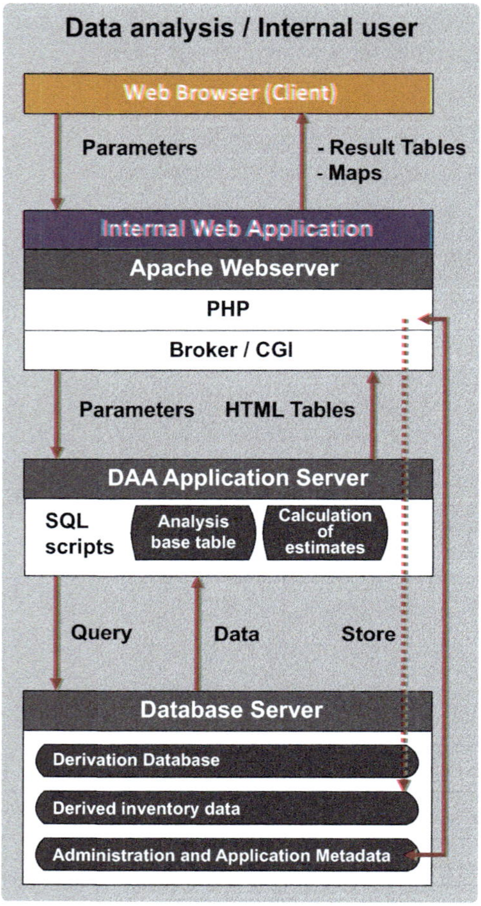

Fig. 20.7 Multi-tier client-server architecture of NAFIDAS with a presentation tier, logic tier and data tier. The user access to the system is established by the web client (web browser)

central database. To reduce the number of field plots having to be visited, an aerial-image interpretation process is applied (Sect. 6.3). The data coming out of it is also saved as raw data. Furthermore, several spatial vector data sets are stored in the raw database. They originate not only from earlier NFI-projects, but also from other sources, such as the Swiss federal government or cantons. Other tables with various data sources are also administered and documented in the database, e.g. tariff coefficients, conversion factors and vegetation models.

All data stored in the raw database is used in further processes predominantly for generating derivation data. Since these processes expect raw data to be immutable, no modifications or updates are admitted at a later stage.

Derivation Database Unlike in the raw database, all algorithmic calculations, manipulations, corrections, imputations and data enhancements are executed and documented in the derivation database. The concepts and terms listed in Table 20.1 provide the basis for the relational model of the derivation database. Since most entity types depend on others, the implicit hierarchy is made explicit by building up foreign key relationships between tables. Apart from the concept *variable*, all entity types inherit from the concept inventory. Furthermore, the central concept *derivation* drives the process of generating derived data in the analysis tables. At the same time, it serves as a reference for the parametrisation concepts described in Table 20.3 and depicted in Fig. 20.8.

All metadata about inventories, variables, analysis tables, derivations with their derivation algorithms, analysis tables and further structures and concepts is formalised and stored in the database as well. When combined with several other applications, this leads to a high degree of automation and transparency. All dependent applications from the NAFIDAS web interface, the DAA, and the Massimo application (Chap. 17) rely on the interfaces and the semantic specification provided by the derivation database.

20.3.2 Analysis Tables

The analysis tables contain derived data and reflect semantic concepts like *recorded plots*, *recorded trees* or *geographical location* (Fig. 20.9). The records (entities) are structured based on location and timeframe (inventory) and hence always have a unique primary key (PK) consisting of:

1. Location ID, which identifies the geographical location (represented by x- and y-coordinates) and
2. Inventory ID, i.e. a unique time period for the analysis. In addition,
3. Key columns, such as tree ID to identify objects in the table as recorded trees, may be included. Each tree keeps its tree ID throughout later inventories as long as it can be re-identified in the field.

The records contain further entity attributes, such as the dbh of recorded trees stored to (derived) variables. Table 20.5 shows the semantic and structural aspects of variables, which are the basis for the columns in the analysis tables defined by variables. These columns are filled with data from derivations in a downstream process.

Since a derivation specifies a variable in its inventory context, derivations particularly important for data analysis. Its underlying semantic meaning must be concise and applicable to all trees that could be part of the analysis table. A derivation contains an inventory specific algorithm, which is executed to update the data on the respective column in the analysis table. Only records of the corresponding inventory are treated in the calculation. Derivation metadata covers information such as the inventory ID, variable name (available in four languages), scale (interval, discrete), name of lookup table in case of discrete scale, time of calculation, validity of the algorithm and the level in the hierarchy based on dependencies in the calculation algorithm (Sect. 20.3.3).

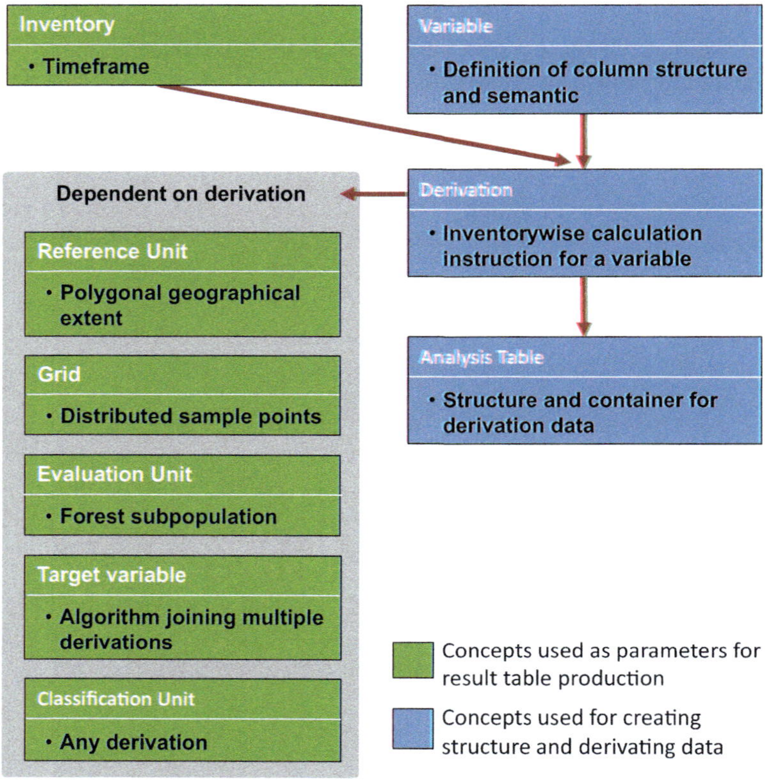

Fig. 20.8 Hierarchy and brief descriptions of the main concepts implemented in the derivation database (see also Table 20.1 and Table 20.2)

20.3.3 Processing Variables and Derivations

The complete processing of the variables and derivations in the analysis tables consists of three steps as depicted in Fig. 20.9.

1. Adding a column to a previously defined analysis table with information about a variable's metadata. This information is then used automatically in a parametrised database script. For example, column S is added to the inventory tree table to show integer values. The variable is semantically specified as the tree diameter in centimetres of a monitored tree.
2. Run record-inserting scripts – stored as metadata in the database, i.e. scripts that add all records of a specified inventory to the analysis tables. For example, all tree records of inventory no. 92 are inserted into the inventory tree table.
3. As a final step, this column can be updated with the values calculated for a derivation. A user-defined script calculates the values of a specific column for the defined number of records as described in step 2. For example, column S is

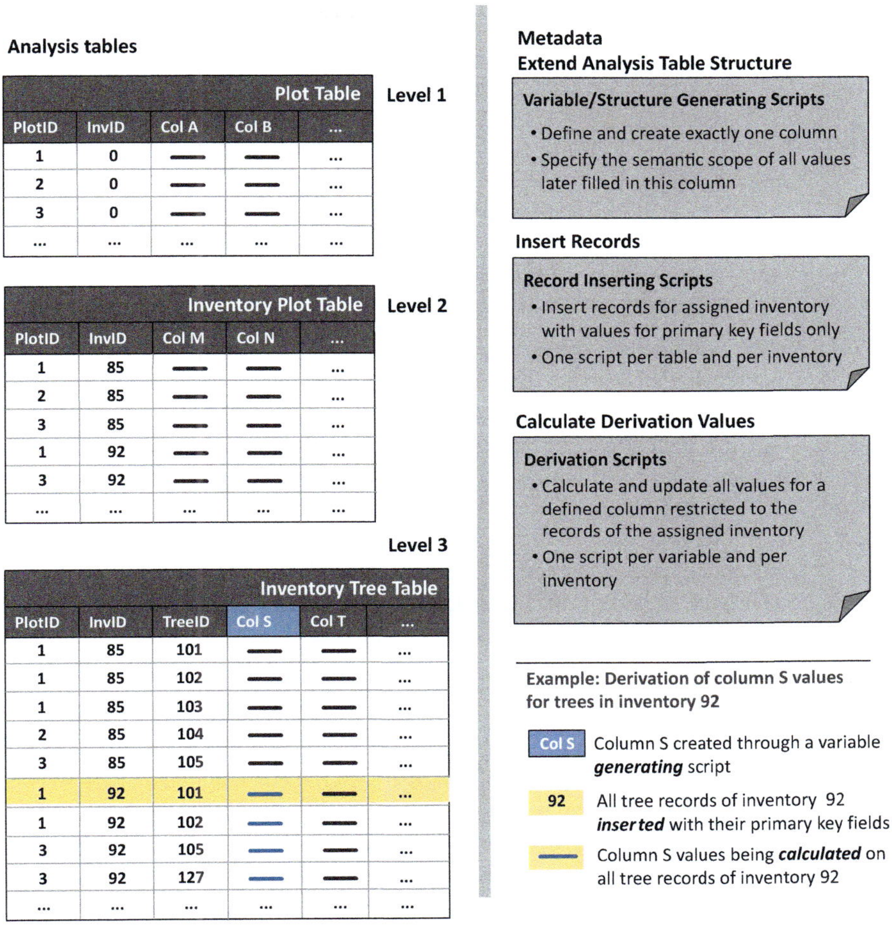

Fig. 20.9 Three main levels of granularity for major NAFIDAS analysis tables. (Adapted from Traub et al. (2017))

updated on all trees of inventory no. 92 using the corresponding script, which contains the procedures to extract source data, transform these values and loads the results in the tree diameter column (column S).

20.3.4 Specific Features

Dependency of Derivations and Recalculation Derivation algorithms recorded in the system refer either to raw data or to other derivations. The dependency vectors are manually saved for each derivation using a web application. As the algorithms and the calculated data are deterministic, a dependency tree can be built. This tree with its nodes (derivations) always represents a directed acyclic graph (DAG)

Table 20.5 The two aspects of a variable

Semantic concept	Structural concept
Defines the semantics of the data to be added to this column, such as the definition of dbh	A column, i.e. a structural feature of an analysis table
Metadata: description of semantic meaning (short and long versions)	Metadata: database column name with storage precision, unit, scale (metric, discrete) and affiliation to analysis table

(Thulasiraman and Swamy 1992), which allows an automatic recalculation of all algorithms to update the derived data.

Long-Term Availability Derivation algorithms and the resulting derived data are updated and recalculated several times a year. An algorithm to freeze certain states of data and metadata was therefore developed. All analysis tables are archived automatically in the database once a month. Additionally, the derivation algorithms, parameter sets, saved result tables and most relevant metadata are archived continuously. This highly formalised process allows these different states of derived data to be used as immutable input such as for MASSIMO (Chap. 17). All NFI inventories are subject to continued maintenance. Even NFI1 data collected in 1983 can be accessed using the same algorithmic structures, derivations and statistical methods as those developed for the most recent inventory Project.

Multilingual Translations of database entities used in result tables, such as names of derivations and target variables, are stored as metadata alongside the data in the database. Hence result tables can be automatically calculated in any of the available languages (German, French, Italian, English). All measurements and units are European metric regardless of the language chosen. The NAFIDAS web application is available in German only, whereas documentation is mostly written in English. The NFI glossary and dictionary on the public website is multilingual, but is not yet driven by the database.

Executing Derivations Deriving data is usually started on the top level of the dependency tree, where the nodes depend only on raw data. As soon as all appropriate derivation nodes have been calculated, the next child level can be tackled. This process continues from top to bottom until no more nodes are available. Note that all derivations on the same level can be calculated independently, which will allow for parallelisation in future versions of NAFIDAS. Since derivations in inventories may depend on derivations from other inventories, a semantically defined group of inventories is usually selected for recalculation. The process of derivation calculation can be characterised as follows:

- Prerequisites:
- Dependency vectors for all derivations are available
- The dependency vectors form a valid acyclic tree
- Inventories are selected for inclusion

- Levels are processed serially, because lower levels in the hierarchy must be processed after higher levels. Derivations consuming a lot of runtime may be excluded from recalculation if neither the algorithm nor the input data has been changed (e.g. complex algorithms using raw GIS data).
- The algorithms of derivations may theoretically be written in any programming language. NAFIDAS predominantly uses SQL. Error-handling routines are enabled to stop the calculation process should errors occur.
- For each derivation metadata is provided, e.g. time and date of last recalculation.

20.3.5 Quality Assurance and Quality Control

Data quality is highly correlated with risk management (McDowall 2005; Sect. 21.2). The resources spent on avoiding risks must, however, be balanced against the risk's impact; efforts to mitigate risks should focus on those risks with high priority. Computerised system validation (CSV) is an effective measure to mitigate and control risks in a system such as NAFIDAS. CSV focuses on the specification and testing of the computerised system itself (McDowall 2005). The concepts of good programming practice and the four eyes principle are applied. Derivation programming in particular and validation are supported by various internal web application tools to access, compare and visualise data or metadata. In addition, it is crucial to validate the processes needed to securely run and maintain the system itself, which includes their documentation and communication to users. Examples of such processes and security features are:

- Inspection of large sets of result tables calculated at different points in time to display differences, e.g. before and after a full recalculation of derivations
- Dependency trees (DAGs) of derivations allow users to trace back calculations from the resulting data to the initial raw data and corresponding metadata.
- Tasks related to user administration or automated table creation are restricted to the administrator to protect the database structure and consistency.

20.4 Data Analysis Application

20.4.1 General Approach of Estimate Calculation

The DAA specifications cover the features *analysis and result tables*, as described in Table 20.2. Using the methods described in Chap. 2, the DAA calculates reproducible and consistent estimates, such as basal area per ha of the Swiss forest, and their corresponding standard errors. The final output is an html file, which serves as the basis to publish tables and maps on the NFI website (Figs. 20.4 and 20.5). The generic architecture of the DAA supports high quality standards and the requirements necessary for ensuring maintainability, extensibility, and scalability. The executable code is created from generic modules, controlled directly by parameters and indirectly by metadata. Data processing is exclusively based on SQL.

Once invoked, the DAA reads metadata and data from the derivation database and processes the calculation of estimates controlled by parameters posted to the DAA (Fig. 20.7). The analysis steps are based on the two-phase sampling for stratification method described in Chap. 2. The internally implemented estimator algorithms apply to state and change analyses in the same way. A complete list of all equations applied is provided in section "Equations for Point Estimators and Standard Errors" in Appendix.

The parameter *reference unit* derived from GIS data (Table 20.3 and Table 20.7) defines the sampling frame F of size λ_F, which is divided into 1-L non-overlapping regions that form post-strata[3] k of surface area λ_k. $\lambda_F = \sum_{k=1}^{L} \lambda_k$ and L is determined by the number of sub-regions of the reference unit. Within each of these strata k, two-phase (double) sampling for stratification is applied to estimate the means per stratum (Y_k) and the ratios of the two means per stratum ($R_k = Y_{X.k} Y_{Z.k}^{-1}$).

Two-phase sampling for stratification requires assigning sample plots to first-phase strata e and second-phase strata k, and calculating stratum size ($\dot{n}_{k.e} \ddot{n}_{k.e}$), which is carried out during the DAA's runtime.

First-Phase Sampling The first-phase auxiliary variable is taken from the sample $\dot{S}$ of $\dot{n}$ systematically distributed sample plots on a 100 m × 100 m grid (Sect. 2.4.4 and Chap. 7). $\dot{S}$ is partitioned into a set of 1-H disjoint first-phase strata e and the number of sample plots falling (by chance) into stratum e is denoted by $\dot{n}_e$. The exact size (surface area) of λ_e is not known, but $\dot{n}_e \dot{n}^{-1}$ provides unbiased estimates of stratum size λ_e. The sampling error, which arises in estimating the stratum size, is considered in the algorithms of the standard error of ratios and totals accordingly.

The number of first-phase strata H is determined by the first-phase stratification algorithm. The stratification model was redeveloped for NFI4 (Pulkkinen et al. 2018). The model is generally based on: (a) a 'forest/non-forest' classification derived from a forest cover map of Switzerland, and (b) information on vegetation height derived from a vegetation height model (Ginzler and Hobi 2015), which splits the forest stratum into five additional strata. The stratification process, i.e. the allocation of sample plots to strata, is applied to each sample plot of $\dot{S}$ on the 100 m × 100 m grid. The sample plots are allocated to strata simply according to where the sampling grid intersects with the perimeter of the forest cover map, and thus directly allows $\ddot{n}_{e.k}$ and $\dot{n}_{e.k}$ to be determined. A critical sample size of 10 terrestrial sample plots per first-phase stratum ($\ddot{n}_{e.k}$) was determined. Small sample sizes typically occur if second-phase strata are derived from forest districts. In this case, the first-phase stratification must be repeated for the critical districts by stepwise application of more simplified models meeting the condition $\ddot{n}_{e.k} \geq 10$.

[3]Splitting into subunits is carried out only during analysis, i.e. after sampling.

Second-Phase Sampling The terrestrial field sample $\ddot{S}$ of $\ddot{n}$ systematically distributed sample plots is based on a 1.4 km × 1.4 km sub-grid of $\dot{S}$ ($\ddot{S} \subseteq \dot{S}$ and $\ddot{n} \leq \dot{n}$). The basic attributes (y_i) of the target variables are measured on these sample plots.

Simple random sampling is assumed to be applied in both phases. The equations of the subsequent sections refer to section "Equations for Point Estimators and Standard Errors" in Appendix and reflect this assumption and are exhaustive and valid for all combinations of base parameters, including target variables, classification units, inventory and sampling grids. This allows for completely generic implementation of algorithms. The computational process for calculating estimates is built on a stepwise aggregation of the target variable values, calculated by the algorithm of each target variable.

20.4.1.1 Aggregation Steps of Target Variable

1. **Local density values on plot level (j)** This step is applied for all target variables that describe elements with multiple occurrences/observations per plot, such as trees. The single subject values y_i are aggregated into local density values on the plot level y_j. Usually the target variable algorithm combines derivations, such as single-tree basal area, with an extrapolation factor. For target variables of type area (one value per plot), the constant local density value $y_j = 1$ is assigned to each plot in the reference domain.
2. **Mean local density on level first-phase strata (*e*)** The mean local densities in the first-phase strata *e* within the second-phase stratum *k* are calculated according to Eq. 20.99. The sample size $\ddot{n}$ is derived from the entire set of plots in post-stratum k, which is exclusively determined by the sampling grid. The mean local density in stratum *e* is the basis for calculating any total or a ratio estimator.
3. **Summarising mean local densities on level second-phase strata (*k*)** Values are calculated according to Eq. 20.98. The stratum allocation and stratum weights $\dot{n}$ are derived from the first-phase sample.
4. **Summarising mean local densities to results for the whole country** Estimates are derived from aggregated (area weighted) values on level k according to Eq. 20.96.

The total $\hat{T}$of any target variable Y is based on its estimated global mean $\hat{Y}$ multiplied by the surface area of the second-phase strata λ_k, and is calculated by: $\hat{T} = \sum_{k=1}^{L} \lambda_k \hat{Y}_k$ (Eq. 20.96 modified). The variance of $\hat{T}$ is calculated by: $\widehat{var}\langle\hat{T}\rangle = \sum_{k=1}^{L} \lambda_k^2 \widehat{var} \langle\hat{Y}_k\rangle$ (Eq. 20.97 modified).

20.4.1.2 Estimating Ratios

The estimates are calculated by $\hat{R} = \hat{Y}_X \hat{Y}_Z^{-1}$ (Eqs. 20.103, 104a, 104b), covering the initial sampling strata *k* and strata *e* within stratum *k*. The variance of $\hat{R}$ is calculated using Eq. 20.105, which is based on the variance of $\hat{U}_k$ (Eq. 20.110) and on the variate $u_j = x_j - z_j \hat{R}_k$ (Eq. 20.114). If the second-phase post-strata is used as a CU such as production regions by definition only one stratum exists per CU category such as 'Plateau'. In this case $L = 1$ and $k = 'Plateau'$, i.e. only $\hat{Y}_{X.Plateau}$ and $\hat{Y}_{Z.Plateau}$ values are considered for stratified estimation, and all other strata values are set to zero by means of indicator variables (Sect. 20.4.2). Furthermore, Eq. 104b reduces to $\lambda_k \hat{Y}_{X.k} \lambda_k \hat{Y}_{Z.k}^{-1} = \hat{Y}_{X.k} \hat{Y}_{Z.k}^{-1} = \hat{R}_k = \hat{R}_{Plateau}$, i.e. the correct ratio estimator is created automatically for any type of domain estimation (Sect. 20.4.2) and the exclusive use of $\hat{R}$, also for variance calculation, is correct.

The combined ratio estimator works the same way for all types of ratios, but certain types of ratio estimates must be distinguished in terms of data processing.

For all target variables on the sub-plot level, such as basal area, distinctions between two implicit and one explicit selection case of target variable 2 must be made:

1. **Implicit** % analysis, whereby the target variable 1 is divided by the sum of target variable 1 over columns, rows or the entire result table and then multiplied by 100 [%].
2. **Implicit** per ha ratio estimation. Target variable 2 refers to the (estimated) area of the reference domain (Table 20.6). The principle of basal area per ha calculation can be derived from Fig. 20.10. In the example, a region with area $\lambda_F[ha]$ and $n = 48$ sample plots is shown, each having a local density of basal area (Yxj) with unit $[m^2]/[ha]$. The means of $\hat{Y}$ are calculated as $\hat{Y}x = \sum_1^{48} iYxj/n = 51/48\ [m^2]/[ha]$ and $\hat{Y}z = \sum_1^{48} iYzj/n = 16/48\ []/[]$, where $i = 1$ if the sample plot is part of the reference domain (sample plot lies within the green patch) and $i = 0$ otherwise. The totals are calculated as $\hat{T}x = \lambda\hat{Y}x$ with unit $[m^2]$ and $\hat{T}z = \lambda\hat{Y}z$ with unit $[ha]$. The ratio estimator is calculated as: $\hat{R} = \hat{Y}x/\hat{Y}z = \hat{T}x/\hat{T}z = \lambda 51n/\lambda 16n = 51m^2/16ha$.

Table 20.6 Types of domain as used in the DAA

Type of domain/description	Area known	Known at level	Included in result tables
Reference domain (Sect. 2.8.1): all plots assigned to a certain type of forest, such as accessible forest without shrub forest Determined by DAA parameter reference domain	No	Plot	No
Domains (Sect. 2.5.3):			
1. The surface area is known (area domain of interest) such as production regions. Determined by DAA parameter reference unit	Yes	Plot	Optional
2. The surface area is unknown (such as altitudinal vegetation zones). Determined by DAA parameters CU1–CU3	No	Plot	Yes
Subpopulation (Sect. 2.3.3): a subset of trees to be analysed, such as broadleaved trees determined by DAA parameters CU1–CU3	No	Tree	Yes

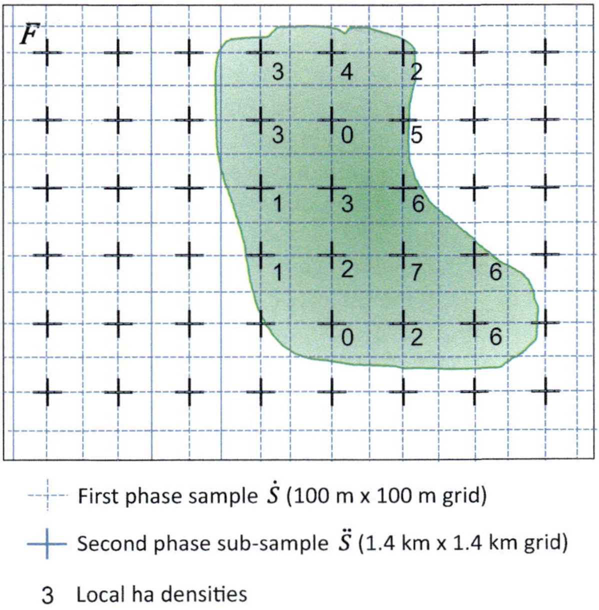

Fig. 20.10 General principle of the two-phase sampling design of the NFI. The green patch defines the reference domain. The local ha densities of sample plots outside this unit are zero (0). F denotes the (known) area of the entire region. The values within the reference domain denote fictive ha densities and must be understood as based on the single tree basal area multiplied by an extrapolation factor

3. **Explicit** ratio estimation. Target variable 2 is selected by the user from the associated analysis table of target variable 1 or a table that has a less granular primary key structure. For example, if target variable 1 is the basal area, the associated analysis table is the tree table. Valid choices for target variable 2 would be *number of stems* (also associated with the tree table) or *forest area* (associated with the plot table).

20.4.1.3 Estimating Domains

The values of result table cells and margins (Fig. 20.4) are built by estimates of domains and subpopulations. The statistical principles of domain estimation are explained in Sect. 2.5.3. The calculation process is mainly controlled by the (non-overlapping) categories of the selected CUs which define the single domains and subpopulations. The domain terminology applied in the DAA is explained in Table 20.6.

In successive inventories, plot or tree attributes may shift between the two inventory occasions from one to another domain, i.e. a subject is allocated to different domains in successive inventories. As an example, the value of attribute *ownership* may shift from domain *public* to domain *private*. This affects change estimation and is termed *domain-shift* in the subsequent sections.

20.4.1.4 General Rules

- Each subject (tree, plot,...) is allocated to exactly one domain per CU, e.g. to one altitudinal vegetation zone (Fig. 20.4).
- Result tables usually consist of single domain estimates in the table margins, such as (a) results on altitudinal vegetation zones OR (b) regions; and (c) estimates built by domain intersections in the table cells, such as altitudinal vegetation zone x region (Fig. 20.4).
- Domain estimation is controlled by indicator variables $i_{d1}, i_{d2}, \ldots, i_{dn}$, which are set to $i = 1$ if the subject belongs to a certain domain d_i and $i = 0$ otherwise (Köhl 2001; Sect. 2.5.3).
- CUs on metric scale must be assigned to classes (supported by tools in the parameterisation interface).
- When estimating domains, all domains are handled as if they were CUs and thus are not distinguished into domains with known vs. unknown area in subsequent paragraphs.
- Domain estimation must be conducted independently for $\hat{Y}_X$ and $\hat{Y}_Z$ as the example in Fig. 20.4 illustrates where basal area per ha of forest area is analysed according to the CUs altitudinal vegetation zone AND tree species. Whereas basal area (target variable 1) can be evaluated for all combinations of both CUs (species x altitudinal vegetation zone), forest area (target variable 2) can only be evaluated according to altitudinal vegetation zones because it does not make sense to classify forest area by tree species.

20.4.1.5 Processing Steps in Estimating Domains

1. **Plot level:** If the local density value y_j is assessed at tree level, such as basal area, the y_i values must first be aggregated to plot-level values by $y_j = \sum_{i \in \ddot{S}_{k.e.j}} i_{sub}\, y_i$, where i_{sub} denotes the indicator variable for the subpopulation, such as oak trees, and y_i denotes single tree basal area values multiplied by an extrapolation factor (Sect. 2.3.3 and Fig. 20.11).
2. **Domain level** Mean local density estimates of domains on level first-phase strata ($\hat{Y}_{k.e}$) are calculated by: $\hat{Y}_{k.e} = \ddot{n}_{k.e}^{-1} \sum_{j \in \ddot{S}_{k.e}} i_d\, y_j$, i.e. the value of y_j is considered in the summary term only where $i_d = 1$.

20.4.2 Estimating Change

20.4.2.1 Analysis of Change Components

Change components typically include growth, yield and mortality target variables, and are based on measurements of identical tree individuals on two different occasions. Because the resulting change figures are assigned to one combined inventory ID, the change components are analysed analogously to state estimates. Estimation of domains and handling of domain-shifts is straightforward: as there is only one observation per plot/tree per (change)-inventory, by convention all change figures are allocated to the domain according to the most recent inventory. Following the example in Sect. 20.4.2, an imaginary shift of all plots from *private* to *public* ownership would mean that all change component figures are allocated to the domain *public forest.*

20.4.2.2 Analysis of Change

General Changes in target variables can be calculated between any two state inventories. The user may choose the target variables interactively from a selection list generated by the NAFIDAS web interface according to the two state inventories selected. In NFI4, change estimates are based on paired samples only. Due to the co-variance between the subjects, a lower standard error than with an independent analysis of the results of two inventories is yielded. Paired sample analysis requires the target variables to be defined in the same way in both inventories, selected on a common sampling grid and on a common reference domain such as *common accessible forest without shrub forest in NFI3 and NFI4.* The changes are calculated as differences between the total values of the target variable ($\hat{T}x$ *current inv* – $\hat{T}x$ *previous inv*) during DAA runtime.

The calculation of mean local densities of change is divided into two steps:

1. Local density values for each inventory occasion per sample plot are built
   Aggregation steps:
   (a) All target variable values (local densities) of the first occasion are treated as losses and assigned an inverted sign. All values from the second occasion are treated as gains and keep their sign.
   (b) The change is obtained by summing up all target variable values of the two inventories.
2. The mean change is calculated according to Eq. 20.99.

Any domain-shifts between two inventory occasions are treated following the gains and losses principle; subjects are allocated to the domains as recorded in the particular inventory.

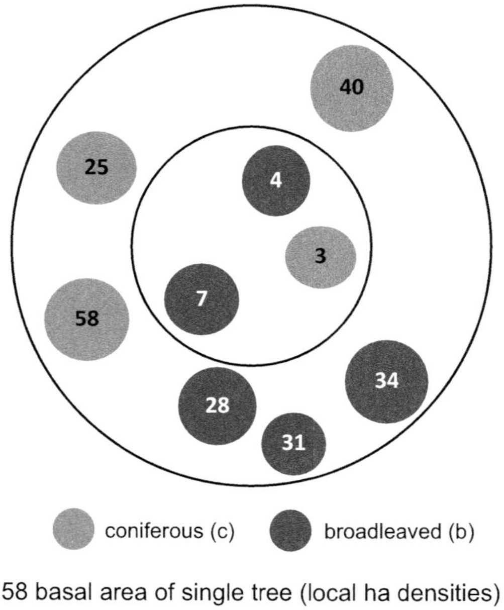

Fig. 20.11 Aggregation of single tree values into local densities on the plot level considering subpopulation estimates. The values are fictive ha densities and must be understood as based on the single tree basal area multiplied by an extrapolation factor of 50 for the small (inner) 0.02 ha circle and 20 for the large (outer) 0.05 ha circle. $i_c = 1$ for coniferous trees and zero otherwise, $i_b = 1$ for broadleaved trees and zero otherwise. The ha densities are calculated as $y_{jc} = \sum_{i=1}^{4} i_c Y_i = 126m^2/ha$ and $y_{jb} = \sum_{i=1}^{5} i_b Y_i = 104m^2/ha$ for coniferous and broadleaved trees, respectively

Analysis of changes in forest area To obtain meaningful results, the target variable algorithms and reference domains must be combined appropriately. Two main types of analysis are available:

1. **Analysis of relative changes:** this analysis is based on reference domains including a commonality definition such as *common accessible forest*. It uses a dummy target variable (value =1) in combination with a CU, such as the plot attribute *ownership*, which may change between inventories. The results are based just on the numbers of plots per domain in the previous and current inventories and enable direct analysis of the domain-shift. The balanced total is always zero due to the commonality definition.
2. **Analysis of absolute change:** here the target variable is the reference domain itself assessed on the entire grid including 'non-forest' plots. It provides absolute change figures in combination with any CU.

20.4.2.3 Changes Per Year

The equations applied are based on the methods described in Sect. 2.8.3. Change and change component results may be calculated either over the entire inventory period between two occasions (periodical change), or as annual[4] figures (change per number of vegetation periods).

Any comparisons between NFI3 and NFI4, in particular, should generally be provided as annual figures since the number of vegetation periods (Δ_v) between the times of recording may vary greatly because in NFI4 fieldwork was changed from periodic to permanent (Sect. 2.4.3). For the calculation of annual changes, the periodic change results must be divided by Δ_v. The mean number of vegetation periods a in domain d is calculated by: $a_d = \sum_{k=1}^{L} \lambda_k \ddot{n}_k^{-1} \sum_{j=1}^{n_k} i_{d.j} \Delta_{v.j} \left(\sum_{k=1}^{L} \lambda_k \ddot{n}_k^{-1} \sum_{j=1}^{n_k} i_{d.j}\right)^{-1.}$
where λ_k is the surface area, $\ddot{n}_k$ the number of terrestrial sample plots in stratum k and $i_{d.j}$ the domain allocation of plot j. Note that the mean number of vegetation periods is calculated per domain, which yields more realistic (smoothed) results than deriving the mean annual values directly from the local densities per plot ($\sum y_j \Delta_{v.j}^{-1}$).

Any total of Y_x or Y_z in domain d is denoted by T_d, and the differences in totals between two occasions by Δ_{Td}. The annual change $\Delta_{Td/a}$ is estimated by $\Delta_{Td/a} = \Delta_{Td} a^{-1}$. When calculating annual ratio estimates, a distinction again needs to be made for per-ha, percent and for explicit target variable 2 selection. In particular, the effects arising from the additivity of results should be taken into consideration.

1. **Percent analysis over plot-domain estimation** The annual change values of target variable1 $\Delta_{Td}\, a_{d.i}^{-1}$ are divided by $\sum \Delta_{Td} a_{d.j}^{-1}$, where i denotes the table cell domain and j the domain in the column/row margin of the result table. In this context we have: $\ddot{n}_{d.i} \subseteq \ddot{n}_{d.j}$, $\Delta_{vd.\ i} \neq \Delta_{vd.j}$ and $a_{d.\ i} \neq a_{d.j}$ since both domains represent a different set of sample plots. This means that any annual results cannot be expected to be additive, and hence percent figures will also not be additive, i.e. they will not add up to 100%. The problem can be directly translated into percentages of annual per-ha results or other ratio estimates, such as a growth and yield ratio.
2. **Percent analysis within subpopulations** Additivity is given as long as a common $a_{d.\ j}$ value exists, which is the case if the subpopulation percentage is calculated within a common domain.
3. **Per ha analysis** $Y_x = \Delta_{Td}$ and $Y_z = T_d^{(r)}$ (the total forest surface area in domain d defined by the reference domain). The ratio estimator in this domain is defined by $\Delta_{T/r} = \Delta_T T^{(r)\ -\ 1}$ and the annual change of $\Delta_{T/r}$ by $\Delta_{T/r\ \cdot\ a} = (\Delta_T a^{-1}) T^{(r)\ -\ 1}$. Here the term a is unique and unambiguous because it is only applied to the difference of Y_x.
4. **Explicit selection of target variable 2** If both Y_x and Y_z are determined by change components, such as the ratio between growth and yield, the ratio is

[4]The two inventory occasions under consideration.

built by $\Delta_{Tx \cdot a/Tz \cdot a} = (\Delta_{Tx}a)(\Delta_{Tz}a)^{-1} = \Delta_{Tx}\Delta_{Tz}^{-1}$. Hence division by a is obsolete because the mean of a cancels to 1.

The constant a must also be considered in variance calculation. As the mean of Δ_v does not estimate a population parameter, it is statistically treated the same way as a constant such as λ. Hence, the variance is calculated by: $\mathrm{Var}(\Delta_{T/a}) = \mathrm{Var}\ \Delta_T\ a^{-2}$ (totals), $\mathrm{Var}(\Delta_{T/r \cdot a}) = \mathrm{Var}\ \Delta_{T/r}\ a^{-2}$ (per ha ratio) and $VAR(\Delta_{Tx \cdot az/Tz \cdot ax}) = Var(\Delta_{Tx/Tz})a_x^{-2}a_z^{2}$ (percent ratio).

20.4.2.4 Data Sources and Metadata

The data of the first-phase sample $\dot{S}$ (auxiliary data) is collected from the vegetation height information and the forest cover map (Ginzler and Hobi 2015, Pulkkinen et al. 2018, Waser et al. 2015; Sects. 7.1–7.2).

The data of the second-phase sample $\ddot{S}$ (terrestrial sample) originates from the field survey, the GIS data (area of regions), the road survey and the forest service interview survey. All data from both phases is exclusively retrieved from the derivation database. The DAA never accesses raw data in any processing or data-retrieval step.

Any data-retrieval, data-flow and analysis processing of the DAA is primarily controlled by parameters (Section "Parameterisation of Data Analysis" in Appendix), stored in tuples with their values in a temporary dictionary during runtime. In combination with other metadata, they exhaustively define the type of analysis and the layout of the result tables. In addition to the dictionary, other meta-information, such as names of tables and columns or target variables, are selected dynamically from external metadata. Hence changes in the derivation database or changes in the attributes of the metadata never require an adaptation of DAA modules.

20.4.3 Software Design Principles

In technical terms, the DAA is designed as a generic software application. After invocation, generic elements are transformed into executable code according to defined parameters and metadata. Similarly to the web applications, the generic modules of the DAA follow the design principle *separation of concerns* (Laplante 2007), i.e. tasks are separated into modules and redundant code is avoided as much as possible. The software code is only as generic as needed to enable a good balance between code redundancy and maintainability.

Each module may be invoked either in a controlled mode by handing over keyword parameters or in direct mode without parameters. Three types of modules can be distinguished:

1. **Top-level module:** this module is invoked by the web server and calls all steering modules sequentially, i.e. each steering module is invoked once after the preceding module has terminated.
2. **Steering modules:** contain functions that cover: (a) management of internal metadata and (b) call(s) of operational modules.
3. **Operational modules:** which are called from steering modules. They can be divided into: (a) preparation modules such as for data-retrieval tasks and (b) execution modules such as for estimate calculation. Calls of other operational modules within such modules are not allowed. Operational modules may be executed repeatedly at the time of call, but not at any later stage in the program flow.

To keep maintainability and quality assurance at a high level, certain basic rules are applied to the software design of the modules:

1. The code style is guided by the principles: (a) keep it as simple as possible and (b) use the same language/programming techniques for the same purpose.
2. Data retrieval and manipulation is exclusively based on SQL and restricted to the minimum number of records and derivations needed to create user-defined result tables.
3. Mean or variance terms are never calculated using internal functions provided by mathematical or statistical libraries, but rather by algebraic code, implemented within an SQL SELECT statement. Sample size is always retrieved separately to avoid problems with the handling of NULL and 0 (zero) values. Estimate figures are not truncated or rounded up or down by the DAA, apart from for display in the final result tables.

20.4.4 Data Management

Metadata The initial management module establishes the meta-information base and enhances the metadata dictionary with new and transformed parameters and values.

Rules for handling metadata:

- Metadata should only be manipulated within the module they are created in.
- The dictionary should be enhanced and updated during application flow, but basis parameters are immutable.
- The semantics of the metadata is predefined and should never be changed in other modules or processing steps.

Data Import The import module consists of data queries, built from generic elements. For translation into valid SQL, the code elements are read from the dictionary and/or external metadata and are assembled into several preparation modules. Each module starts by calling a database connection module, continues with the retrieval of metadata from the derivation database and proceeds thereafter with the retrieval of the inventory data. The number of records retrieved is based on the base parameters *inventory*, *sampling grid*, *reference domain* and optionally by additional user-specific restrictions. The retrieved data is stored in temporary tables. These tables constitute an essential component of the DAA information flow and are persistent through the whole DAA run. Data and external metadata retrieval is exclusively conducted in the import module. Immediately before termination, the connection to the derivation database is closed and should not be opened at a later stage in the application flow.

Handling of NULL Values The value of a target variable is zero should there be no data on a sample plot. Thus, statistically correct calculation of estimates requires NULL values to be transformed into zero values in SQL algorithms.

Calculation of Domain Estimates The computational implementation of domain estimation is based on the combination of CU(s) within the first-phase strata of a certain region. The number of categories of the CU determines the number of domains and how they are combined, and thus the structure of the result table. These structural details are analysed in a statistical steering module, which calls the operational modules for estimate calculation in nested loops. In the example in Fig. 20.4, any analysis according to altitudinal vegetation zones (5 categories) and production region (5 categories) results in 5×5 (= 25) domain combinations or table cells plus $5 + 5$ (= 10) separate domain results (table margins), plus the overall value (intersection of the table margin, i.e. the cell furthest to the right and lowest down). Thus, the example finally results in a total of 36 estimates (35 domains +1 overall estimate). All domains, domain combinations and so on are calculated using the same sampling algorithms as those in the estimate calculation modules.

20.5 Discussion

20.5.1 General

NAFIDAS has proven to be a scalable and stable system for storing and analysing NFI data. Security, stability and maintainability, but also efficiency and performance, characterise NAFIDAS. Key features are:

- Separate data storage concept for raw and derived data. The derivation database serves as a single point of truth for both data and applications. Derivation calculation is based on the concept of dependency trees, which is of high relevance for the consistency of the database.
- All DAA processes run completely metadata-controlled without user interaction needed, which ensures statistically correct calculation of estimates. The DAA

parameterisation and the other tools available on the internal web application can be mastered within a few days by any user with a solid knowledge of the NFI's methodology.
- All NAFIDAS components are integrated and interact via the internal web applications based on an Apache web server.
- The public website of NAFIDAS is very intuitive to use. It has gradually developed into a comprehensive information system, which currently provides access to more than 57,000 tables and maps per language. User satisfaction relates to the huge amount of general and specific information about the NFI; Tables and maps can be found quickly thanks to an efficient querying and filtering system.
- Any results from inventories or single annual panels can be generated within a few weeks after the field data collection is completed.

The system also supports **regional** inventories. If they are conceptually based on the NFI methods, it just involves adapting a set of already existing derivation scripts; the database structure and the DAA modules remain unchanged. Thus, regional inventories can benefit greatly from using the NFI's well-established and tested methods. Any reporting to **national** stakeholders, such as Switzerland's Greenhouse Gas Inventory (FOEN 2016), and to **international** protocols, such as the Global Forest Resources Assessment (FAO 2015) and the State of Europe's Forests Ministerial Conference on the Protection of Forests in Europe (FOREST EUROPE 2015), is supported as well, either by executing standard analysis definitions or by querying the derivation database directly. The latter requires, however, in-depth knowledge of the data model and the derivation algorithms.

20.5.2 Quality Assurance

NAFIDAS users can rely on a high level of data quality, which is even more important than the sheer amount of information. QA in NAFIDAS is mainly ensured by making data and processes highly transparent and by clearly defining the interfaces between the components. The process chain, from generating single raw data entities to final forest statistics, is supported by highly automated and metadata-controlled processes in combination with high-performance computer systems. Interactions with humans are kept to a minimum to reduce the risk of failures. The technical specifications of the system are well documented and user requirements are reviewed regularly.

The following measures are particularly important in maintaining NAFIDAS' high data quality:

- Modularising and applying the *separation of concerns* principle (Laplante 2007) on all levels.

- Defining secure user interfaces including secure background processes.
- Supporting users with comprehensive meta-information and allowing the system to check user input, e.g. to prevent any data analysis that could lead to statistically incorrect estimates.
- Ensuring that, should any risks occur, they are quickly detected.
- Applying the four eyes principle, e.g. by independently re-programming statistical analysis algorithms and verifying calculated estimates (plausibility checks).

Although many risks are covered by the applied QA measures, residual risks, mainly connected with the misinterpretation of estimates, may, however, remain. Documenting statistical methods accordingly should help to keep this type of risk to a minimum.

Future improvements in terms of efficiency and risk mitigation should focus on making the process for change requests more formal to avoid misunderstandings and inefficient cyclic discussions between programmers, administrators and users. Implementing a professional test suite for integral unit testing is another example of an important QA requirement to be met in the near future.

20.6 Conclusions and Outlook

NAFIDAS has been continuously improved and extended over the past 10 years. It has developed into a reliable and sustainable information and decision support system and as such has become an indispensable backbone of the NFI. The system covers the complete data-processing chain from field-data collection to the final tables and maps for long-term and large-scale inventory forest statistics. It enables a broad range of users to access the results of the NFI in a timely and transparent manner. With its multilingual query and filtering techniques, the public NFI website provides a wide range of users quick access to a large number of tables and maps.

Further developments of NAFIDAS will focus on both quality and performance, for example analysing how sensitively results react to small changes in a certain derivation. Other examples include methods for making faster calculations through parallelisation, where possible, and for integrating GIS data optimally into the NAFIDAS system. At the same time, keeping the costs and benefits in balance will be given high priority for each measure under consideration.

Two main issues will remain relevant in future, as described by Traub et al. (2017, p. 105): "...(1) What is the proper level of interactivity for certain user groups, and (2) what is the optimal design of the user interface aimed at simplifying matters as far as possible...". In the light of these considerations, the focus in future should be primarily on stability, security and maintainability rather than on extensive revisions, e.g. to allow more interactivity or provide real-time analysis features.

Appendix

Parameterisation of Data Analysis

On the first parameterisation page, the parameters *type of analysis*, *inventory period (s)*, *target variable*, *sampling grid(s)* and *reference domain* are determined by the user (Fig. 20.3). Valid combinations are queried from an administrative database table, which has to be maintained manually.

On the following page, one to three classification units can be selected. Which of these is valid depends not only on the parameters already selected but also, in particular, on the hierarchy of the database tables the target variable is stored in. In some cases, a specific classification unit is also mandatory. Furthermore, if the difference between two inventories is to be analysed, only classification units are considered where the lookup values are the same.

On the third page, the method of double sampling stratification can be selected, as well as the *reference unit*, which depends on the type of analysis and the inventory period. The user can select a second target variable and other optional parameters, such as the number of decimal places. The choice of the second target variable has to follow certain rules because not every kind of ratio makes sense or is even valid (Table 20.7).

Table 20.7 Enhanced information on the DAA parameter (minimum set required). IDs refer to a derivation variable (column in an analysis table). Further information about the parameters based on derivation is given in Table 20.3

Parameter	Description
Basic	
Target variable 1	The algorithm is exhaustive for total estimates. It constitutes the numerator part in ratio estimates
Target variable 2	Optional (with few exceptions); it has generally the same properties as target variable 1, but its algorithm constitutes the explicit denominator in ratio estimates
Restrictions	
Sampling grid	Defines the size of the second-phase sample
Reference domain	Defines the forest domain, such as common accessible forest in NFI3 and NFI4
Restrictions explicit	Allows specific domain(s) or subpopulations to be excluded from analysis

(continued)

Table 20.7 (continued)

Parameter	Description
Control of denominator for estimating ratios (implicit target variable 2 selection)	
Area reference	Defines whether the total or a per-ha ratio estimator is calculated for target variable 1. The per-ha analysis type implies that target variable 2 is identical to the reference domain.
Percent calculation	Reference types: total, row, column and sub-column. If the analysis type is *change*, the change value of target variable 1 is expressed as the percentage of the state value of the previous inventory (target variable 2)
Control of domain estimation	
Classification units (CU)	Basis for domain estimation
Control of statistical analysis	
Reference unit	Determines post-strata by splitting Switzerland or a certain canton into disjoint non-overlapping regions. Special type of CU of which area is known from GIS
Type of first-phase stratification	Determines the model of first-phase stratification (no stratification, forest/non-forest stratification, and additional strata within forest)
Time reference	Determines whether annual or periodical change figures are calculated

Equations for Point Estimators and Standard Errors

All DAA equations and explanations are taken from Lanz (2017, Chap. 7), with double sampling for stratification under simple random sampling, with stratified sampling of first-phase plots.

General Equations

(96) $\hat{Y} = \lambda_F^{-1} \sum_{k=1}^{L} \lambda_k \hat{Y}_k$

(97) $\widehat{var}\langle \hat{Y} \rangle = \lambda_F^{-2} \sum_{k=1}^{L} \lambda_k^2 \widehat{var}\langle \hat{Y}_k \rangle$

(98) $\hat{Y}_k = \dot{n}_k^{-1} \sum_{k.e=1}^{H} \dot{n}_{k.e} \hat{Y}_{k.e}$

(99) $\hat{Y}_{k.e} = \ddot{n}_{k.e}^{-1} \sum_{j \in \ddot{S}_{k.e}} y_j$

(100) $\widehat{var}\langle \hat{Y}_k \rangle = \dot{n}_k^{-1} (\dot{n}_k - 1)^{-1} \left[\sum_{k.e=1}^{k.H} \dot{n}_{k.e} (\dot{n}_{k.e} - 1) \ddot{n}_{k.e}^{-1} \hat{\sigma}_{k.e}^2 \langle y \rangle + \sum_{k.e=1}^{k.H} \dot{n}_{k.e} \left(\hat{Y}_{k.e} - \hat{Y}_k \right)^2 \right]$

(101) $\hat{\sigma}_{k.e}^2 \langle y \rangle = \left(\ddot{n}_{k.e} - 1 \right)^{-1} \sum_{j \in \ddot{S}_{k.e}} \left(y_j - \bar{y}_{k.e} \right)^2$ with $\bar{y}_{k.e} = \hat{Y}_{k.e}$

(103) $\hat{R} = \hat{Y}_X \hat{Y}_Z^{-1}$

(104a) $= \left(\lambda_F^{-1} \sum_{k=1}^{L} \lambda_k \hat{Y}_{X.k} \right) \left(\lambda_F^{-1} \sum_{k=1}^{L} \lambda_k \hat{Y}_{Z.k} \right)^{-1}$

(104b) $= \left(\sum_{k=1}^{L} \lambda_k \hat{Y}_{X.k}\right)\left(\sum_{k=1}^{L} \lambda_k \hat{Y}_{Z.k}\right)^{-1}$ (which corresponds to $\hat{R}_k$ if $L = 1$)

(105) $\widehat{var}\langle\hat{R}\rangle = \hat{Y}_Z^{-2}\lambda_F^{-2}\sum_{k=1}^{L}\lambda_k^2\widehat{var}\langle\hat{U}_k\rangle$

(106) $\hat{Y}_{X.k} = \dot{n}_k^{-1}\sum_{k.e=1}^{k.H}\dot{n}_{k.e}\hat{Y}_{X.k.e}$

(107) $\hat{Y}_{Z.k} = \dot{n}_k^{-1}\sum_{k.e=1}^{k.H}\dot{n}_{k.e}\hat{Y}_{Z.k.e}$

(108) $\hat{Y}_{X.k.e} = \ddot{n}_{k.e}^{-1}\sum_{j\in\ddot{S}_{k.e}} x_j$

(109) $\hat{Y}_{Z.k.e} = \ddot{n}_{k.e}^{-1}\sum_{j\in\ddot{S}_{k.e}} z_j$

(110) $\widehat{var}\langle\hat{U}_k\rangle = \dot{n}_k^{-1}(\dot{n}_k - 1)^{-1}\left[\sum_{k.e=1}^{k.H}\dot{n}_{k.e}(\dot{n}_{k.e} - 1)\ddot{n}_{k.e}^{-1}\hat{\sigma}_{k.e}^2\langle u\rangle + \sum_{k.e=1}^{k.H}\dot{n}_{k.e}\left(\hat{U}_{k.e} - \hat{U}_k\right)^2\right]$

(111) $\hat{U}_{k.e} = \ddot{n}_{k.e}^{-1}\sum_{j\in\ddot{S}_{k.e}} u_j$

(112) $\hat{U}_k = \dot{n}_k^{-1}\sum_{k.e=1}^{k.H}\dot{n}_{k.e}\hat{U}_{k.e}$

(113) $\hat{\sigma}_{k.e}^2\langle u\rangle = \left(\ddot{n}_{k.e} - 1\right)^{-1}\sum_{j\in\ddot{S}_{k.e}}\left(u_j - \bar{u}_{k.e}\right)^2$

(114) $u_j = x_j - z_j\hat{R}_k$

(115) $\bar{u}_{k.e} = \ddot{n}_{k.e}^{-1}\sum_{j\in\ddot{S}_{k.e}} u_j$ with $\bar{u}_{k.e} = \hat{U}_{k.e}$ and, in general, $\bar{u}_{k.e} \neq 0$

Variation Between Sampling Units All variance estimators contain a term $(n-1)^{-1}\sum_{j=1}^{n}\left(u_j - \bar{u}\right)^2$. For the variance of a mean, one finds $u_j = y_j$ and $\bar{u} = \hat{Y}$ (with $\hat{Y} \neq 0$), the sum term of the variance can be written as:

(146) $\sum_{j=1}^{n}\left(u_j - \bar{u}\right)^2 \doteq \sum_{j=1}^{n}\left(y_j - \hat{Y}\right)^2 = \sum_{j=1}^{n} y_j^2 - n^{-1}\left(\sum_{j=1}^{n} y_j\right)^2$

For the variance of a ratio, u_j needs to be pre-calculated at the plot level. Alternatively, the variance terms can be expressed as a function of the original local densities Yx and Yz. Then $u_j = x_j - z_j\hat{R}$, $\bar{u} \neq 0$ because $\hat{R}$ is the global ratio. The sum term of the variance of a ratio of means estimator can then be written as:

(148) $\sum_{j=1}^{n}\left(u_j - \bar{u}\right)^2 \doteq \sum_{j=1}^{n}\left(\left(x_j - z_j\hat{R}\right) - \left(n^{-1}\sum_{j=1}^{n}\left(x_j - z_j\hat{R}\right)\right)\right)^2 =$

$$\left(\sum_{j=1}^{n} x_j^2 - n^{-1}\left(\sum_{j=1}^{n} x_j\right)^2\right) - 2\hat{R}\left(\sum_{j=1}^{n} x_j z_j - n^{-1}\left(\sum_{j=1}^{n} x_j\right)\left(\sum_{j=1}^{n} z_j\right)\right)$$
$$+\hat{R}^2\left(\sum_{j=1}^{n} z_j^2 - n^{-1}\left(\sum_{j=1}^{n} z_j\right)^2\right)$$

Variation Between Double Sampling Strata Under two-phase sampling schemes, the variance estimators contain a term in the form $\sum_{e=1}^{H}\alpha_e\left(\hat{U}_e - \hat{U}\right)^2$. A shorthand computation of the term is:

(152) $\sum_{e=1}^{H}\alpha_e\left(\hat{U}_e - \hat{U}\right)^2 = \left(\sum_{e=1}^{H}\alpha_e\hat{U}_e^2\right) - \hat{U}^2\sum_{e=1}^{H}\alpha_e$

because $\sum_{e=1}^{H}\alpha_e\hat{U}_e = \hat{U}\sum_{e=1}^{H}\alpha_e$.

Under single-plot double sampling with stratified sampling of first-phase sample plots, $\alpha_e = \dot{n}_{k.e}$ and $\sum_{k.e=1}^{k.H} \dot{n}_{k.e} \hat{Y}_{k.e} = \hat{Y}_k \sum_{k.e=1}^{k.H} \dot{n}_{k.e}$

The sum term can then be written as:

$$(155)\ \sum_{e=1}^{H} \alpha_e \left(\hat{U}_e - \hat{U}\right)^2 \doteq \sum_{e=1}^{H} \dot{n}_e \left(\ddot{n}_e^{-1} \sum_{j \in \ddot{S}_{e.j}} x_j\right)^2 - \dot{n}^{-1} \left(\sum_{e=1}^{H} \dot{n}_e \left(\ddot{n}_e^{-1} \sum_{j \in \ddot{S}_{e.j}} x_j\right)\right)^2$$

$$+\hat{R}^2 \left\{ \sum_{e=1}^{H} \dot{n}_e \left(\ddot{n}_e^{-1} \sum_{j \in \ddot{S}_{e.j}} z_j\right)^2 - \dot{n}^{-1} \left(\sum_{e=1}^{H} \dot{n}_e \left(\ddot{n}_e^{-1} \sum_{j \in \ddot{S}_{e.j}} z_j\right)\right)^2 \right\}$$

$$-2\hat{R} \left\{ \sum_{e=1}^{H} \dot{n}_e \left(\ddot{n}_e^{-1} \sum_{j \in \ddot{S}_{e.j}} x_j\right) \left(\ddot{n}_e^{-1} \sum_{j \in \ddot{S}_{e.j}} z_j\right) - \dot{n}^{-1} \left[\left(\sum_{e=1}^{H} \dot{n}_e \left(\ddot{n}_e^{-1} \sum_{j \in \ddot{S}_{e.j}} x_j\right)\right) \right.\right.$$

$$\left.\left. \left(\sum_{e=1}^{H} \dot{n}_e \left(\ddot{n}_e^{-1} \sum_{j \in \ddot{S}_{e.j}} z_j\right)\right) \right] \right\}$$

References

FAO (2015) Global forest resources assessment 2015 desk reference. Food and Agriculture Organization of the United Nations, Rome

Fielding RT (2000) Architectural styles and the design of network-based software architectures. Doctoral Thesis, University of California

FOEN (2016) Switzerland's Greenhouse Gas Inventory 1990–2014. National Inventory Report including reporting elements under the Kyoto Protocol. Federal Office for the Environment FOEN, Climate Division, Bern

FOREST EUROPE (2015) State of Europe's Forests 2015. Ministerial Conference on the Protection of Forests in Europe, Forest Europe, Liaison Unit Madrid, Madrid. Spain

Ginzler C, Hobi ML (2015) Countrywide stereo-image matching for updating digital surface models in the framework of the Swiss National Forest Inventory. Remote Sens 7 (4):4343–4370. https://doi.org/10.3390/rs70404343

Köhl M (2001) Inventory concept NFI2. In: Brassel P, Lischke HS (eds), Swiss national forest inventory: methods and models of the second assessment. Eidg. Forschungsanstalt WSL, Birmensdorf, pp. 19–44

Lanz A (2017) Forest inventory estimation. Unpublished report, WSL, Birmensdorf

Laplante PA (2007) What every engineer should know about software engineering, vol 40, 1st edn. Taylor & Francis, Boca Raton

McDowall RD (2005) Effective and practical risk management options for computerised system validation. Qual Assur J 9:196–227. https://doi.org/10.1002/qaj.339

Pulkkinen M, Ginzler C, Traub B, Lanz A (2018) Stereo-imagery-based post-stratification by regression-tree modelling in Swiss National Forest Inventory. Remote Sens Environ 213:182–194. https://doi.org/10.1016/j.rse.2018.04.052

Thulasiraman K, Swamy MNS (1992) Directed graphs. In: Graphs: theory and algorithms. Wiley, pp 97–125. https://doi.org/10.1002/9781118033104.ch5

Traub B, Meile R, Speich S, Rösler E (2017) The data storage and analysis system of the Swiss National Forest Inventory. Comput Electron Agric 132:97–107. https://doi.org/10.1016/j.compag.2016.11.016

Waser LT, Fischer C, Wang ZY, Ginzler C (2015) Wall-to-wall forest mapping based on digital surface models from image-based point clouds and a NFI forest definition. Forests 6 (12):4510–4528. https://doi.org/10.3390/f6124386

Part VII
Data Quality

Chapter 21
Quality in Aerial-Image Interpretation and Field Survey

Berthold Traub, Fabrizio Cioldi, Christoph Düggelin, Markus Keller, and Christian Ginzler

Abstract In this chapter, we discuss several types of quality assurance (QA) and quality control (QC) measures, which are described for the three main stages in the field survey in sequence. This comprises the process of aerial-image interpretation, which is crucial for ensuring the complete coverage of the Swiss forest in the field sample. We also discuss the QA activities applied during fieldwork, including the training of field teams, the maintenance of the field manual, and several organisational measures to prevent mistakes. Finally we describe the additional surveys conducted to assess the quality (reproducibility) of the field assessments in terms of precision (uncertainty) and accuracy, and summarise the quantitative results of selected attributes from the Swiss National Forest Inventory (NFI)'s repeat survey.

21.1 Introduction

The quality of the estimates and conclusions from empirical data depends on the quality of the original input data to the analyses. The data for national forest monitoring is largely derived from experts' observations of numerous variables recorded over large areas in the field or through remote sensing on a large number of sample plots. Given this complex setting, data quality is of paramount relevance in all forest monitoring endeavours, including the Swiss NFI (NFI). Extensive quality assurance (QA) measures have therefore been implemented in NFIs and continuously developed with the goal to avoid systematic observation errors and reduce random errors as far as possible.

B. Traub (✉) · F. Cioldi · C. Düggelin · M. Keller
Scientific Service NFI, Swiss Federal Research Institute WSL, Birmensdorf, Switzerland
e-mail: berthold.traub@wsl.ch

C. Ginzler
Remote Sensing, Swiss Federal Research Institute WSL, Birmensdorf, Switzerland

C. Fischer, B. Traub (eds.), *Swiss National Forest Inventory – Methods and Models of the Fourth Assessment*, Managing Forest Ecosystems 35,
https://doi.org/10.1007/978-3-030-19293-8_21

Studies of data quality in forest monitoring programs have focused particularly on the reliability and comparability of the collected data (e.g. Ferretti et al. 2009; Ferretti 2009; Gasparini et al. 2009; Kaufmann and Schwyzer 2001; Kitahara et al. 2009; Pollard et al. 2006). Drawing on experience in collecting inventory data over many decades, precise definitions and consistent fieldwork instructions seem to be especially important. The effect of measurement errors has been addressed by several authors (Berger et al. 2012; McRoberts et al. 1994; McRoberts and Westfall 2016; Westfall et al. 2016). In the framework of the development of the statistical methods of the NFI, Gertner and Köhl (1992) and Köhl 2001 investigated several types of errors. The authors developed error-budgets and showed that systematic (inaccurate) measurements have a strong influence on the overall sampling error (standard errors), which is greater the larger the sample size. Random errors due to imprecise measurements have, in comparison, comparably less influence on the total error.

In the following sections, we discuss several types of QA and quality control (QC) measures, which are described for each stage in the field survey in sequence. Beginning with Sect. 21.3 we focus on the process of aerial-image interpretation, which is crucial for ensuring the complete coverage of the Swiss forest in the field sample. During this process, the sample plots to be visited in the field are selected. Thus a lack in quality in deciding what counts as forest in the aerial images could lead to: (a) re-forested or newly forested areas being overlooked, which is a major issue from a statistical point of view, and (b) the field teams being sent too often to non-forested areas, which is expensive. The focus of Sect. 21.4 is particularly on the QA activities applied during fieldwork. This includes the training of field teams, the maintenance of the field manual, and several organisational measures to prevent mistakes. In Sect. 21.5 we describe the additional surveys conducted to assess the quality (reproducibility) of the field assessments in terms of precision (uncertainty) and accuracy, and summarise the quantitative results from the NFI's repeat survey, as they may also be relevant for other national forest inventories.

21.2 Quality Assurance and Quality Control

The collection of high-quality data is based on several preparation steps, as well as on highly standardised methods applied during fieldwork. The basis of the field survey is the aerial-image interpretation to identify clearly the field plots to be visited. Those assessing the aerial-image and field-plot data require regular training and stable definitions of attributes over time. Pollard et al. 2006 state: "...The goal of a quality assurance (QA) program is to provide a framework to assure the production of complete, accurate, and unbiased forest information of known quality...". In this sense, QA measures were implemented at all stages during the field survey, as well as for the aerial-image interpretation in the NFI. Table 21.1 summarises the QA and

Table 21.1 Examples of QA and QC activities in the field survey process chain

Task	QA	QC
Identify field plots to be visited	Highly automated processes and state-of-the-art hard- and software technology (Chap. 6)	Repeat survey of aerial-image interpretation (Sect. 21.3)
Field-plot assessment	Comprehensive field manual, instructions and training (Sect. 21.4), MAIRA plausibility checks (Sect. 23.7)	Repeat and control survey (Sect. 21.5)
Data transfer and upload	Manual crosschecks, notes from field survey	Plausibility checks (Sect. 23.7)

QC measures for the main activities in the survey process-chain to ensure the highest possible data quality.

The risk of committing systematic, as well as random, observation errors plays an important role in forest inventories and has specific causes. Some random observation errors cannot, because of the nature of the data, be completely eliminated or avoided. The most critical failures are those arising from misinterpreting instructions, which lead to systematic deviations and thus introduce bias. Thus, it is straightforward to think about how the risks of committing these types of errors can be managed. Risk management, which is important in planning and deciding on QA and QC activities (McDowall 2005), consists of three major components:

- **Risk assessment**: Identifying (assessing) those processes, assessment techniques, measurement devices and soft- and hardware with a high risk and prioritizing (evaluating) them in QA measures. Risk management in field surveys involves being able to carry out these assessments at any time so that the resources to be spent on avoiding risks can be continuously balanced against their risk priority.
- **Risk treatment**: Treating risks to reduce the risk of failure. Here well-documented instructions and processes introduced with regular training, for example, are very important.
- **Risk acceptance including risk communication**: Accepting and communicating the residual risks that remain after implementation of QA activities, i.e. those risks which cannot be avoided without excessive investment of resources into QA and therefore have to be accepted. Even tree diameter measurements may not be exactly reproducible depending on calliper placement and bark irregularities. Assessments of (forest) stand attributes are even more difficult for different field teams to reproduce. QC measures are used to monitor and verify the success of all QA measures. In particular, they help to reveal the residual risks and indicate the achievable reproducibility of the assessment under given conditions.

Risk management is an ongoing process that aims to find reasonable ways to obtain the achievable degree of quality, i.e. to minimise the residual risks wherever possible. The actual degree of reproducibility should always be considered in the interpretation of forest statistics, especially if they refer to change.

21.3 Aerial-Image Interpretation

As several people interpret the aerial images, differences between their interpretations and measurements are likely, but are assumed to be small. Interpreters may differ in assessing the land-use class, in determining the middle of the tree top for measuring the forest boundary line and in measuring the height of the raster points. To obtain an idea about how accurate classifications of predefined sample points in the category 'forest' are according to the NFI definition, we carry out quality assessments periodically to determine how well different interpreters' classifications agree.

Before beginning the interpretation process, interpreters attend training and discussion sessions to give them a platform for training, discuss important topics and inspect difficult plots on-site. The most critical tasks in interpreting aerial images is classifying the land use on the plots, e.g. determining whether a plot covered with any trees should be categorised as managed by forestry or as another category such as a cemetery or park, and digitising the forest boundary line.

On 3% of the sample points, the interpretations are carried out at least twice, although the actual number of repetitions depends on how many interpreters are involved. In NFI4, two individuals conducted the stereo-image interpretation so that each interpretation was performed twice. The double interpretation is organised as a 'blind check', i.e. the interpreter does not know whether the assessment of the plot is of type repeated or single. This ensures that all interpretations are carried out independently in the same way.

All measurements taken during the interpretation are documented and compared. This applies to classifications of the land cover, height measurements, forest boundary line mapping and the final 'forest/non-forest' decisions. The differences are discussed after each field season. Any descriptions that appear to be missing are added to the manual and the definitions are revised accordingly.

The results of the comprehensive double interpretation in the inventory in Liechtenstein may serve as an example of the current state of quality level. The image data and methods were the same as in the NFI, and two experienced NFI interpreters assessed the aerial images (n = 1019 plots). The first interpretation resulted in 56.7% 'non-forest', 36.2% 'forest', and 0.7% other forest land ('shrub forest'), while for five plots no interpretation was possible. The repeated interpretation resulted in an agreement of 96.8% regarding the forest decision. On 33 plots (3.2%), the forest decision was different. The disagreements arose because of confusion in classifying the land use (three plots; e.g. classifying a tree nursery as 'forest' rather than 'non-forest') and different digitising of the Forest Boundary Line (FBL), and thus different calculations of the forest width (30 plots) (Fig. 21.1). Given the uncertainties in delineating the forest boundary line, all 'non-forest' plots with their centre within 7 m of the forest boundary line are visited in the field for verification.

Fig. 21.1 Example of a field plot location with forest boundary lines selected by two interpreters (light and dark green lines). The forest width is calculated as the shortest distance between the forest boundary lines through the plot centre (red circle). For the interpretation with the light green forest boundary lines, the width of the tree group is too small to fulfil the forest definition, and this resulted in a 'non-forest' decision for the plot. Using the delineation of the other interpreter (dark green), the width fulfils the forest definition and the plot qualifies as a forest plot

21.4 Field Work

21.4.1 Personnel, Working Plan and Equipment

An NFI field team consists of a forest engineer and a forester or forest ranger. This combination of professional skills ensures each team has the necessary practical and theoretical background, and was found to work well in the previous inventories. The same criteria were therefore applied in recruiting the NFI4 field teams.

The 'human factor' is of decisive importance for the quality of the data collected in the field. Not only do the field team members need to be familiar with the Swiss forestry sector and have detailed expert knowledge about the NFI methods, but they also need to work together harmoniously. The NFI's team of instructors pays great attention to the compatibility of the field team members. Normally, they spend long periods of time working together without breaks. In summer, working days exceeding 8 h are the rule and not the exception. To ensure good data quality, it is essential to take every measure possible to keep the teams' motivation high and to avoid personal conflicts. For this reason, criteria for recruitment are not only length of experience and technical skills, but also very much the personalities of the candidates.

Field team members are recruited, if possible, from all geographical and language regions in Switzerland. German is the main language used in general for communication and instruction. In addition, team members should have good knowledge of another national language, namely French, Italian or Romansh. Normally at least one member of the team speaks the language of the area they are working in fluently.

This is particularly important for contact with the local population, e.g. with forest owners, and for the interviews with the forest services. In addition to those who work for almost a whole season in a field team, NFI also has a group of four to five back-up people who can replace, at short notice, members of the regular team who are absent because of e.g. holidays or illness. In general, former NFI team members who prefer to work for just short periods are considered for this replacement function. They participate in all training courses and have the same level of expertise as the regular field team members.

Field team members are encouraged to continue to be involved in NFI surveys to maintain the continuity of the fieldwork and build on prior knowledge and experience. Several measures have been taken to make the work as attractive as possible even though it is physically extremely demanding, including improving the efficiency of the organisation of fieldwork and generating a generally motivating working environment. Further measures are:

- Paying appropriate salaries with generous expenditure allowances.
- Allowing team members freedom in organising their work time, personal professional equipment, training in safety and security, as well as team-building events.
- Having no hierarchical structures within the field teams. Both field employees use the field computer and measure the attributes. They are encouraged to take it in turns typing and measuring to make the work more varied and interesting, and to discuss any critical decisions. If they cannot decide after consulting the field manual, they can contact the NFI instructors for advice or instruction (Sect. 8.2).
- Moving field team members two to three times per year. These exchanges provide opportunities for the field staff to get to know each other better and to learn from each other, which also improves the standardisation of the work flow between the different field teams.
- Allowing field team members to work without time pressure. They are paid on an hourly basis and not per sample plot so that they can assess each sample plot properly and not in a hurry. The working plan is optimised for three field teams, so that there is not too much pressure to work quickly.
- Allocating the sample plots to the survey teams (working regions) so that team members can return home regularly. This means, on average, daily or every other day during the spring and autumn and at least once a week during the summer months when working days in alpine regions may be long (Sect. 8.2.2). In principle, each field team works in all regions of Switzerland and the sample plots in the larger cantons are assessed by at least two different field teams. This should balance out any effects of one team potentially making inaccurate assessments. Working away from home for several months without interruption can be a problem, especially for employees with family duties, but the allocation of working regions takes this into account so that team members remain motivated.
- Letting field teams organise their work individually and independently. They can, for example, be flexible in choosing which out of the pool of sample plots assigned to them to assess when. This allows them to choose the region each week with the best possible weather conditions to be as safe as possible and

obtain data with the best quality possible. Their work progress is continually monitored by the headquarters at WSL. The field teams are required to send their data to the WSL database several times a week and to complete a time sheet every 2 weeks.

- Having the functionality of the equipment regularly checked by the instructors, as the measurement devices and tools are of crucial importance for the quality of the fieldwork. The instructors also test new measurement instruments and other tools (Sect. 9.2) that could improve the quality and efficiency of the fieldwork. Special attention is paid to the calibration and maintenance of measurement devices. This is done once at the beginning of each field season (compass), once per week (calliper), once per day, or even more frequently (vertex). The field teams are also encouraged to give regular feedback on the equipment and the software MAIRA for data collecting (Sect. 23.2.1) so that they can be improved as needed. Moreover, all the NFI instructors and the whole NFI team are always glad to receive feedback from the field teams and to discuss with them a wide range of topics along the entire data-processing chain. This leads to a better team spirit and makes the field team even more aware that they constitute a substantial part of the NFI Project. Their work involves considerable responsibility, and is essential for high-quality data assessment, analysis and reporting.

21.4.2 Field Manual

The NFI4 manual for the field survey is published annually (Düggelin and Keller 2017) and is mandatory for guiding the work of the field teams on a sample plot. It defines the sequences and contents of the various work steps, as well as the aims, definitions and exact procedures, including the measuring instruments to be used for each attribute assessed. Having clear, unambiguous and conclusive descriptions of the attributes is of absolute importance to ensure the collected data is reliable. In order to avoid confusing real changes in plot attributes with definitional changes, any changes in the definitions of attributes must be kept to a minimum and only made if really necessary. In most cases, such changes are slight and concern details about an attribute definition that need to be adjusted to take into account special circumstances. Any refinements of attribute definitions or assessment techniques are communicated to the field teams in newsletters during the ongoing field season and the most recent version of the field manual adapted accordingly. New issues of the field manual include the contents of the newsletters and are published at most once per season and handed over at the beginning of the next field season. The manual is, however, never replaced or updated during the field season.

The field teams need to make sure that the field manual is always at hand. Together with the training they receive, it provides the basis for building up a common understanding of the field survey standards.

21.4.3 Training

Every year, at the beginning of the field season, all field teams – including the very experienced ones – spend a whole week having theoretical and practical training for the fieldwork. Particular attention is paid to any changes in the field manual, field computer software or field equipment. Important basic skills, such as knowledge of woody species, are revised. In the practical part, the participants are trained in using the field computer software and the measuring procedures. They are given complicated tasks to solve and situations to handle, and shown how to follow the field manual strictly and record information on the field computer. They are tested both as individuals and as a team. All the special cases presented in the course are discussed thoroughly with the whole group of field teams until a common understanding is reached, e.g. about how to interpret the course of the various boundary lines to be specified on a forest margin plot. After the instruction days, an NFI instructor accompanies each field team member, including those with considerable experience, for about 3–4 days during the first 2 weeks of the field season to ensure they collect good quality data, focusing in particular on accuracy, and to clarify any remaining uncertainties directly on the sample plot ('hot checks').

In addition to the instruction at the beginning of the field season, about two so-called 'training days' (lasting 2–3 days) per field season are conducted where specific and emerging topics are taken up, taught and trained. A particular goal here is to teach and practise using assessment standards for qualitative attributes such as *regeneration cover* or *stand stability*. During the training days, the field teams also receive feedback on the quality of their recent work, mainly based on the evaluation of the control survey (Sect. 21.5). Furthermore, particularly challenging field situations, such as when individual field teams need to make tricky ad hoc decisions in the field, are presented and discussed in plenary. In addition to the inventory-specific training, courses in first aid, with a particular focus on outdoor accidents and basic life support, and in driving are given regularly during the training days and described in more detail under safety aspects in Sect. 8.2.2. As with the instruction week at the beginning of the season, all field team members are required to attend these training days.

During the field survey, the field teams can contact the NFI instructors via hotline at any time. They can send, as additional information, photos showing the situation, e.g. to illustrate difficult 'forest/non-forest' decisions, immediately via smartphone. If technical problems with the field computer software occur, the technical support team at headquarters is able to directly access the computer via the mobile network and provide support.

21.5 Repeat and Control Surveys

Both repeat and control surveys are conducted in many national forest inventories, where a percentage of 5–10% of remeasured plots is usual (Berger et al. 2012; Gasparini et al. 2009; Kitahara et al. 2009; Pollard et al. 2006; Tomppo et al. 2010).

Since the terminology of control and repeat surveys varies between countries, these surveys are, however, rarely comparable. They e.g. differ in whether the re-visited plots are remeasured completely, whether the QA teams are expert teams and whether the QA teams work under the same conditions as the regular field teams with regards to time pressure. In the subsequent paragraphs we briefly describe the two types of inventories as applied in the NFI.

Control Survey The control survey allows the success of training to be checked immediately, and focuses on how well the fieldwork complies to the NFI definitions and instructions. An instruction team visits about 2% of the field plots, i.e. about five per field team, directly after the field measurements (FM) to check on particular attributes of interest. Because this team consists of instructors, they should uncover deviations from an assumed true standard and may identify misunderstandings, unclear or different interpretations of definitions between groups, as well as any malfunctioning of measuring devices. Giving the teams immediate concrete feedback helps improve accuracy as it reduces the risk of introducing systematic errors. The sample plots are selected according to topic or assessment phase. The control survey is referred to as a 'cold check' (Pollard 2005), i.e. it is carried out after the field teams have assessed the plot. The control survey crew have all the data of the FM crew at hand, and may repeat and compare the assessment completely or just focus on special topics, such as changes in the 'forest/non-forest' decision. The results of control surveys are communicated regularly and provide topics for the training courses during the field seasons.

Repeat Survey This survey enables us to assess the actual level of reproducibility by conducting quality measurements (QM) by some of the regular field teams (then called QA teams). It is conducted on a random sub-sample of the annual panels (QA plots) as a 'blind check' (Pollard 2005), i.e. the QA team has no access to the data assessed by the FM team to assure an independent remeasurement of the plot. For the sake of comparability, the QA teams remeasure the plot completely using the same methods and equipment as the regular field teams.

The repeat survey covers about 8–10% of a team's approximately 280 plots per year. It is conducted during three 1-week periods in June, September and November for efficiency reasons. To avoid field teams being influenced, they do not know which of their plots will be remeasured. In the ideal case, also the QA teams would not know that they were **re**-measuring a plot. It is not, however, possible to prevent the QA teams from finding out about their role because there will inevitably still be traces from the first FM assessment on and around the sample plot. The QM results are exclusively used to evaluate the overall quality of the survey, i.e. the repeat survey data is never used to correct the original field survey data.

The NFI2 repeat survey is described by Kaufmann and Schwyzer (2001), together with the statistical methods used. These include descriptive statistics and visual interpretation of differences, as well as association measures (Gamma, Kappa) and significance tests (sign test, Wilcoxon and McNemar test for concordance and marginal homogeneity). The NFI4 repeat survey was compared with the NFI3 repeat survey in Traub et al. (2016), who noted that observer agreement was better in NFI4 than in NFI3, although the improvement was slight. They attribute the improvement to the change from temporary short-term fieldwork to permanent fieldwork throughout the year with far fewer field teams employed.

21.6 Analysis of the NFI4 Repeat Survey

21.6.1 Attributes Examined and Database

Out of about 280 attributes taken on the field plots, we selected 12 attributes that are very relevant for estimating volumes and biodiversity (Table 21.2). For some

Table 21.2 List of attributes examined. The abbreviations in brackets are used in subsequent sections to identify the attributes

Attribute	Definition	Scale
dbh [cm]	Stem diameter at breast height (1.3 m aboveground) of tally tree measured with calliper. Restricted to standing living trees	i
d7 [cm]	Upper stem diameter 7 m aboveground measured with Finish calliper. Restriction: measurement limit = 60 cm	i
Tree height [m]	Height of the top of the tree aboveground	i
Stand age [years]	Age of the reference stand[a] in years (age of the dominating trees in the reference stand without standards and preparatory planted trees)	i
Crown length [3 categories]	Length of the green crown (to the lowest green branch without epicormic branches) in relation to total tree height	o
Layer [3 categories]	Position of the crown of sample trees with respect to the mean height of the 100 strongest trees/ha. Restriction: without free-standing trees	o
Stand stability [10 categories]	Expected resilience of the reference stand to disturbing influences for a 10-year or 20-year period, depending on region	o
Degree of mixture [4 categories]	Proportion of basal area of conifers and deciduous trees in percent referred to the reference stand	o
Regeneration cover [7 categories]	Degree of established regeneration cover in the reference stand. (Height >0.1 m and dbh <12 cm)	o
Development stage [5 categories]	Determined with the mean dbh of the 100 strongest trees of the reference stand. Restriction: without mixed in order to keep ordinal scale	o
Stand structure [4 categories]	Reflects the proportion of the different layers	n
Crown closure [8 categories]	The intensity of interaction of the tree crowns in a stand	n
Woody species	Trees and shrubs with a completely lignified plant axis at least 40 cm high	o

[a]In case of one or more stand borders crossing a sample plot, only the reference stand is to be considered for stand attribute assessment

Scale: *i* interval, *o* ordinal, *n* nominal. Note: d7 and Tree height are measured for tariff trees only (Sect. 2.3.4)

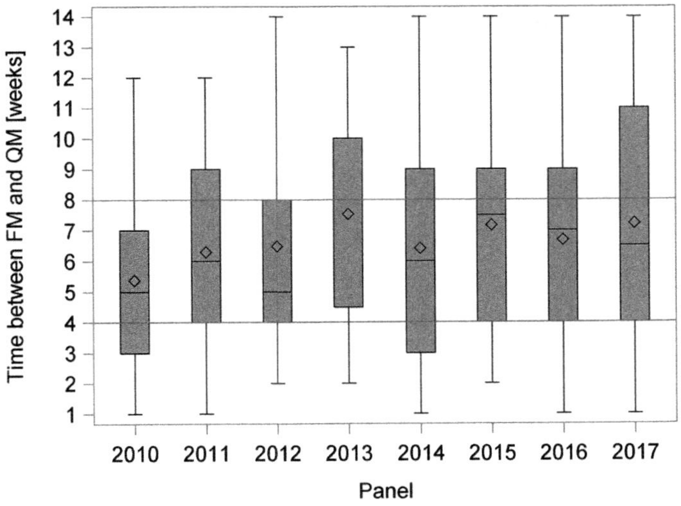

Fig. 21.2 Time between field and repeat surveys in weeks for each NFI4 annual panel

attributes, the full range of measurement or assessment classes was reduced to avoid mixing measurement scales. The definition and measurement instructions of the attributes are described in detail by Düggelin and Keller (2017). For the sake of better readability, attributes are set to italics when referred to in the context of the examination.

The analysis of the attributes is based on the repeat survey data from NFI4 (annual panels 2010–2017). The repeat survey was conducted about 1–2 months after fieldwork (Fig. 21.2).

Several filter and standardisation steps were applied to the database of the repeat survey (447 sample plots) to make it as comparable as possible with the data from the FM. Highly comparable pairs of data in terms of repeated measures were selected and the following restrictions applied:

- On 442 of the 447 sample plots assessed in the repeat survey, the FM and QM groups agreed that the plots were accessible, but disagreed about four plots. They agreed that, of these, 429 were 'forest without shrub forest' according to the NFI forest definition, but disagreed about five plots. Shrub forest plots were excluded because most of the attributes selected for analysis had not been assessed on these plots. Finally, three more plots were excluded where the QM groups did not find the plot centre but re-established it, resulting in a database of **426 sample plots** valid for the analysis of tree attributes.
- A subset of **259 sample plots** was selected to compare stand attributes, where the FM and QM groups agreed that no stand border crosses the sample plot, i.e. that the plot covers only one stand. They disagreed in 67 plots. Plots with stand borders should be filtered out because the course of this border, determined by each group individually, is used to allocate the plot centre to the reference stand, to which the plot attribute assessments refer to. Different border courses thus could lead to incomparable assessments of stand properties.

- Only those trees identified by both the FM and QM groups were included. For ingrowth trees to be identified as matching, the following threshold values were applied: centre-tree distance <0.5 m, azimuth <15° and dbh <5 cm. Certain trees were only measured by one of the two teams and were therefore excluded (170 ingrowth trees and 81 trees from the previous inventory). This filter resulted in a total of **5139 trees** valid for analysis.
- The *dbh* and *d7* comparisons refer to standing, living trees measured with the calliper (up to 60 cm) to ensure, with high probability, that the diameters were measured at the same clearly defined location and with the same device.

Data cleaning was only applied to a few clearly mixed-up measurements. Outliers and extreme values were not eliminated because they reflect the inherent reality of measurement uncertainty in fieldwork. They will, however, be considered in interpreting the results because they play an important role in the context of normal approximation and significance tests on the interval scale (Cochran 1977, p. 44).

21.6.2 Analysis Methods

Data Types and Statistical Methods The data analysis methods applied to assess reproducibility included descriptive statistics, significance tests and observer agreement measures, such as Kappa statistics. The analysis methods used are listed in Table 21.3 with the scale of the attributes, and explained in the following sections. The data were processed and statistically analysed with PC- SAS© Release 9.4 (SAS Institute 2013; Walker 2002).

Descriptive Analysis of Data Reproducibility The repeated assessments of field sample attributes provides a set of paired pre/post data which enables reproducibility quantification. Because a small amount of uncertainty cannot be avoided, rules must be established to evaluate and interpret any deviations found.

Table 21.3 Methods to analyse reproducibility

	Scale		
Method	i	o	n
Calculation of differences = QM – FM	x	x	
Pairwise t-test/Standard deviation of differences	x		
Wilcoxon signed rank test for pairwise observations	x	x	
Symmetry tests (2 categories. McNemar, >2 categories. Bowker test)		x	x
Pearson coefficient of correlation (r_p)	x		
Spearman coefficient of correlation (r_s)		x	
Kappa statistic		x[a]	x

[a]Weighted Kappa
i interval scale, *o* ordinal scale, *n* nominal scale

An intuitive method based on data quality requirements will first be introduced. It was designed to take into account specific *measurement quality objectives* (MQO) which provide a window of performance that is allowable for individual observations also called measurement tolerance level or expected level of precision. The MQOs were established as the "…best guess of what experienced field teams should be able to consistently achieve…" (Pollard et al. 2006). The NFI instruction team assigned such tolerance levels to the attributes assessed (Table 21.2). These levels were based on their assumptions of field crews' abilities to make repeatable measurements or observations. Their thresholds thus reflect the quality of measurements that can be achieved on average in the long term, rather than optimal results under ideal conditions.

MQOs have also been used in other forest inventories (e.g. Gasparini et al. 2009; Kitahara et al. 2009), as well as in assessing and monitoring the effects of air pollution on forests (Allegrini et al. 2009; Ferretti 2009; Ferretti and König 2013). These studies use a slightly different terminology and define data quality objectives (DQOs) as a combination of MQOs and data quality limits (DQLs) which determine the minimum acceptable frequency of observation within the MQOs. Following this notation, the DQO of the *dbh* in the NFI was denoted as $98\ \%\ \ @\ \pm 1cm$ (to be read as: all measurements are expected to lie within $\pm 1cm$ (MQO) in 98% (DQL) of all cases).

Measures of Relationships and Observer Agreement To evaluate the DQOs, the DQL part is contrasted with the actual quality results (DQR) derived from the repeat survey. The DQR is calculated as

$$DQR = \frac{number\ of\ observed\ differences\ within\ MQO}{number\ of\ all\ observed\ differences} * 100\% \qquad (21.1)$$

Correlation coefficients can be used to describe the relationship between the two FM and QM measures. The Pearson (r_p) coefficient is used for interval scale data and the Spearman (r_s) coefficient for nominal scale data (range:± 1). Values close to one indicate a strong positive correlation and negative values an inverse relationship. The interpretation of the correlation is based on the assumption that the intercept and slope of the linear regression between the FM and QA values does not differ significantly from zero (intercept) or from one (slope).

Observer agreement may generally be estimated by calculating the proportion of the measurement pairs that shared the same category (actual agreement) in relation to the pairs that did not. However, for skewed distribution with many observations, the actual agreement may be misleading because the marginal totals may not be taken into account (Wulff 2004). This problem is compensated for by the Kappa statistic, which summarises the agreement in a single index. It compares the probability of agreement to the agreement expected from pure chance, i.e. if the observations were independent (Agresti 2013). The Kappa statistic is given as

$$Kappa = \frac{(P(A) - P(E))}{(1 - P(E))} \qquad (21.2)$$

where *P*(*A*) denotes the proportion of observed / actual agreements and *P*(*E*) denotes the proportion of random agreements. With complete observer agreement $Kappa = 1$, with complete disagreement $Kappa = -1$, and with chance agreement $Kappa = 0$ (Siegel and Castellan 1988, p. 285). For attributes on an ordinal scale, the weighted Kappa is calculated to take into account the order of the categories (Agresti 2013, p. 435; Siegel and Castellan 1988).

The symmetry of discordant agreements can be evaluated using the Bowker or McNemar test. These tests consider non-agreements only and evaluate the relationship of observations above and below the diagonal in the contingency table (Siegel and Castellan 1988, p. 75)

$$\chi^2 = \frac{(b-c)^2}{b+c} \ (1 \ degree \ of \ freedom) \tag{21.3}$$

where b = number of occasions where the FM category of attribute shifted from category 1 to 2, and c = number of occasions where the QC category of attribute shifted from category 2 to 1.

As the Kappa statistic focuses on agreements between raters only, the symmetry statistic together with Kappa provides a more complete picture because agreement is not necessarily correlated with symmetry. Moreover, if the disagreements deviate substantially from symmetry, the Kappa statistic should be interpreted extremely cautiously because it is strongly dependent on: (a) the number of categories the attribute in question has, and (b) the distribution of the categories. If there are many categories and/or skewed marginal distributions, Kappa values tend to be smaller, which should be considered when comparing the performance of different attributes. A comparison of values over time selected from the same attribute, however, is assumed to be stable and appropriate. It is therefore essential to examine the marginal distributions, which are very relevant for interpreting Kappa and which help explain how attribute categories are filled.

To evaluate accuracy, the mean or the median of differences is examined. The paired t-test and Wilcoxon rank sum test may indicate a significant deviation of the differences from zero. Additionally, we have provided the 95% confidence interval (CI) to support the interpretation of arithmetic means. As long as zero is included in the CI, a significant bias is not likely.

Species Richness The quality of species richness assessments (attribute *woody species*) between field teams was examined by the pseudoturnover (PT) of species composition (Nilsson and Nilsson 1985), which may be interpreted as an index that aggregates agreement ratios. It assumes that no true turnover happened between the points in time of the assessment, i.e. the turnover rate can only be accredited to observer disagreement.

$$PT = \frac{A+B}{S_A+S_B}100 \tag{21.4}$$

where

A= number of species exclusively found by team A and B = number of species exclusively found by team B.
S_A = all species found by team A, and S_B = all species found by team B.

For the PT analysis, one PT value per plot was calculated. A separate analysis by recorded height categories (height 40–130 cm and height >130 cm; dbh = 12 cm) was waived. Depending on the season and time between the measurements, the height class of small plants may change, and thus introduce an unknown amount of noise in the results.

Notes on the Interpretation of Test Results For attributes on the interval scale, parametric tests about the null hypothesis (H0: the deviation of the differences from zero is not significant) can be applied. For the sake of simplicity, we use the confidence interval (CI) of the differences to interpret the results and if the CI includes zero the H0 will not be rejected. For the proper interpretation of confidence intervals, see Cumming and Finch (2005). It is good practice to confirm significant results from the test using its nonparametric equivalent, the Wilcoxon test. However, if the parametric test rejects H0 and the non-parametric does not, it does not mean something is necessarily wrong because the non-parametric test has less statistical power if assumptions for the parametric test are met.

The sample size must always be considered when interpreting results from significance tests and confidence limits. With increasing sample size, smaller differences become statistically significant. Thus significance may never be interpreted without having a clear understanding of the (practical) relevance of the differences found for the attribute examined. The determination and evaluation of the DQO are related to this topic. The required sample size to detect a certain difference as significant to significance level alpha may be determined by experimental design analysis or post-hoc power analysis (Cohen 1988). This is, however, beyond the scope of this evaluation.

21.6.3 Analysis Results of Tree Attributes

The results of the statistical analysis for interval scale attributes are presented in Table 21.4 and Fig. 21.3 (left and central columns). Results for the ordinal scaled attributes *crown length* and *layer* are given in Table 21.5. As *d7* and *tree height* are measured on tariff trees only, far fewer measurements are available than for *dbh*. The *tree height* could be measured on almost all trees, but *d7*-measures are frequently missing because the tree is forked below 7 m, because *d7* is located within the crown or covered by plants such as *Hedera h*elix, or for some other reason.

FM and QM correlations and the distributions of the QM-FM differences for the interval scale attributes are presented in Fig. 21.3. The *dbh* measures correlate most, followed by *d7*, *tree height* and *stand age* (left column), with similar results for the distribution of differences (central column). The uncertainty for all attributes is generally higher for broadleaved trees. The *dbh* differences are slightly positively

Table 21.4 Analysis of tree attribute differences (QM-FM) on interval scale

Attribute	DQO	Type	DQR[%]	*s*	r_p	Mean [CI 95%] t-test	*n*
dbh	98%@	c	98.65	0.56	0.99	0.04 [0.02;0.07]***	2294
	±1 cm	b	98.22	0.63	0.99	0.04 [0.00;0.07]***	1573
d7	95%@	c	92.08	1.64	0.99	0.04 [−0.13;0.20]	379
	±2 cm	b	**82.47**	2.93	0.97	−0.07 [−0.54;0.40]	154
Tree	85%@	c	85.48	1.53	0.98	0.03 [−0.10;0.15]	606
height	±2 m	b	**68.54**	2.61	0.94	0.42 [0.10;0.73]*	267
Stand age	75%@ ±20 y	c+b	80.12	28.78	0.88	−4.48 [−9.38;-0.42]*	135

DQO: (Sect. 21.6.2), type: *c* coniferous, *b* broadleaved, *c+b* coniferous and broadleaved combined DQR: (Eq. 21.1), *s* standard deviation, r_p Pearson correlation coefficient, *CI* confidence interval, *n* number of trees. Level of significance (paired t-test): *** $\alpha = 0.001$, ** $\alpha = 0.01$, *$\alpha = 0.05$

correlated with the time between measurements if both were measured within the vegetation period from April to August (Fig. 21.3, right column), with a more marked trend for coniferous trees than for broadleaved trees, but no such trends for the other attributes.

The quality results (DQO) for *dbh* and *stand age*, *crown length* and *layer* correspond to the expected values, while those for *d7* and *tree height* of broadleaved trees are more than 10% below the limit. The positive signs of the mean differences (*dbh, d7* and *tree height*) mean that the QM values tended to be higher than the FM values. The *dbh* differs highly significantly ($\alpha < 0.001$) from zero, as does the *tree height* ($\alpha < 0.05$) of broadleaved trees. The attributes of broadleaved trees varied more than those of coniferous trees. Acceptable results for (weighted) Kappa were obtained for the attributes *crown length* and high Kappa values for the attribute *layer* (>75). Results for broadleaved trees were again slightly weaker than for those for coniferous trees.

21.6.4 Plot Attributes

The results of the plot attribute analysis are listed in Table 21.6.

DQO The attributes assessed on the sample plots performed differently. Four of them (*stand age*, *stand stability*, *regeneration cover* and *development stage*) exceeded the defined DQOs. The results for *degree of mixture* showed fair results (approx. 5% below DQO). The attribute *stand structure* and *crown closure*, assessed on the nominal scale, however, had DQO results substantially below the defined threshold (18%).

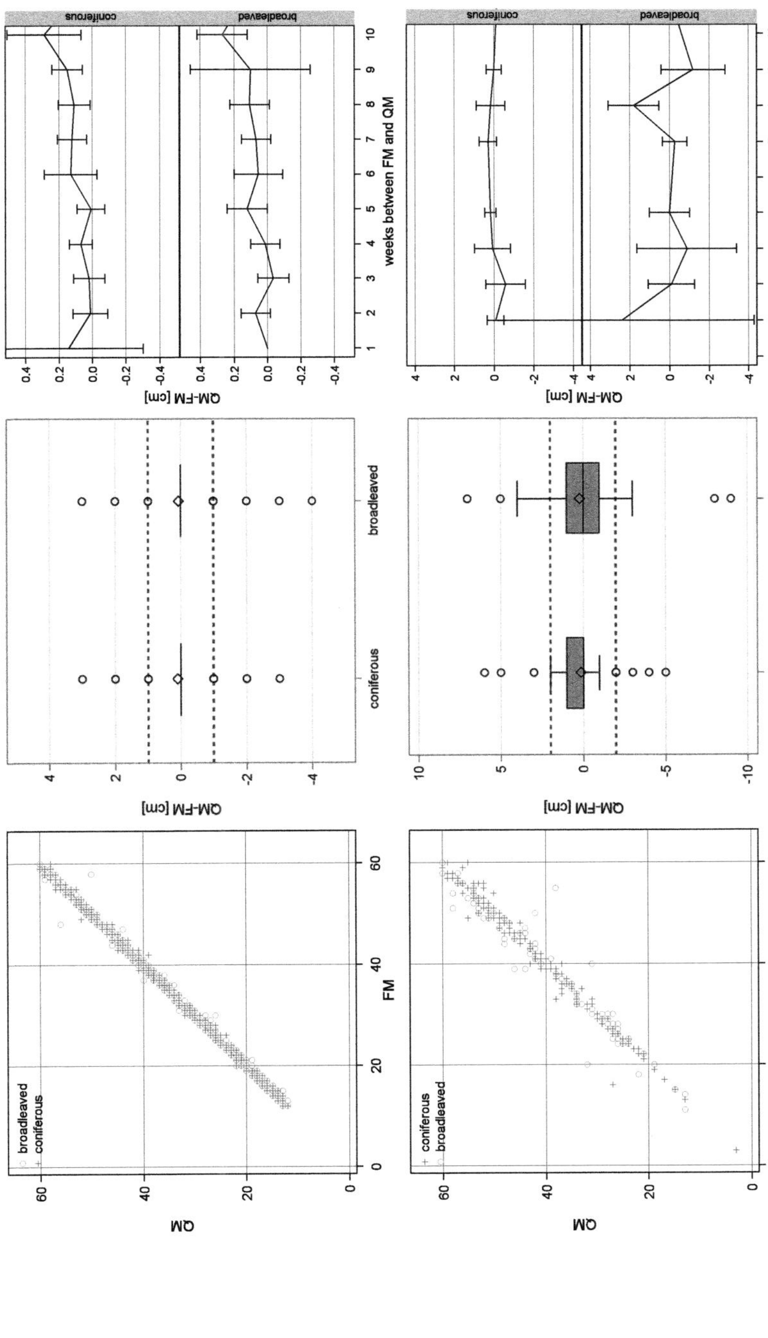

Fig. 21.3 Correlation and QM-FM difference distribution of attributes on interval scale. The *dbh* is measured on all trees, but *d7* and *tree height* on tariff trees only. All tree-based attributes are given separately for coniferous and broadleaved trees. *Stand age* is assessed only on plots without stand borders. The left column shows the correlation between FM and QM, and the middle the distribution of differences, where the dashed lines mark the defined DQO standards. The right column shows mean differences and 95% confidence limits of the means over the number of weeks between QM and FM (both measurements were taken between April and August)

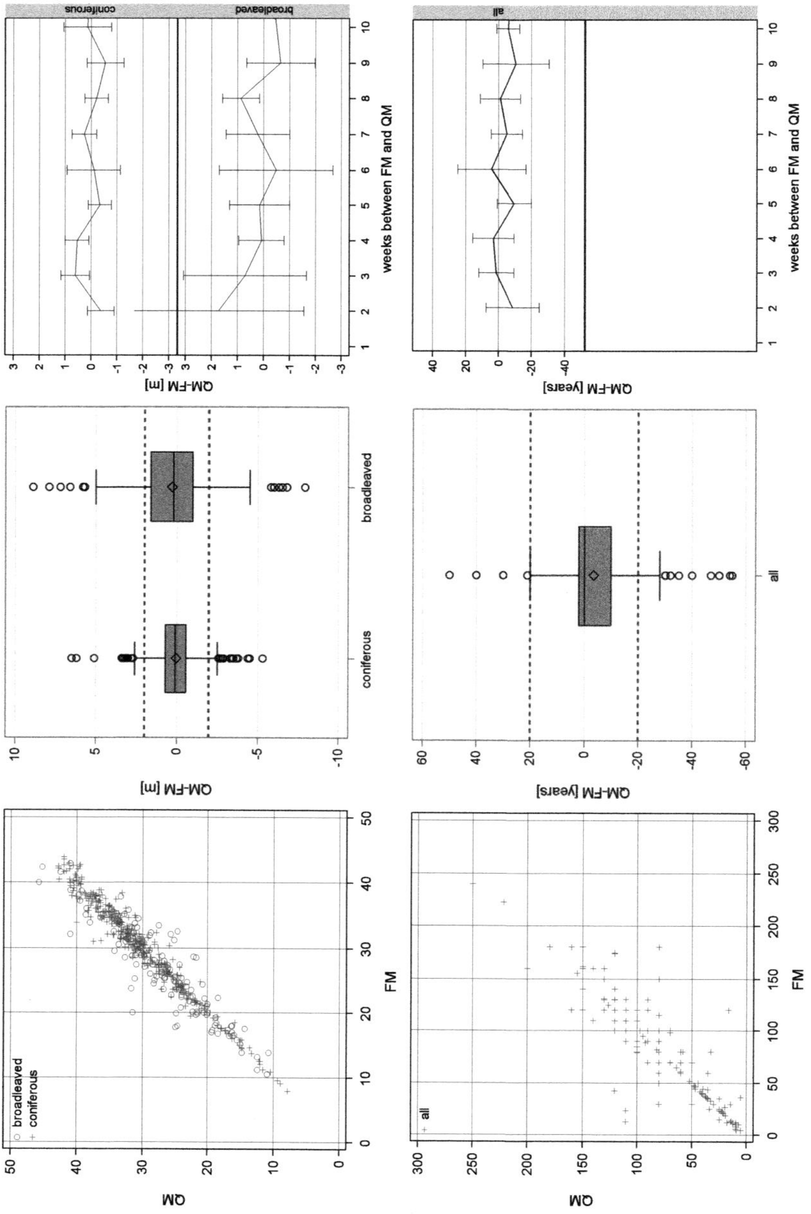

Fig. 21.3 (continued)

Table 21.5 Analysis of tree attribute differences (QM-FM) on ordinal scale

Attribute	DQO	Type	DQR[%]	r_s	Kappa [CI 95%]	*n*
Crown	85%@	c	89.41	0.68	0.64 [0.60;0.68]	2568
length (3)	±0 categories	b	82.31	0.54	0.51 [0.47;0.56]	1679
Layer (3)		c	89.79	0.86	0.80 [0.78;0.83]	2663
		b	87.83	0.79	0.75 [0.72;0.78]	1742

DQO: (Sect. 21.6.2). Type: *c* coniferous, *b* broadleaved, DQR: (Eq. 21.1), r_s Spearman rank correlation coefficient, *CI* confidence interval, *n* number of trees, () number of categories of the attributes

Table 21.6 Analysis of plot attributes on ordinal and nominal scale

Attribute	DQO	type	DQR [%]	r_s	Kappa [CI 95%]	*n*
Stand stability (10)	75%@±1 Kl	c+b	79.85	0.42	0.26 [0.19;0.33]	268
Degree of mixture (4)	90%@±0 Kl	c+b	84.33	0.92	0.87 [0.83;0.91]	268
Regeneration cover (7)	80%@±1 Kl	c+b	90.08	0.76	0.59 [0.52;0.65]	242
Development stage (5)	90%@±1 Kl	c+b	98.76	0.94	0.88 [0.83;0.92]	161
Stand structure (4)	75%@±0 Kl	c+b	**66.42**		0.44 [0.36;0.53]	268
Crown closure (8)	65%@±0 Kl	c+b	**45.90**		0.35 [0.28;0.42]	268

DQO: (Sect. 21.6.2), type *c* coniferous, *b* broadleaved, *c+b* = coniferous and broadleaved combined, DQR: (Eq. 21.1), r_s = Spearman rank correlation coefficient, *CI* confidence interval, *n* number of trees, () number of categories of the attributes

Kappa The attributes *degree of mixture* and *development stage* had equally high Kappa values (0.87 and 0.88). The Kappa value for the attribute *stand structure* was 0.44 and for *crown closure* 0.35. Note that the attribute *stand stability*, which had the most categories (10), had the lowest value for Kappa (0.26) but still exceeded the DQO expectations, although a low Kappa is to be expected according to the rules for interpreting Kappa (Sect. 21.6.2). Including the contingency tables provided below for interpreting the Kappa values is moreover recommended.

Marginal Distribution and Symmetry The contingency table for the attribute *stand stability* (Fig. 21.4) shows the actual assessment was more-or-less restricted to the four categories 6, 7, 8 and 9, with differences in one class being quite frequent (framed cells), and the categories 1 and 2 never occurring during FM or QM. However, as frequencies are symmetric to the main diagonal, no systematic trend is apparent.

The marginal distribution of the *degree of mixture* in Fig. 21.5 is different because marginal frequencies are more evenly distributed, with the majority of assessments falling into the pure categories 1 (coniferous) and 4 (broadleaved). The deviations between FM and QM are symmetric to the main diagonal. The relative uncertainty is generally higher in the mixed categories (2, 3) than in the pure categories (1, 4). The marginal distributions of both the QM and FM attributes reflect the distributions of the attribute across the entire NFI4 FM sample.

Fig. 21.4 Contingency table of attribute *stand stability*. *FM* field measurement, *QM* quality measurement. Framed cells: most of the observations were in categories 6–9. QM-FM deviations of one class are quite frequent

FM	QM								
	3	4	5	6	7	8	9	10	Total
3	**0**	0	0	0	0	0	0	0	0
4	0	**0**	0	2	2	0	0	0	4
5	1	0	**0**	2	5	3	0	0	11
6	0	1	4	**9**	10	9	0	0	33
7	1	2	3	3	**36**	38	7	0	90
8	0	0	1	3	22	**42**	19	2	89
9	0	0	1	2	9	13	**11**	3	39
10	0	0	0	0	0	0	2	**0**	2
Total	2	3	9	21	84	105	39	5	**268**

Fig. 21.5 Contingency table of attribute *degree of mixture*. *FM* field measurement, *QM* quality measurement. Framed cells: mixed categories 2 and 3 with higher uncertainty

FM	QM				
	1	2	3	4	Total
1	**100**	6	1	1	108
2	7	**31**	7	0	45
3	1	9	**26**	5	41
4	1	0	4	**69**	74
Total	109	46	38	75	268

Fig. 21.6 Contingency table of attribute *stand structure*. *FM* field measurement, *QM* quality measurement. Framed cell: category with the highest uncertainty (compare with QM and FM margin totals)

FM	QM				
	1	2	3	4	Total
1	**64**	21	7	1	93
2	**21**	**100**	14	0	135
3	6	16	**12**	1	35
4	1	1	1	**2**	5
Total	92	138	34	4	268

The frequency results of the nominal scaled attribute *stand structure* (Fig. 21.6) contrast the frequency of agreements, with the total frequency of a category at the margin, which is the basis for Kappa. Owing to the nominal scale, category numbers cannot be put into a meaningful sequence in terms of higher/lower or larger/smaller, i.e. it does not make sense to interpret the neighbourhood of a single frequency

Fig. 21.7 Contingency tables of attribute *crown closure*. *FM* field measurement, *QM* quality measurement

FM	QM								
	1	**2**	**3**	**4**	**5**	**6**	**7**	**8**	**Total**
1	**49**	15	1	1	1	3	1	0	71
2	13	**31**	5	5	0	2	1	2	59
3	3	9	**6**	8	0	0	1	1	28
4	0	2	8	**16**	5	1	2	2	36
5	0	1	1	5	**7**	2	2	3	21
6	6	1	1	2	0	**4**	2	0	16
7	1	0	2	6	3	3	**5**	1	21
8	0	0	3	5	1	0	2	**5**	16
Total	72	59	27	48	17	15	16	14	**268**

figure. Category 3 of (all-aged/all-sized) has apparently the most uncertainty, the share of matches (12) is about one-third of the marginal values (34, 35).

The attribute *crown closure* had the worst results regarding failure to meet the DQO and the Kappa value (Fig. 21.7), where the attribute categories refer to different characteristics of the crown layer. The categories 1–5, which are separated by dashed lines in the table, form a group which describes the number of gaps in the crown layer, followed by categories 6–7, which reflect the extent of crown grouping and category 8 for stepped heterogeneous crowns. The frequency counts are quite heterogeneous, with FM and QM clearly disagreeing within as well as between these groups of categories, which indicates that the content and definition of this attribute leaves room for interpretation.

No attribute deviated significantly from symmetry, and visual assessments of the marginal distributions indicated that all FM and QM frequency distributions were similarly shaped apart from *crown closure* and *stand stability*, which are either substantially skewed or unevenly distributed. Hence the Kappa values of these attributes must be interpreted with special care.

Pseudoturnover In Fig. 21.8 the distribution of PT values found from the NFI4 annual panels is presented. Though exact DQOs have not yet been defined for this derived measure, it is expected that most of the PT values are between 10%-30% and the frequency of higher PT values decreases substantially. The results found from NFI4 data confirm this expectation in general. However, in the course of the annual panels no clear trend towards a continuous improvement is visible. The quality level decreased slightly from the 2010 values, with the worst results of the mean PT in 2013 (30%), and then improved to a constant level of around 20% up to the last annual panel in 2017.

21.6.5 Summary

- The *d7* and *tree height* measurements on single trees and the assessment of the plot attributes *stand structure* and *crown closure* were not as easily reproducible

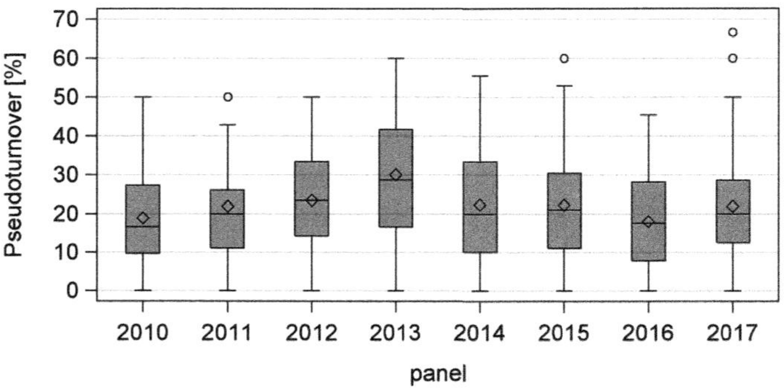

Fig. 21.8 Distribution of Pseudoturnover (PT) values for the eight NFI4 annual panels from 2010 to 2017

as expected in terms of the DQO. Intermediate levels for achieving DQO results were found for the *degree of mixture*.

- The low Kappa values for the *stand stability* may be influenced by the comparably large number of categories (10) and the skewed marginal distribution.
- The results of the DQO analysis were not necessarily congruent with the Kappa and correlation measurements.
- If the statistical significance of the systematic deviation in *dbh* measurements is interpreted in the light of the time between measurements during the vegetation period, the differences found are plausible. The significant deviation from zero of the mean differences of *tree height* for broadleaved trees and *stand age* were not, however, expected and are difficult to interpret.
- The Pseudoturnover, a measure of the reproducibility of the attribute *woody species*, has not improved substantially over approximately the past 3 years. No NFI-specific DQO benchmarks have, however, been defined so far for this attribute.

21.6.6 Discussion

21.6.6.1 Results

Tree Attributes Most of the tree attributes examined proved to be sufficiently reproducible, with generally better results for coniferous trees. The *dbh* measurements fully met the DQO defined by the instructors' team. The overall standard deviation of 0.59 cm was close to the deviation of 0.57 cm found for NFI2 (Kaufmann and Schwyzer 2001). The significant deviation of the mean difference of 0.04 cm was, however, only half that in NFI2 and is certainly of minor practical relevance. The factual increase in the *dbh* between FM and QM could be expected, particularly as both measurements were taken within the vegetation period (April to

August). The reproducibility of the *d7* measurements were rather different for coniferous and broadleaved trees. In fieldwork it seems that *d7* frequently (in about 40% of cases) cannot be measured at all and that broadleaved trees are in general more difficult to measure. The quality of *d7* measurements in NFI4 was, however, generally less biased and more precise than in NFI3, and has improved. Thus, setting the DQO at 85%@± 2 cm is apparently too ambitious for broadleaved trees. The instructors' team are continuing to try to improve *d7* measurement training. The quality results of the *tree height* measurements are acceptable for coniferous trees, but the results for broadleaved trees were more than 15% below the DQO and positively biased (0.41 m). Unlike for *dbh*, the differences in *d7* and *tree height* did not seem to correlate with the time differences between QM and FM. The assessments of the attributes *crown length* and *layer* met the defined DQO in general. The observer agreement was also sufficient for both attributes, but again somewhat less for broadleaved trees.

Stand Attributes The quality of the *stand age* assessment was sufficient, but negatively biased (about 4–5 years). The reasons for this bias have so far not been clarified. The results of the attributes *stand stability* and *regeneration cover* meet the DQO, but the Kappa value of *stand stability* is quite low. This can be at least partly explained by the marginal distribution, which is strongly concentrated in about three out of ten categories. The results of the *degree of mixture* analyses are 5% below the DQO and show that the measurements of the categories pure coniferous or pure broadleaved perform better than those of the two mixed categories. The attributes *stand structure* and *crown closure* were more than 10% points below the DQO. The attribute *crown closure* could be reproduced better in the ordinal scale categories 1–5 (different type of crown closure) than in the nominal scale categories 6–8 (different type of grouped structure). The median and maximum PT values for *woody species* presence show a quite stable median around 20%, which is acceptable. It can obviously not be improved even though the field teams are trained intensively and regularly. The exercises in *woody species* assessment during the training days still show a markedly varying quality performance between individual team members.

The results for stand attributes from the NFI4 repeat survey can unfortunately not be compared directly to the according NFI2 results (Kaufmann and Schwyzer 2001). Since information about stand border was not recorded in NFI2, the authors included all plots in the study, whereas the NFI4 results refer to only a subset of plots where no stand border crosses the plot area (Sect. 9.7). This kind of standardisation, however, was found to be important in order to create comparable conditions for the observer agreement evaluation.

The time between QM and FM measurements apparently has no influence on the observer agreement. We suspected this factor is particularly relevant for the agreement on the number of woody species. However, in none of the annual panels examined did we find any evidence that the time between QM and FM influenced the PT frequency distribution (Fig. 21.9). The same holds true for all other attributes examined apart from *dbh*, which is expectable. This important finding allows more flexibility in the planning of organisational changes in the repeat survey in NFI5.

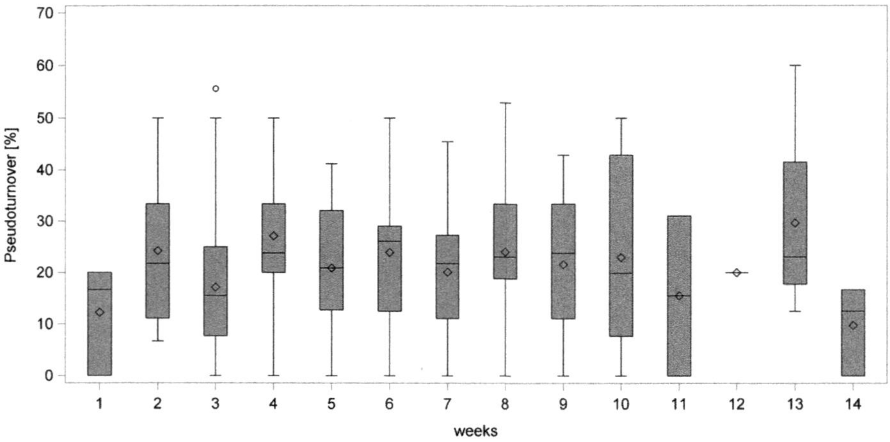

Fig. 21.9 Distribution of Pseudoturnover (PT) values as a function of weeks between FM and QM. Database: all measurements of QM and FM were taken between April and August. Sample sizes for weeks 1–14 were: 3, 19, 31, 28, 20, 14, 25, 28, 17, 10, 2, 1, 4, and 3

21.6.6.2 Methods

General Various analysis methods were employed to evaluate data quality, which enabled the reproducibility of the attributes to be interpreted from different perspectives. The DQOs proved to be very helpful, as they are clearly understandable and reflect the reality of fieldwork very closely. The definitions of the DQOs could probably be improved in two ways: (a) using different DQLs for coniferous and broadleaved trees and (b) defining relative versus absolute measurement tolerances should help to cope with the fact that the reachable absolute precision is correlated with the magnitude of the measurement. For example the *dbh* tolerance could be expressed in percent of the measured *dbh,* such as a threshold of ±5%, instead of being a fixed value to allow more measurement tolerance for larger trees.

Regularly switching field team members between the field teams helps to improve communication between them and standardise interpretations of definitions and instructions. However, some attributes are difficult to assess, which means a mean deviation of plus or minus one category has to be accepted. Although the number of categories could be reduced to cope with this problem, we recommend not doing so because attributes should be assessed in as much detail and as granularly as possible. Moreover, any changes in the definitions of attributes in permanent forest inventories should be kept to a minimum to ensure long-term comparability.

Statistical Analysis Parametric significance tests are important if a potential bias is relevant. Nonparametric tests should, however, always be involved or even prioritised if the data does not meet the requirements for the parametric test perfectly. Using a single number or an index such as Kappa to represent the observer agreement has the drawback of severely reducing the amount of information (Siegel and Castellan 1988). Ideally, the interpretation of these values should be supported by evaluations of the underlying contingency tables and graphical depictions.

The significance tests applied rely on the null hypothesis (H0), in which it is assumed that there are no differences between the results of the FM and QA teams. The alternative hypotheses (H1) of these tests always contain what it is hoped to demonstrate. Should the p-level be greater than α, the H0 is rejected, but if it is lower than α, this is not a proof for H0, i.e. it does not mean that the repeat survey results are equivalent to the original survey, but merely that there is not enough evidence to say they were significantly different. Equivalence tests, specifically developed to address this issue, would probably provide more information on the extent to which both teams obtained the same results. These tests use a reversed null and alternative hypothesis to try to demonstrate equivalence, e.g. applied by Robinson and Froese (2004) in model validation. However, applying equivalence tests requires defining the size of a practically important and relevant difference margin between the two results beforehand. In future, the already defined DQOs will be adapted to allow for transformation into corresponding difference margins.

Repeat Versus Control Survey The repeat survey yields important information about the overall reproducibility of attributes. The main drawback of the repeat survey is that only differences between groups can be revealed because neither team can claim to provide reference measurements which are assumed to be the 'true' value.[1] Thus conclusions about absolute quality are not possible. Moreover, a seemingly correct observer agreement could also be based on a commonly misinterpreted assessment instruction or definition. The control survey has the advantage that the FM results are compared to a true reference measurement. Table 21.7 provides an overview of the strength and weaknesses of both types of QC assessments. As a consequence, more resources will be spent on the control survey in future so the instructors can rapidly give specific feedback to the field teams. Control survey results will also be recorded and evaluated more systematically in NFI5. The increased efforts put to the control survey may be compensated by raising the efficiency of the repeat survey, mainly by replacing the random selection of repeat survey plots with a clustered design. This redesign is likely to save considerable fieldwork time in the repeat survey.

21.6.6.3 Conclusions

The repeat survey provides important information on the achieved data quality in terms of the overall reproducibility of attributes. It reflects the level of precision achievable under current NFI fieldwork conditions, but it cannot attribute any systematic deviations to a specific field group or specific person. In the control survey, however, the instruction teams can take reference measurements to identify

---

[1] In terms of the best value which is available.

Table 21.7 Strengths and weaknesses of the repeat and the control survey

	Strength	Weaknesses
Control survey	Enables: Detection of group-specific issues and deviations from a standard Thorough investigations of selected topics Specific and prompt feedback for the groups	Can only be conducted by a team of instructors, who have very limited resources. This fact makes the control survey quite 'expensive'
Repeat survey	Provides an indication of accuracy (how well the two independent observations match)	Time consuming The results do not provide any deviations (bias) from true reference values
	Reveals topics for improvement in instruction and attribute definition	

the source of such deviations. Both surveys help to continuously improve the quality of fieldwork and thus the quality of the collected data. The field teams are generally keen to improve the quality of the field survey and thus welcome feedback from the repeat and control surveys. Knowing that ground staff are monitoring quality seems to be motivating.

Based on the efforts made so far, we can be confident that the assessments of the same field plots recorded by different field teams would yield almost identical results. The field teams' great experience and high motivation has a positive effect on the reproducibility of observations and thus on data quality. Small differences are likely and acceptable, and can be assigned to random measurement errors. What seems crucial in terms of quality is having stable attribute definitions, clear assessment instructions in the field manual and regular training of the field teams.

References

Agresti A (2013) An introduction to categorical data analysis. Wiley series in probability and mathematical statistics, 3rd edn. Wiley-Interscience, Hoboken

Allegrini MC, Canullo R, Campetella G (2009) ICP-forests (international co-operative programme on assessment and monitoring of air pollution effects on forests): quality assurance procedure in plant diversity monitoring. J Environ Monit 11:782–787

Berger A, Gschwantner T, Gabler K, Schadauer K (2012) Analysis of tree measurement errors in the Austrian National Forest Inventory Austrian. J For Sci 129:153–181

Cochran WG (1977) Sampling techniques. Wiley, New York

Cohen J (1988) Statistical power analysis for the behavioral sciences, 2nd edn. L. Erlbaum Associates, Hillsdale

Cumming G, Finch S (2005) Inference by eye – confidence intervals and how to read pictures of data. Am Psychol 60:170–180. https://doi.org/10.1037/0003-066x.60.2.170

Düggelin C, Keller M (2017) Schweizerisches Landesforstinventar Feldaufnahme-Anleitung 2017. Eidgenössische Forschungsanstalt für Wald Schnee und Landschaft WSL, CH 8903 Birmensdorf

Ferretti M (2009) Quality assurance in ecological monitoring-towards a unifying perspective. J Environ Monit 11:726–729. https://doi.org/10.1039/b902728a

Ferretti M, König N (2013) Chapter 20: Quality assurance in international forest monitoring in Europe. In: Ferretti M, Fischer R (eds) Developments in environmental science, vol 12, pp 387–396. https://doi.org/10.1016/B978-0-08-098222-9.00020-0

Ferretti M, König N, Rautio P, Sase H (2009) Quality assurance (QA) in international forest monitoring programmes: activity, problems and perspectives from East Asia and Europe. Ann Forest Sci 66(2009):403. https://doi.org/10.1051/forest/2009025

Gasparini P, Bertani R, De Natale F, Di Cosmo L, Pompei E (2009) Quality control procedures in the Italian national forest inventory. J Environ Monit 11:761–768. https://doi.org/10.1039/b818164k

Gertner G, Köhl M (1992) An assessment of some nonsampling errors in a national survey using an error budget. For Sci 38:525–538

Kaufmann E, Schwyzer A (2001) Control survey of the terrestrial inventory. In: Brassel P, Lischke H (eds) Swiss National Forest Inventory: methods and models of the second assessment. Swiss Federal Research Institute WSL, CH 8903 Birmensdorf, Birmensdorf, pp 114–124

Kitahara F, Mizoue N, Yoshida S (2009) Evaluation of data quality in Japanese National Forest Inventory. Environ Monit Assess 159:331–340. https://doi.org/10.1007/s10661-008-0632-8

Köhl M (2001) Error sources and their influence on the NFI results. In: Brassel P, Lischke H (eds) Swiss National Forest Inventory: methods and models of the second assessment. Swiss Federal Research Institute WSL, CH 8903, Birmensdorf, pp 297–305

McDowall RD (2005) Effective and practical risk management options for computerised system validation. Qual Assur J 9:196–227. https://doi.org/10.1002/qaj.339

McRoberts RE, Westfall JA (2016) Propagating uncertainty through individual tree volume model predictions to large-area volume estimates. Ann Forest Sci 73:625–633. https://doi.org/10.1007/s13595-015-0473-x

McRoberts RE, Hahn JT, Hefty GJ, Vancleve JR (1994) Variation in forest inventory field-measurements. Can J For Res 24:1766–1770

Nilsson IN, Nilsson SG (1985) Experimental estimates of census efficiency and pseudoturnover on islands: error trend and between-observer variation when recording vascular plants. J Ecol 73:65–70

Pollard J (2005) Forest inventory and analysis. Quality Assurance. https://www.fia.fs.fed.us/library/fact-sheets/data-collections/QA.pdf

Pollard JE, Westfall JA, Patterson PL, Gartner DL, Hansen M, Kuegler O (2006) Forest inventory and analysis national data quality assessment report for 2000 to 2003, vol 181

Robinson AP, Froese RE (2004) Model validation using equivalence tests. Ecol Model 176:349–358

SAS Institute (2013) SAS Stat, 2nd edn. SAS Institute, Cary

Siegel S, Castellan NJ (1988) Nonparametric statistics for the behavioral sciences, 2nd edn. McGraw-Hill, New York

Tomppo E, Gschwantner T, Lawrence M, McRoberts RE (eds) (2010) National Forest Inventories. Springer Netherlands, Dordrecht

Traub B, Cioldi F, Düggelin C (2016) Wiederholungsaufnahmen als Instrument zur Qualitätssicherung im Schweizerischen Landesforstinventar. Schweiz Z Forstwes 167:118–127

Walker GA (2002) Common statistical methods for clinical research with SAS examples, 2nd edn. SAS Institute, Cary

Westfall JA, McRoberts RE, Radtke PJ, Weiskittel AR (2016) Effects of uncertainty in upper-stem diameter information on tree volume estimates. Eur J Forest Res 135:937–947. https://doi.org/10.1007/s10342-016-0985-4

Wulff S (2004) Nonsampling errors in ocular assessments, Swedish experiences of observer influences on forest damage assessments. In: Wiersma GB (ed) Environmental monitoring. CRC Press, pp 337–346

Chapter 22
Compilation of a High-Quality Forest Road Data Set

Marielle Fraefel and Christoph Fischer

Abstract A high-quality data set for forest truck roads is an important prerequisite for road infrastructure management and transport route selection, as well as for the analysis of the ecological impact and recreational use of these roads. To support management decisions, the data set needs to fulfil various quality requirements, such as accuracy, relevance (fitness for use), consistency and completeness. In this section, we introduce the steps necessary to produce a reliable forest road data set, following a clearly defined procedure to ensure reproducibility, as well as consistency for the entire country and between inventory cycles. Furthermore, we point out potential error sources and how the Swiss National Forest Inventory (NFI) tries to avoid them, choosing appropriate measures for error prevention as well as error detection and adjustment of methods.

22.1 Introduction

A high-quality data set for forest truck roads is an important prerequisite for road infrastructure management and transport route selection, as well as for the analysis of the ecological impact and recreational use of these roads. To support management decisions, the data set needs to fulfil various quality requirements, such as accuracy, relevance (fitness for use), consistency and completeness (Brackstone 2003; Herzog et al. 2007). In mapping and spatial analyses, different types of accuracy are normally distinguished:

1. Thematic accuracy: how close is an attribute value to the feature's real value?
2. Positional accuracy: how close are features to their real positions?

M. Fraefel (✉)
GIS, Swiss Federal Research Institute WSL, Birmensdorf, Switzerland
e-mail: marielle.fraefel@wsl.ch

C. Fischer
Scientific Service NFI, Swiss Federal Research Institute WSL, Birmensdorf, Switzerland

C. Fischer, B. Traub (eds.), *Swiss National Forest Inventory – Methods and Models of the Fourth Assessment*, Managing Forest Ecosystems 35,
https://doi.org/10.1007/978-3-030-19293-8_22

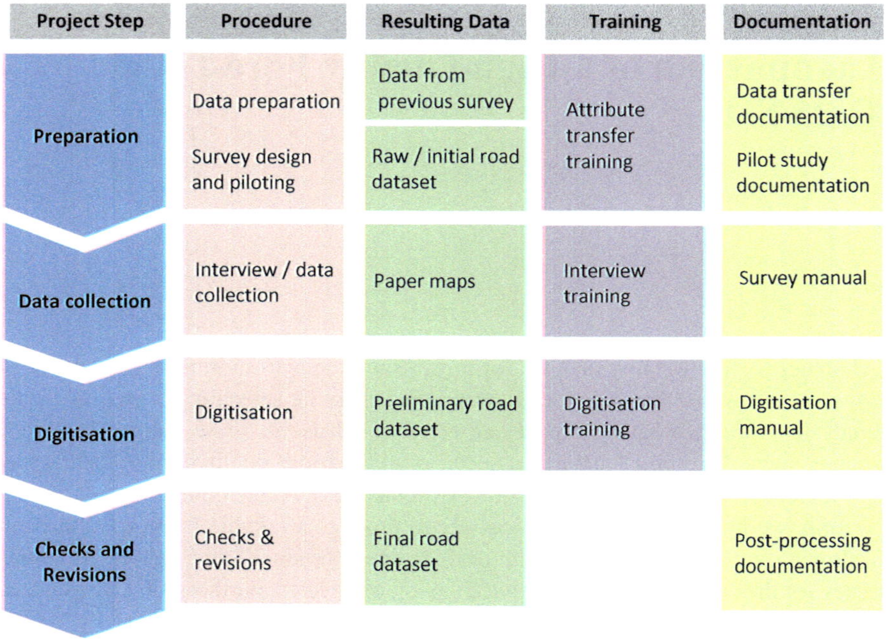

Fig. 22.1 Overview of data preparation, collection and digitisation procedures

3. Temporal accuracy: how up to date is the state that the data set reflects? Does the update interval correspond to the rate at which changes occur?

Data should fulfil the quality criteria of consistency and completeness. For consistency, (a) data entries should not contradict each other, and (b) data from different regions as well as from different inventories must be comparable. Finally, the completeness of the data set ensures that all relevant roads in the entire study area are included. This is important, as the forest road data set also serves as a census.

The forest road data set of the Swiss NFI (NFI) is described in detail in Sect. 10.3. It was first compiled between 1983 and 1985, and has since been updated several times; it is used to calculate road densities (for different road types), and to analyse change and connectivity. We used the geometries of the roads in the swissTLM^{3D} national road and trail data set (swisstopo 2012), which is positionally very accurate, and complemented them with information about trafficability obtained from the local forest services. In this section, we introduce the steps necessary to produce a reliable forest road data set, following a clearly defined procedure to ensure reproducibility, as well as consistency for the entire country and between inventory cycles. Furthermore, we point out potential error sources and how the NFI tries to avoid them, choosing appropriate measures for error prevention as well as error detection and adjustment of methods. An overview of the data generation process, including training and documentation, is given in Fig. 22.1.

22.2 Interview Design and Training

The long and continuous time series in the NFI, which was started in NFI1, allows the comparison of attributes between inventory cycles. This is one of NFI's key strengths. Accordingly, our aim for the forest road survey of NFI4 was to have a data set that is not only internally, i.e. spatially, consistent (between regions), but also temporally (i.e. comparable with previous surveys). Therefore, a large part of the interview set-up was predefined and could not be changed. However, some methodological adjustments and additions were made. In NFI4, the mapped data were digitised directly from georeferenced maps and additional information on road dimensions was gathered for the first time (Sect. 10.3.3).

When defining the new complementary road categories, we faced a trade-off between the goal of having a simple, spatially uniform system with a small number of categories and that of adapting the categories to regions with diverse forest road systems. To make sure the newly defined road categories: (a) were applicable to the entire country, (b) corresponded to the foresters' practices, and (c) could be classified without any additional survey using only the foresters' knowledge, a pilot study was conducted with six forest services in the cantons of Zurich and Grisons covering a wide range of topographic conditions. The pilot study enabled us to adjust the weight and width categories and also to evaluate the digital data collection in the field (i.e. at the forest service offices). Field evaluation, however, proved to be too time-consuming for the interviewers and respondents, and required the presence of a GIS specialist. It was therefore not pursued further, and collecting data from paper maps, as in previous surveys, with subsequent centralised digitisation was favoured.

Integrating the collection of additional road attributes into the road survey led to the situation where the mapping of the original and the new attributes had to be carried out on the same maps, with the same roads evaluated. The values for the minimum width and allowed weight in the complementary survey were higher than in the original survey, which meant that any additional attributes could only be assigned to roads from the original survey. As a result, we chose the mapping codes and symbols carefully to ensure that they are readable and unambiguous. It was also important that all interviewers use them consistently so that the mapped information is comparable between teams and the digitisation team have unambiguous information. All interviewers were therefore provided with training specifically for the road survey to make sure they understood the meaning of the codes, the mapping process itself and frequently occurring cases, as well as some special cases. As Switzerland has four national languages, the interviewers were required to have a good command of the respondent's native language.

22.3 Preparing Data for the Survey

As described in detail in Sect. 10.3, information about road properties was collected by filling in the attribute fields in an existing polyline road data set. To allow direct comparisons between the current and the previous survey, the attributes from both surveys have to be assigned to a common set of road geometries. The data from NFI3 was thus transferred from the Vector25 geometries (swisstopo 2006) used in the third road survey to the most current swissTLM^{3D} (swisstopo 2012) geometries. Although the TLM (Topographic Landscape Model) data set (based on aerial images) contains all roads and paths in Switzerland with a (horizontal) geometric accuracy of 0.2–8 m, it does not contain enough information on width, carrying capacity and cover material for the purposes of the NFI. The transfer was accomplished using semi-automatic matching techniques (Sect. 10.3.2). Once the data had been transferred to the TLM geometries, it could be used as input for the forest road survey, as described in Sect. 10.3.3.

22.4 Quality Assurance During Data Collection

The process of interviewing respondents with the local forest services is described in detail in Chap. 10. Since the interview represents the "point of entry" of the road classification information, great care must be taken to ensure that no errors enter the data during the interview. It should be noted that the information in the survey comes only from the local forest service staff and that none is collected directly in the field.

To avoid systematic classification errors, it is crucial that the interviewers carefully explain to the respondents the criteria, the classification and the reason for collecting the road information. The interviewers also make sure that every potential forest truck road is mapped, i.e. labelled, and that the entire forest district area is covered in each interview.

22.5 Quality Assurance During Digitisation of the Mapped Road Data

After the interviews, the maps with hand-written notes taken during the interviews on them were returned to the office for centralised data entry (Sect. 10.3.4). These were then scanned and geo-referenced to add the newly mapped attributes to the corresponding roads on-screen.

During digitisation, various possible types of error may occur, e.g. typographical errors, misinterpretation, or mix-ups. To minimise the possibility of committing data entry errors, a range of quality assurance mechanisms were applied. These were incorporated into a purpose-built add-in for ArcMap, developed by the NFI, with a

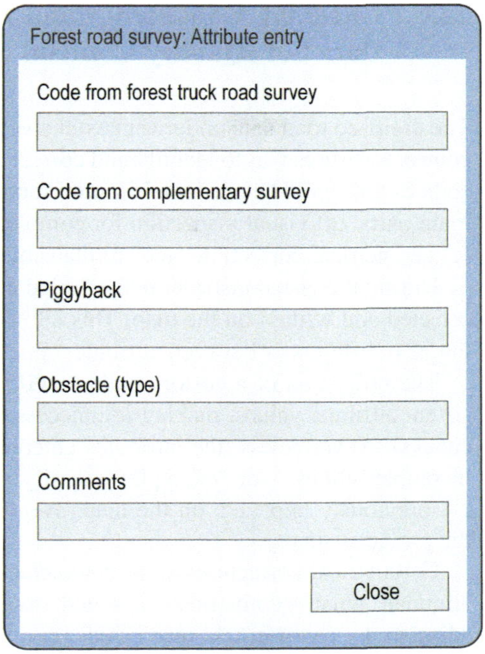

Fig. 22.2 Attribute-entry window with predefined values for road types presented in drop-down lists

drop-down list of available codes to choose from to help eliminate the risk of typographical errors (Fig. 22.2). Furthermore, if the attributes were interdependent, after the first entry the remaining options were adjusted automatically, preventing the entry of inconsistent data. To prevent, as far as possible, the digitisation team making misinterpretation errors, an extensive digitisation manual was written. To prevent small road sections being overlooked, when the first and the last segments of a longer road section were selected, all segments in-between were automatically selected along with the end segments. Using snapping, sticky-move tolerance and warning messages prevented the inadvertent deletion or moving of road segments. In addition, separate buttons for digitising a connection point or a new road allowed quick switching from one task to another. Thus, the tailor-made data entry window helped reduce errors, while at the same time increasing data entry efficiency.

In the case of unanticipated problems, a member of the project management team was consulted. If necessary, open questions were clarified together with the interview team, especially when systematic misclassifications were suspected.

The digitisation manual contained detailed descriptions of all procedures and attributes to ensure reproducibility. The manual was repeatedly updated to incorporate complicated or conflicting cases to ensure consistency during digitisation. The whole digitisation team consisted of six people, who all received thorough instructions and information about the survey and the future use of the data.

22.6 Checking Consistency and Plausibility of the Road Data Set (Quality Control) and Post-processing

The digitised road data underwent extensive post-processing. The aim of the quality-control measures was to identify and correct errors, as well as to detect potential error sources that could be relevant for future surveys. The quality control applied has four main parts: (a) visual inspection for completeness, (b) attribute consistency checks, (c) geometrical corrections and (d) plausibility checks.

During the visual inspection, the digital data were compared with the information selected and written on the map. This allowed not only missing data to be detected but also differences between attribute values and handwritten codes.

The utilisation of a custom-built ArcGIS add-in for data entry prevented inconsistent attribute values, making it unnecessary to check for non-valid values (range checks). Nevertheless, the attributes entered were checked for inconsistencies and incompatibilities, for example: classifying a road as 'newly paved road', i.e. previously unpaved, on the map, even though it had been declared 'paved' in the previous survey.

Geometrical corrections were conducted using topology rules to search for and eliminate small geometrical gaps and overlaps. Very small un-attributed objects adjacent to forest truck roads and very small attributed objects were checked specially (Fig. 22.3). As the digitisation was carried out by a team and the study area was split into smaller workable areas, rigorous edge-matching checks were necessary to make sure that the end vertices of lines at boundaries coincide.

Finally, the plausibility of the data was checked by looking at the percentages of roads belonging to different categories in different regions, and by comparing the results with the information from the previous survey. This evaluation was conducted together with field staff and other NFI subject-matter specialists.

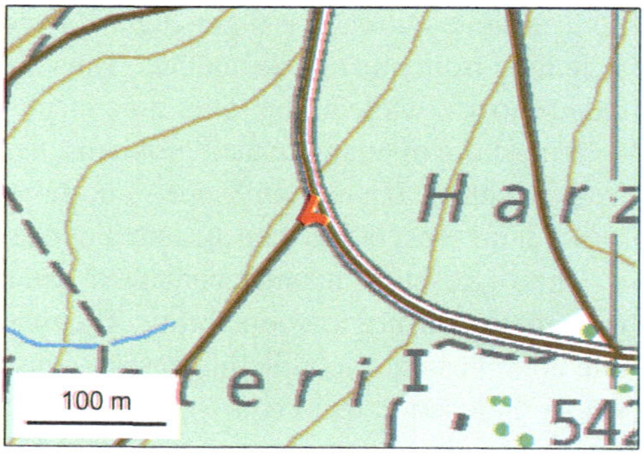

Fig. 22.3 Extract from map used to systematically check all short road segments (red) linking forest truck roads (dark brown)

22.7 Conclusions

Survey data management and quality assurance play a critical role from the very beginning of a survey to the last stages. Usually a balance has to be found between preserving time-series, adapting to technological advances and keeping cost / time expenditure within reasonable limits.

In our data collection process, the main sources of uncertainty turned out to be: (a) how well the foresters knew the roads, e.g. exactly how wide is a specific road element, and (b) how well the digitisation staff interpreted the mapped information, e.g. where does a mapped line end, and how well they handled contradictory information.

We found that running a pilot study to test the applicability of categories and procedures was an important step towards obtaining a relevant and realistic data set. Intensive training proved crucial to ensure comparability between the interview teams.

Our intention to collect information about exact weight limits and widths instead of categories turned out to be unrealistic in our survey, as the foresters usually did not have this information. In some applications, it might make sense to collect this type of data without classifying it.

Recording data directly on computers in the field (paperless interviews) should be given thorough consideration, as it would permit the elimination of errors because the respondents could check that data is being entered correctly. It would also eliminate the need for post-interview digitisation and reduce the editing effort. A good map overview of the area in question should, however, still be provided. More research on methods to implement data entry in the field is needed.

References

Brackstone G (2003) Managing data quality in a statistical agency. UN Economic Commission for Latin America and the Caribbean

Herzog TN, Scheuren FJ, Winkler WE (2007) Data quality and record linkage. Springer

swisstopo (2006) VECTOR25. Das digitale Landschaftsmodell der Schweiz

swisstopo (2012) swissTLM3D Version 1.1 Ausgabe 2012 – Nachführungsinformation

Part VIII
Raw Data Collection Software in the Swiss NFI

Chapter 23
Raw Data Collection Software in the Swiss NFI

Sandro Bischof, Enikő Stüdeli-Fey, and Rolf Meile

Abstract The Swiss National Forest Inventory (NFI) is based upon three main raw data sources: the aerial-photo interpretation, the terrestrial inventory and the interview survey. Once the aerial images have been analysed, all forest sample plots, excluding inaccessible forests, are visited terrestrially. As soon as a forest district is completed, an interview survey is conducted with the responsible forester. The software application MAIRA has been developed for the terrestrial inventory and SILVIS for the interview survey. In the following, we present these two field applications, with special emphasis on the applied data model, the data flow between these two applications and the database, as well as the main tasks of the software during the assessment. We also describe the software architecture and our software engineering approach, and we discuss software quality and data quality considerations.

23.1 Introduction

The Swiss National Forest Inventory (NFI) is based upon three main raw data sources: the aerial-photo interpretation, the terrestrial inventory and the interview survey. Once the aerial images have been analysed, all forest sample plots, excluding inaccessible forests, are visited terrestrially. As soon as a forest district is completed, an interview survey is conducted with the responsible forester. The software application MAIRA has been developed for the terrestrial inventory and SILVIS for the interview survey. In the following, we present these two field applications, with special emphasis on the applied data model, the data flow between these two applications and the database, as well as the main tasks of the software during the assessment. We also describe the software architecture and our software engineering approach, and we discuss software quality and data quality considerations.

S. Bischof (✉) · E. Stüdeli-Fey · R. Meile
GIS, Swiss Federal Research Institute WSL, Birmensdorf, Switzerland
e-mail: sandro.bischof@wsl.ch

C. Fischer, B. Traub (eds.), *Swiss National Forest Inventory – Methods and Models of the Fourth Assessment*, Managing Forest Ecosystems 35,
https://doi.org/10.1007/978-3-030-19293-8_23

23.2 The Field Applications MAIRA and SILVIS

The work of the field staff is supported by two Java applications developed in-house that are used in completely different survey settings. MAIRA is used outdoors on the sample plots and SILVIS is used at the district forester's office. MAIRA helps collect field data and is used every year during the field season, while SILVIS records answers to interview questions and is used once every 5 years.

23.2.1 MAIRA

MAIRA is the principal field application of NFI4 and is used to collect a comprehensive set of various types of data from a sample plot. The application serves as the user interface between the assessment process in the field and the central database hosting the raw data. Therefore, MAIRA has a well-elaborated graphical user interface and a robust transmission service to the database to ensure the plausibility and logical sequence of the assessment. The application runs on a Microsoft Windows tablet PC system.

Data is gathered and processed for multiple purposes: (a) collecting measured or estimated raw data, as well as data in the form of expert evaluations (these form the core data set for subsequent scientific analysis); (b) locating existing plot centres using marking points or establishing new plot centres using reference points; (c) recording the start and end time of certain activities during the assessment to support the preparation of a new inventory cycle; (d) collecting data for documentation reasons, such as the coordinates of the car park or the condition of the path to reach the sample plot; (e) recording binary data, a new category of data introduced in NFI4 that will be post-processed, such as photos and Global Navigation Satellite System (GNSS) data.

The field application LAURO, also a Java-application developed in-house, was used for NFI3. Extensive enhancements and expansions of LAURO led to the development of the present MAIRA software.

23.2.2 SILVIS

Every 5 years, the field staff visit approximately 800 foresters and asks questions about the plots located in their forest districts. All the plots will have been assessed previously in the field, so the aim of this interview survey is to obtain additional information from the foresters concerning forest management. The SILVIS application leads the field staff through the interview questions and records all answers. SILVIS performs plausibility tests and supports the interviewer by supplying information previously recorded in the field by MAIRA.

23.2.3 Hardware and Devices

At the beginning of NFI4, we used a Fujitsu tablet as a field device. Although designed for office work, the tablet was modified with a special screen glass to deal with the variable light conditions in forests. During NFI4, we replaced the Fujitsu tablets with Panasonic Toughpad fz-g1 tablets, which are built for outdoor conditions. We used both types of tablets with the virtual keyboard and the pen as input devices. We have always run the tablets with Windows operating systems. During the 9 years of NFI4, we migrated the operating system from Windows XP to Windows 7 and later to Windows 10. The Java virtual machine makes such migrations quite simple.

The tablet not only runs the field application but also hosts a few non-self-developed programs, for example software to identify plants, to view PDF-files or to fill in various work reports (Microsoft Excel).

Three external devices are currently used via the USB interface: (a) camera, (b) GNSS receiver and (c) backup memory stick. It is possible to communicate with additional devices via Wi-Fi or Bluetooth, for example to automatically transmit measured values, but this has not been a user requirement so far.

To run SILVIS, we use simple Fujitsu notebooks because it is better to have a real keyboard and a larger screen in the interview setting so that information like maps and aerial photographs can easily be shared with the forester when discussing a sample plot.

23.2.4 Distinct Software Purposes

The two field applications are used in fundamentally different circumstances with different work organisations and user settings. With MAIRA a single field team member uses the tablet to record data while another person performs the measurements. In contrast, SILVIS runs on a notebook and the survey is conducted in close collaboration with the interviewed foresters. This setting facilitates discussion and clarification because the interviewer and forester can view a single screen at the same time.

The two field applications also differ considerably technically. MAIRA comprises highly complex business logic, and its attributes have strong visual dependencies. These must therefore be arranged appropriately in the input forms, which means software ergonomics play an essential role (Sect. 23.7.3). The SILVIS data is, however, mainly categorical. Interviews proceed from one question to the next in a linear fashion and SILVIS, therefore, involves much less business logic than MAIRA. The configuration is much simpler as well because SILVIS is only used by NFI and no regional inventories or other users are involved.

Since the conditions and use of the two field applications differ, their development and deployment are completely decoupled, which reduces the complexity of the software engineering and maintenance involved. Nonetheless, the two applications are connected by a common data flow (Sect. 23.3.3), which means that the requirement engineering is simplified because it is limited to just one product.

23.3 Data Model and Data Organisation

23.3.1 Database

The two field applications upload collected raw data to a centralised Oracle database for persistent storage. For this purpose, a relational database model (Codd 1971) was implemented in third normal form, which guarantees a high level of normalisation, referential integrity and atomicity. Concepts like *visited plots*, *recorded trees* or *lichens on recorded trees* are transformed, including their attributes, into separate database tables. Around 50 so-called raw database tables, i.e. base tables, have been implemented, which are connected with primary-foreign key relationships. These tables have been filled with NFI data sets from the inventories of the last 40 years. In addition, about 300 lookup tables have been created to constrain the values for each categorical attribute used in the base tables. The base table data and lookup table content serve as an immutable base for predetermined data in MAIRA, SILVIS and LUBI applications (Sects. 23.3.2 and 4.2). Furthermore, all derivations used in NAFIDAS (Traub et al. 2017; Sect. 20.3.3) continually draw information from this database.

Two of the most important design goals for the database were long-term availability and stability of the base table data. At the same time, flexibility when adding new attributes or concepts had to be guaranteed. The database model implemented met these needs by simplifying all data access and enhancing reliability for later processes. The fact that all data is kept in a single database turned out to be a crucial step for the long-term storage and derivation of NFI time series. A similar approach is taken in the Swedish NFI according to Fridman et al. (2014).

23.3.2 Predetermined Data

When using MAIRA and SILVIS for field data collection, both applications need to be supplied with some initial data. Not only do the module preferences (Sect. 23.4.1) have to be configured, but the applications also have to be driven by predetermined data, i.e. data originating from previous inventories stored in the database. A full data set for the plots to be visited and relevant data collected in all former inventories must, therefore, be available. The predetermined data files must be exported from the database using a specified exchange format. Furthermore, the value ranges of attributes need to be defined in advance, and stored in additional data files.

PDF documents represent one type of predetermined data file. These documents are created from heterogeneous data sources, such as photographs, geographic data sets and pure relational data, using database procedures (PL/SQL) and Python, with data originating from the base tables and GIS web services. Such a PDF-document file serves as an independent information source for the field team. The following files are part of the client repository (Sect. 23.3.4).

- Manually drawn sketches of the sample plot environment digitised from previous inventories. These are used to relocate the plot.
- Sketches generated from the base tables with tree information for each plot, including graphics and tables with basic information about the trees, produced from the predetermined data. These are used for orientation on the plot and for locating and identifying the correct trees.
- Sketch with an aerial image including reference points originating from aerial-photo interpretation. The reference points are used to locate the centre of a new sample plot (Chap. 5).

In addition to the PDF documents, non-binary content files are produced in Extensible Markup Language (XML) format. An associated XML Schema Definition (XSD) is delivered as part of the predetermined data. This definition is used to specify and verify the collected raw data, which is, in turn, stored in an XML file. For each sample plot to be visited in the field, the following three text-based files are passed to the client repository (Fig. 23.1).

- File with all relevant information about the sample plot, including data collected in previous inventories (XML). This provides specific data about the plot and tree history, which is also used for validating newly entered data.
- File with valid lookup values (key-value pairs) for attributes used in the applications (XML). This provides data for filling combo boxes and controlling business logic using selected codes.
- Document for verifying the final result file (XSD). Enables the software to check the final result file, which can then be transferred to the database.

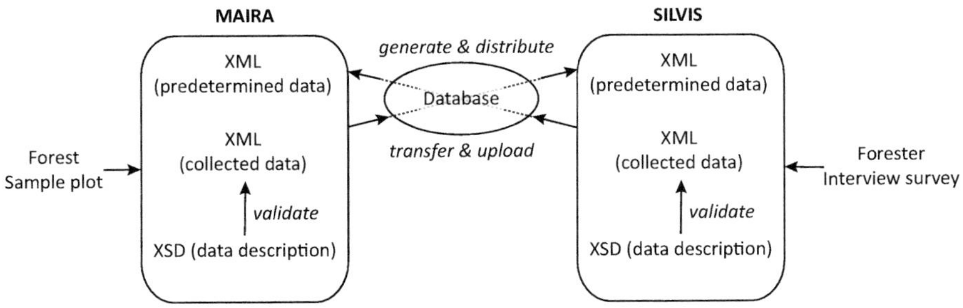

Fig. 23.1 Data flow between the database and the two separate field applications

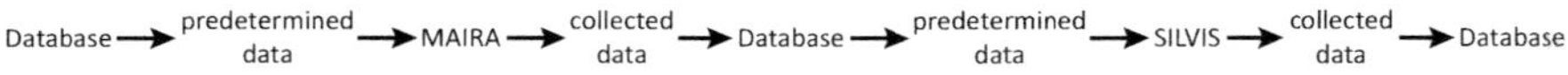

Fig. 23.2 The data flow shown in a serialised form

23.3.3 Data Flow

The two applications MAIRA and SILVIS are technically independent of each other. What connects them, however, is a common data structure and architecture. Both applications are controlled via XML files containing predetermined data, mainly from previous terrestrial inventories or aerial-photo interpretations, if available. In addition, MAIRA, which is used on the sample plots, contains information from the most recent interview with the district forester, while SILVIS contains information on the sample plots (Fig. 23.1). In both applications, this predetermined data is especially important for information on forest management and interventions.

If the data flow is represented in a serialised way (Fig. 23.2), it becomes apparent how easily and uniformly this process can be defined beyond the limits of each application.

23.3.4 Repository

All data related to the field assessment is stored and managed in the MAIRA repository, which is a collection of directories and corresponding files, organised in a well-defined structure. A repository is created on each field tablet as part of the software installation at the beginning of a field season.

The design and implementation of the repository were important for the field application in the initial stage of NFI4 (Fig. 23.3). The MAIRA repository is directory- and file-based, unlike the former field application LAURO in NFI3, which used a locally installed relational database on the tablet. In the LAURO application, the schema of the database was more or less a copy of the schema of the central database with selected sample plots. This concept has several disadvantages: (a) it is inflexible regarding the removal or addition of sample plots during the field season; (b) the relationship between predetermined and collected data is unclear; and (c) a fully implemented relational database on each tablet is inefficient. The new repository therefore had to have greater flexibility and self-containment.

To meet these requirements, each sample plot is stored in a separate directory in MAIRA (Fig. 23.4). Initially, such a directory is equipped with: (a) the task description in the form of an XML file (called the job file), which contains the main predetermined data and defines the assessment; (b) the lookup file containing the codes consisting of key value pairs; (c) the XSD, to check the consistency of the XML file containing the collected data before uploading it to the database; and (d) the sketches and additional handwritten information from former inventories available as PDFs. These files are generated and delivered from the database in work

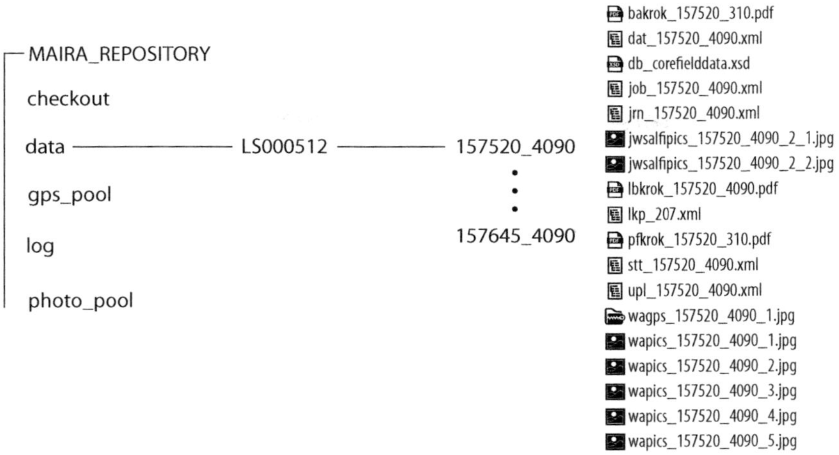

Fig. 23.3 Structure of the MAIRA repository including the full content of the directory from one sample plot

Fig. 23.4 At the end of a plot assessment, the entire collected data as well as any binary data, such as photos or GPS measurements, is stored in the specific subdirectory of the current sample plot

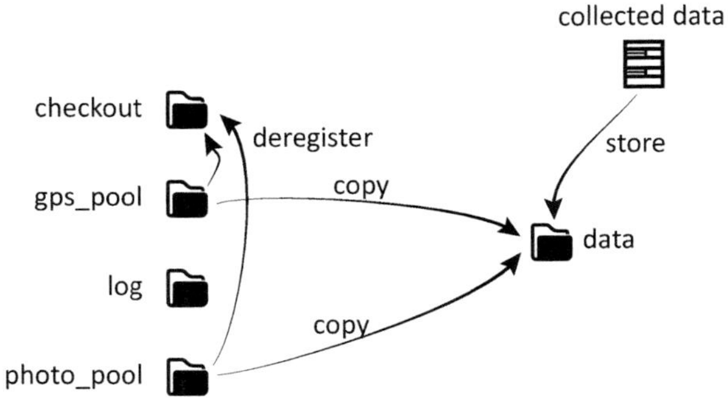

packages. A work package contains a selection of sample plots to be assessed by a field team. At the beginning of the field season, the initial work package defines the main workload for each field team. During the field season, new work packages can be assigned to the team to repeat a survey of randomly selected sample plots for statistical purposes (Sect. 21.5). Each directory representing a sample plot contains all the information needed to undertake an assessment, making it self-contained for our purposes. This approach allows us to reallot sample plots to different teams for organisational reasons throughout the field season simply by moving a directory from one client to another. This gives us greater flexibility.

During a plot assessment, a few files are added to the sample plot directory: (a) the XML data file storing the collected raw data (the most valuable file); (b) the state file containing details about progress in assessing the plot, as well as interim values, which are only relevant during the assessment and thus not permanent; (c) the journal file storing the overall progress in the assessment of a sample plot; (d) all binary files like photos and GNSS measurements; and (e) the upload file, created from the data file at the end of a plot assessment, with an XML structure optimised for easy handling in the database.

23.4 MAIRA at a Glance

23.4.1 Chapters and Modules

The software MAIRA is organised into chapters and modules. A chapter is a visible and selectable component on the user interface and is tightly linked to the chapters in the Manual of the Field Survey (Sect. 9.3). A chapter consists of a sequence of input forms organised as modules.

A module is an abstraction for a delimitable part of the assessment process, which can be activated or deactivated. The activated modules therefore determine the sequence of the assessment. A deactivated module, along with its windows, remains hidden during an assessment. A module covers one topic and is not dependent on other modules, and therefore has no side-effects on the further assessment and the later statistical analysis. A single module covers an entire chapter in some cases.

The capability to activate or deactivate a module makes it possible to configure the application for different customers, such as regional forest authorities or other countries (such as the Principality of Liechtenstein). MAIRA is then used in these forest inventories, which are usually carried out on a condensed grid of sample plots. Compared with NFI assessments, regional inventories typically have fewer attributes to assess on a sample plot so that fewer modules are activated in MAIRA.

23.4.2 Plot Selection

The entry point of the graphical user interface (GUI) in MAIRA is a table presenting the task list that includes all sample plots a field team has to assess during a field season (Fig. 23.5). The table entries correspond directly to the directory structure of the repository, as discussed above in Sect. 23.3.4. MAIRA does not require plots to be assessed in a particular order. Instead, the order is chosen by the field staff and, in most cases, depends on short-term considerations. The application provides extensive information about, e.g. each plot's geographic location, elevation, existing forest boundaries and type of forest (standard vs. shrub forest), which makes it easier for the field team to plan their routes and assessment areas.

The assessment process for a chosen sample plot is more clearly defined than the selection of sample plots. Some of the chapters are inter-dependent, and they have to be completed in a strict order. This applies to the four basic chapters, which have to be completed before the assessment can continue with further chapters.

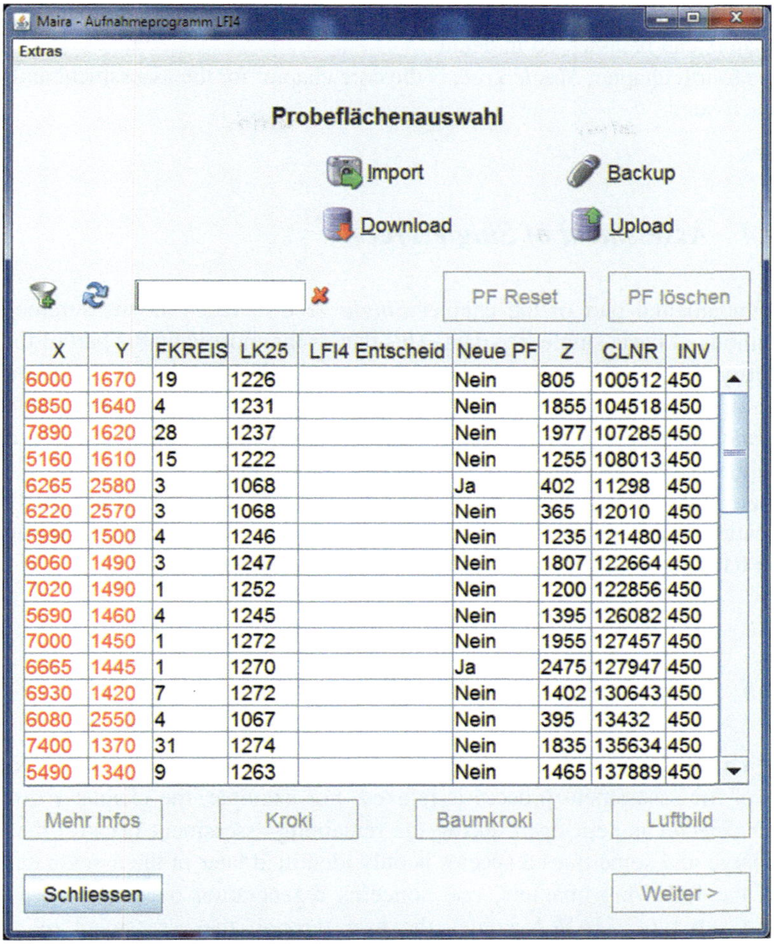

X	Y	FKREIS	LK25	LFI4 Entscheid	Neue PF	Z	CLNR	INV
6000	1670	19	1226		Nein	805	100512	450
6850	1640	4	1231		Nein	1855	104518	450
7890	1620	28	1237		Nein	1977	107285	450
5160	1610	15	1222		Nein	1255	108013	450
6265	2580	3	1068		Ja	402	11298	450
6220	2570	3	1068		Nein	365	12010	450
5990	1500	4	1246		Nein	1235	121480	450
6060	1490	3	1247		Nein	1807	122664	450
7020	1490	1	1252		Nein	1200	122856	450
5690	1460	4	1245		Nein	1395	126082	450
7000	1450	1	1272		Nein	1955	127457	450
6665	1445	1	1270		Ja	2475	127947	450
6930	1420	7	1272		Nein	1402	130643	450
6080	2550	4	1067		Nein	395	13432	450
7400	1370	31	1274		Nein	1835	135634	450
5490	1340	9	1263		Nein	1465	137889	450

Fig. 23.5 Example of the form for selecting sample plots in MAIRA (German)

23.4.3 Compulsory Modules of MAIRA

Two of the basic chapters are: (a) *Identification* and (b) *Sample Plot*, which both consist of modules that must be completed in any case. *Identification* includes some basic documentary questions needed for the preparation of an assessment while *Sample Plot* includes the questions that determine the procedure for the rest of the assessment. These are: (a) sample plot accessibility: if the centre of the plot is unreachable, the assessment is terminated; and (b) 'forest' (forest or shrub forest) or 'non-forest'. If the plot is classified as 'non-forest', or shrub forest for some regional inventories, the assessment is stopped.

The first part of the third chapter, *Boundary*, determines, for example, if the plot is limited by an accessibility boundary or has a stand boundary. Other boundary information concerning the forest edge is grouped in a separate module where these parameters can be assessed in detail.

The fourth chapter, *Single Tree*, is the core chapter for the assessment and cannot be deactivated.

23.4.4 Assessment of Single Trees

The fundamental part of the chapter *Single Tree* includes all measurements and evaluations of basic single tree data. This data is the most valuable part of the forest inventory and plays a central role in the data analysis. The chapter additionally has several modules, such as *Assessment of dead sample trees*, *Assessment of fungi on trees* and *Tariff tree*, all of which depend on basic single tree data. For a detailed description of the assessment of single trees, see Chap. 9.

The single tree assessment is the most complex and extensive part of the application with regards to business logic. Furthermore, nowhere else in the application has such an effort been made to prevent incorrect data input.

23.4.5 Assessment of Further Plot Data

In the chapters concerning the interpretation area (50 × 50 m), the strict sequential order of the assessment process is relaxed. For example, the chapter *Presence of Woody Species* is kept open during the remaining assessment because the area is quite large and sometimes a species is only identified later in the assessment.

Of the following chapters, one concerns regeneration occurring on a clearly defined sub plot. MAIRA guides the user through the assessment of a set of parameters, such as quantifying damage on young trees or counting the young trees in different growth classes. There are also chapters dealing with (a) lying deadwood, (b) damage caused by vehicle tracks during timber harvesting, and (c) preliminary questions for the forester's interview.

Finally, there are two chapters dealing with organisational issues. The *Import* chapter incorporates all binary data, i.e. photos and GNSS data. In the last chapter, *Time Measurement*, the field staff can review and complete the registered time information.

23.4.6 Selected Complementary Functionalities

MAIRA has not only the core function of enabling scientific data collection on forest sample plots, but also some complementary capabilities. Three of these capabilities,

Upload, *Master Copy* and *Download* are described in detail below because they considerably simplify and improve the efficiency of the assessment process.

23.4.6.1 Upload

Uploading is ideally the last working step at the end of each fieldwork day. The aim is to transfer all assessed data to the central database as soon as possible after data collection. The tablets have a SIM card installed that can connect to the mobile network if the user allows explicit access to it. The advantages of this approach are: (a) it provides additional data safety beyond the backup; (b) it can track the overall progress in the field; and (c) initial statistical analyses can be started as soon as the last sample plot has been uploaded at the end of the field season. As a result, the entire process can be more efficient.

23.4.6.2 Master Copy

A so-called *Master Copy* is provided to support a control survey on selected sample plots, which is conducted to improve the reliability of the assessments. Such a survey is undertaken with full access to data collected in the regular assessment and serves to identify general disagreements or misunderstandings. Any discrepancies observed are used to improve future instruction of the field teams. All these sample plots are coloured blue in the plot selection table to identify them as copied and completed (or uploaded) sample plots.

23.4.6.3 Download

Some randomly selected sample plots are re-visited and entirely re-assessed for statistical reasons, i.e. a repeat survey is done (Sect. 21.5). The *download* functionality is used to receive new allotted tasks to conduct such repeat surveys or to organise the workload anew. The advantage of this approach is that the tasks can be downloaded remotely via the mobile network. This means the field team does not have to come to the headquarters to obtain additional tasks and no physical handover of a storage medium is needed.

23.5 Software Architecture and Principles

Having good software architecture reduces, in our opinion, the complexity of understanding the existing code and provides guidelines for new code. It is important that the software architect firmly asserts the architecture and that the developers adhere to it completely. Therefore, in a small software development team, such as the group working with NFI, it is important to involve the developers in the conceptual design of an appropriate and understandable architecture. This section focuses on MAIRA because the difference between SILVIS and MAIRA regarding software architecture is primarily a difference in complexity.

23.5.1 Applied Architecture

Software architecture should be simple and clear. In MAIRA, we distinguish three different aspects: structure, pattern and naming.

23.5.1.1 Structure

The structure includes the organisation in software packages and components. For example, each module or larger functionality (e.g. database upload) is put in a separate package. The components of each package follow a given structure defined by a template, and the name of these components follows a uniform name pattern across the different packages. For the software engineer, these measures lead to a better overview of the package contents and an efficient identification of each component's capabilities.

23.5.1.2 Pattern

An architectural pattern defines the interactions among the software components. The main architectural pattern we use is the Model-View-Controller (MVC) pattern. Each of these three components has a specific and clearly defined task. The advantage of this approach is that we do not have to seek new solutions for problems that have already been solved. Furthermore, the approach is reusable for each module and throughout the whole application.

For over a decade, the Swing framework in Java has been the main framework available for developing graphical user interfaces. In our development environment, where most of the code is developed to handle user input appropriately through a graphical user interface, the MVC pattern is considered best practice, but the Swing framework has no explicit concept to support it. As a result, we developed our own MVC architecture (Fig. 23.6) to separate these three categories, even though the view and controller can actually be completely separated. Nevertheless, the pattern helped to split the problem into smaller parts.

The central part of the architecture presented in Fig. 23.6 is the controller which mediates between the model and the view. The model contains the data and the business logic. The view contains the elements of the graphical user interface and determines the layout. Because the view is associated to the model only via the

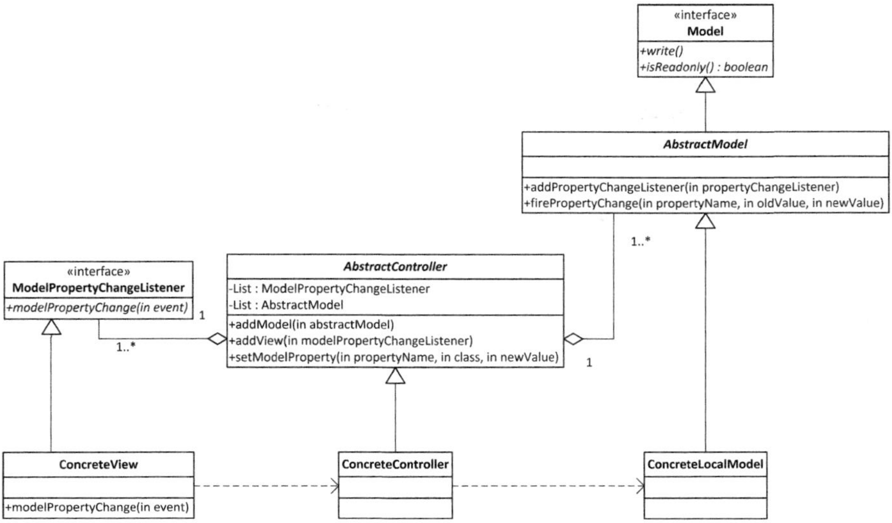

Fig. 23.6 The Model-View-Controller pattern guarantees complexity reduction (UML class diagram)

controller, the view and model are not coupled, which makes the implementation more flexible.

The concrete implementation of the view contains part of the business logic, in that input options are changed according to values entered by the user. The presence of business logic in the view thus makes it impossible to completely separate the architectural building blocks. Our architecture has the additional advantage that several models can be registered at the controller so that different types of data, e.g. predetermined and collected data, can be used.

23.5.1.3 Naming

The third concept is naming, i.e. finding mnemonic names to identify variables, methods, classes, packages and interfaces. This core task is not easy to perform while writing software. One convention is that an attribute stored in the database takes its original name from the database, and this name is propagated unchanged through all layers in the entire code. This convention facilitates efficient communication between different positions, makes the code more readable, and also facilitates finding bugs faster. No translation between different layers of the software is needed because the database is the source of all attribute names.

Besides applying the three architecture areas described above, we aim to develop as many reusable software components as possible and, on a more detailed level, to write clear code. To communicate and implement software architecture, we rely on several design patterns (Gamma et al. 2009). Finally, we feel that good software architecture reduces unnecessary dependencies between the components, and we therefore apply the concept of *separation of concerns* (Laplante 2007).

23.5.2 Languages and Libraries

Both field applications, MAIRA and SILVIS, are written in Java. We used Version 1.6 at the beginning of NFI4 and switched to Version 1.7 later on. Java has a comprehensive system of libraries and frameworks. To develop the graphical user interface, we used the Swing framework.

MAIRA consists of three core libraries: the main library (maira-[version].jar) and two libraries containing the data model, i.e. a predetermined data library (LFICoreFieldReq.jar) and a global data model which leans on the relational database structure (LFICoreFieldDat.jar).

Some of the other technologies used include:

- OJDBC: for connecting to the remote relational Oracle database.
- LOG4J: a logging utility.
- JDOM: for accessing, manipulating and outputting XML data.

23.6 Software Engineering and Software Quality

23.6.1 Guiding Principles

The core guiding principle is that MAIRA exists only as one operative version regardless of whether the application is used in the national or in a regional inventory. The application is written specifically for the needs of the national inventory. Regional authorities are invited to use this application for their own inventories. They usually have the same needs in the core part of the assessment, but if the needs differ too much, modules can be deactivated, as mentioned in Sect. 23.4.1. This principle of having one single but highly configurable version reduces the complexity in software engineering and data management and keeps the IT resources small. However, owing to the continuous inventory in NFI4, the software versions may differ from year to year to take into account new user requirements or bug fixes.

23.6.2 Software Development Process

To describe the software development process, it is necessary to introduce two terms: (a) the field season and (b) the software development cycle. The field season,

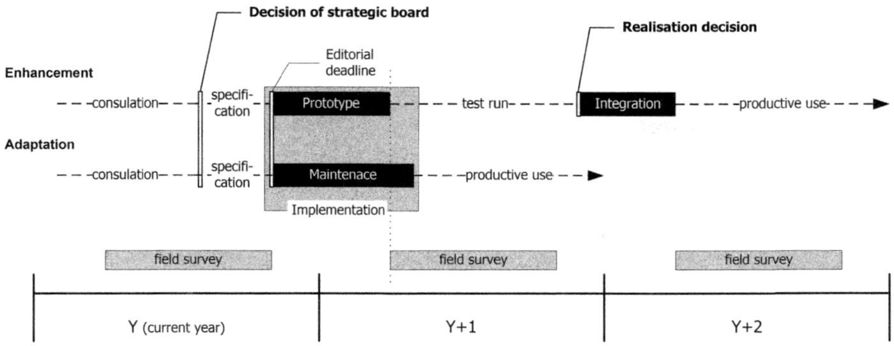

Fig. 23.7 Software development cycle over 3 years

typically April to November, is the working period. NFI4 was designed as a continuous inventory lasting 9 years, which means it is possible to improve, modify or adapt the assessment, and therefore the field application, each year (Fig. 23.7). Not every change has a major impact on the assessment, software or database. Bug fixes are typically necessary and should not affect the existing system, usually they should be solved anyway, depending on their importance.

NFI has established a process to steer the ideas and prioritise upcoming needs and requirements in the so-called software development cycle. A software development cycle is usually completed between two field seasons, but preparation for a software development cycle already starts during the field season. No changes are made to the software during the field season, except for necessary bug fixes, which are usually performed at the beginning of the season. This approach leads to a trouble-free field season and prevents unnecessary software updates.

The software development cycle recurs each year and consists of a set of activities and milestones (decisions). We distinguish between two forms: enhancement and adaptation, where an enhancement requires a test run in the field before it can be implemented and therefore leads to a prototype. An adaptation is maintenance work that, after a specification, is directly implemented and put into productive use. The consultation and the specification phases are times of intense communication between the developers and the ordering party.

Changes generally affect the data structure or the extent of the attributes. Changes in the business logic, graphical user interface, background processes, general features or other parts of the software can be complex and expensive, but they normally have a smaller impact on the data.

Removing or adding one or more attributes, or modifying the meaning of an attribute, always affects the core of NFI. The longitudinal data, i.e. repeated observations of the same variables, and its consistency over the last 40 years is primarily what makes NFI so valuable. To preserve the significance of this longitudinal data, changes have to be carefully analysed and implemented.

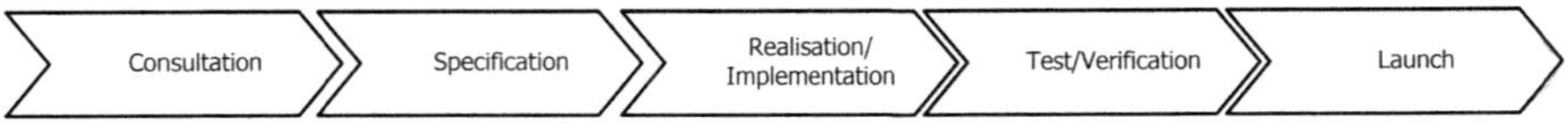

Fig. 23.8 The phase model applied in one software development cycle

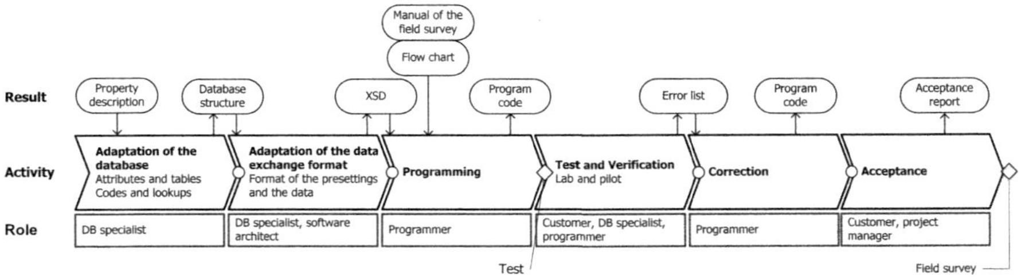

Fig. 23.9 Detailed view of the realisation phase as an example

The development cycle is illustrated in Fig. 23.8. The cycle is a phase model inspired and downsized from HERMES, the project management method for information technologies of the federal administration (Swiss Confederation 2016).

Each of the project phases has a more detailed structure that includes a sequence of activities. Figure 23.9 shows an example of the detailed structure of the realisation phase.

Depending on the type of a change, some or all layers of the software architecture are affected. In general, the insertion or removal of a new attribute to the database passes through all layers. For example, introducing a new clinometer with a scale in degrees instead of in gon, the scale units used previously, changes the method for measuring the inclination of deadwood pieces. This means that the meaning of an existing attribute has been changed because only values are stored and not the measurement unit, and a new attribute in addition to the old attribute has to be added to the same table to mark the change. If this is not done, values with different units become mixed in the same attribute and it will only be possible to identify the correct measurement unit if detailed information about the different inventories available. If the database table is altered or, in this case, extended with a new attribute, the change has to be implemented and propagated through several layers of the application Fig. 23.10.

23.6.3 *Software Quality*

Software quality can be described with various characteristics depending on the focus (Elmir et al. 2013). In addition to the academic definitions, ISO 25010 (ISO/IEC 2010) has established a standard for software quality. In this section, we discuss some quality criteria relevant for the development of the two NFI field applications.

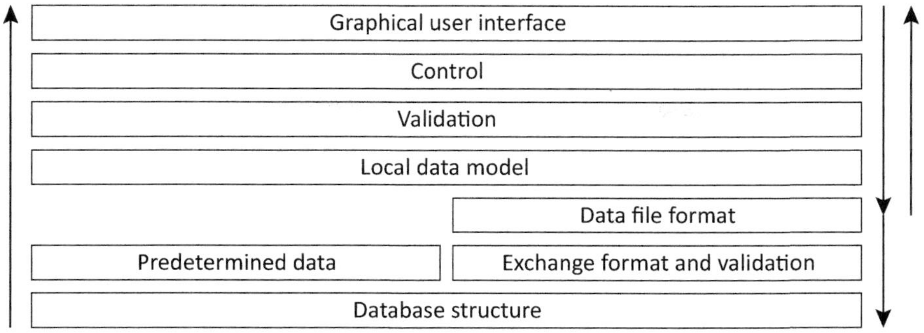

Fig. 23.10 Each change type affects a different number of application layers and thus affects the effort to make changes. The arrows indicate the possible scope of change types

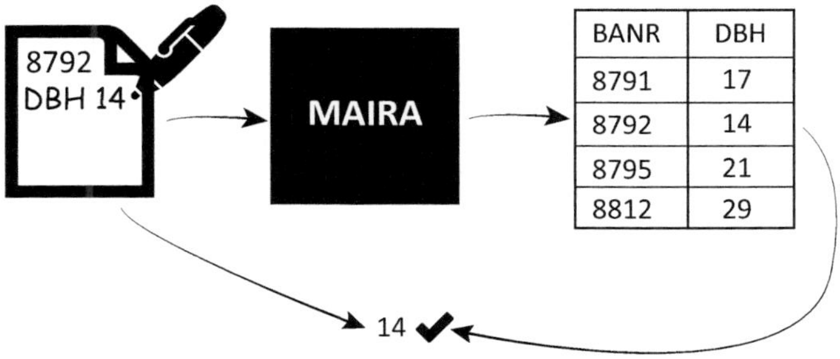

Fig. 23.11 The black-box verification method for testing data correctness

23.6.3.1 Correctness

Correctness means the input–output behaviour is correct with respect to the specification. This is clearly the main criterion for any data collection software. To test correctness of MAIRA, we use a black-box method in which the entire field application is treated like a black box (Fig. 23.11). The user works through the whole application and notes every entered value on a paper print-out of the input forms. After the data has been uploaded into the database, all uploaded values are compared with the noted values. This test is performed only once, at the very end of a broader testing and verification phase, because it requires a considerable amount of effort. Nonetheless, it is highly effective for testing the essential data correctness.

23.6.3.2 Extensibility

Extensibility means that future growth of the software is taken into account. This is an important criterion in our case since NFI is a continuous inventory and may potentially be changed or extended every year. The software architecture has therefore been specifically designed, through its structure, templates and conventions, to support further extension (Sect. 23.5.1.1).

23.6.3.3 Re-usability

Re-usability is a criterion for software quality that concerns all outcomes of the software development process, the code, software components, test suites, designs and documentation. Our software libraries have a large number of components that can be re-used, especially GUI components. In addition, the use of sophisticated software configuration features means that the entire MAIRA application can be re-used in other forest inventories.

23.6.3.4 Compatibility

Compatibility allows an application to run on multiple platforms, including operating systems and hardware. Compatibility is achieved in our software because Java is used as the programming language. The code can therefore run in any Java virtual machine, regardless of the underlying operating system. During the past 10 years, we have used different hardware and Microsoft Windows systems (XP, 7 and 10) and have never had a problem.

23.6.3.5 Usability

Usability concerns how intuitive and responsive the user interface is. This criterion is highly relevant to software whose main purpose is to help with manual scientific data collection, as opposed to collecting data with automatic measuring devices. In MAIRA, warnings and error messages are displayed in an unambiguous way. Additionally, the graphical components are presented with a meaningful and logical layout, taking into account the best practices of the field teams. In Sect. 23.7.3 we discuss further details of usability from the point of view of data quality.

23.6.3.6 Maintainability

Maintainability is an important overall quality criterion for any long-living software. In our software, this criterion is met by ensuring the software architecture is appropriate (Sect. 23.5), the code quality high, the versioning suitable, the documentation clear and the personnel competent. Importantly, these features also decrease the cost of software development.

23.6.4 *Development Tools*

Development tools help us manage the software implementation and deployment. Specifically, they help organise repetitive tasks and establish a defined development process and a common framework. The tools increase the overall quality of our work and the software reliability, and they support cooperation among the developers.

The main development tool is an integrated development environment (IDE). All of our developers use Eclipse, which is the most widely used IDE for Java. The IDE is capable of integrating further development tools. We use the following such tools:

- Apache Subversion to control and document the software evolution and to support collaboration and sharing of the source code.
- Sonatype Nexus to organise, store and distribute software components and to integrate a common external repository.
- Maven to automatize the building process in a software project and to manage all dependencies. Maven ensures that all the different software components, such as configuration files, icons, images, Java classes and libraries, are located, formatted and named correctly after the building is completed.
- NSIS (Nullsoft Scriptable Install System) to make the application built by Maven installable. The NSIS installer enables the installation of the application on an arbitrary Windows client.
- Bugzilla, to record all new features, change requests and bugs appearing during the field season. The processed records can be used to quantify progress, and recent records give an indication of the future workload. At the beginning of a development cycle, a list is made and each unprocessed record is prioritised and assigned to a work package to be completed.

The above tools cover important domains in software management and support a well-maintained code base, leading directly to higher data quality.

23.7 Ensuring Raw Data Quality with MAIRA

Raw data collected in the field should be not only of high quality, but also accurate and consistent. For the NFI, raw data represents the beginning of the whole data chain, as all subsequently derived data, analyses and results are based on the raw data collected in the field. Consequently, any imprecision at this early stage may lead to much larger errors later on. It is therefore essential that the tools used for data assessment prevent the collection of erroneous data. Raw NFI data of excellent quality can be achieved by emphasising software quality strongly (Sect. 21.5) and especially by using the multilevel checks and messages built into the implementation of MAIRA.

23.7.1 Multilevel Checks and Messages

The goal of MAIRA is to guide the field staff step by step through the process of raw data collection. Multilevel checks and extensive input data validation are conducted to support and assist the field staff in collecting complete and correct data.

The recorded data is checked in multiple contexts. The value of the currently assessed attribute is checked, along with the attribute's value in combination with the values of other attributes. In terms of the GUI, this means checks are made on: the current text input field, the current input form, the current working step (roughly corresponding to a field manual chapter), and the entire plot survey. These multilevel checks, along with sophisticated business logic, ensure data consistency, plausibility and completeness.

Following such checks, GUI messages may be shown to the user. These messages also serve to support the field crew in assessing attributes correctly. Attention was paid to enabling as many GUI messages as necessary to be shown during data recording, but also as few as possible in order not to interrupt the workflow. These messages have different levels of severity:

- Hints: in cases where data is probable but no data was entered, e.g. "Fungi on dead sample trees".
- Reminders: in cases where data is expected but no data was entered, e.g. "Presence of woody species" on the survey plot.
- Blockers: in cases where data must be recorded but was not, e.g. 'dbh measurement value'.

The final save action of a completed plot survey involves further consistency checks: the XML file generated from the recorded data is validated against an XSD definition, which contains the constraints of the database schema where raw data will be uploaded.

All checks conducted by MAIRA are an integral part of the software. It is not possible to add external, user-defined scripts or validation rules, which are a feature of some other field-data collecting tools such as Field-Map (2017).

23.7.2 Predetermined Data at the Attribute Level

Predetermined data, i.e. values of a certain attribute previously recorded in former inventories (NFI1, NFI2 and NFI3), plays an important role in the data value checks.

Predetermined data values are used in various ways:

- Visible: the previous value might be presented directly on the GUI if an attribute value is expected to stay constant over time. In such cases, the predetermined data helps the user to correctly assess the new attribute value.
- Hidden: unbiased recording of the new attribute value is indispensable if it is likely to depend on values recorded in former inventories but is expected to have

changed since the last inventory. The predetermined data is not presented to the user in such cases, but is used inside business logic to validate the newly recorded value.

As an example of the visible use of predetermined values, boundaries like accessibility borders are not expected to change over time, so the attribute values describing such border lines are considered to be constant and the corresponding data is presented on the GUI, while still allowing the user to change the value. Hidden predetermined values are used, for example, for the correct and unbiased re-measurement of dbh on single trees. The new value is compared with a calculated estimate of the expected value. This expected value is determined according to a small model, which also incorporates different site-specific parameters that could influence tree growth. If there is a substantial deviation between the measured and the expected value, a GUI message is presented.

Predetermined data values are additionally used to simplify the identification of assessed objects, e.g. single trees. For example, attribute values, such as the distance and angle from the plot centre of an individual tree, enable its precise identification, thereby ensuring that the measurements in subsequent inventories are carried out on the same tree. This approach is similar to the recovering procedure of the tree identification numbers used in Field-Map (2017).

More generally, using predetermined data to check current values ensures the consistency and comparability of the assessed data in NFI1, NFI2, and NFI3 and NFI4.

23.7.3 Ergonomics and Usability Issues with MAIRA

The field of ergonomics deals with refining the design of products to best adapt them for their users. With this aim in mind, MAIRA was designed and implemented to optimise user friendliness, reduce error sources, and thus obtain better data.

In MAIRA, specific input forms are used to help the user enter correct data. Additionally, special attention has been paid to creating GUI messages that are unambiguously formulated and present the user with only two options as possible answers, e.g. *Yes/No*.

In NFI4, field data was recorded on tablet PCs with a pen as an input device. In our experience, this hardware combination facilitates efficient data collection. The combination of tablet and pen gives the GUI a specific look and feel, which clearly differs from that of software tools designed for data input with a keyboard and mouse, such as the *Open Foris* tool (Open Foris 2016). The tablet and pen design makes it much more comfortable and efficient for the user to select items from a selection list than to type words into an input field. If the number of possible choices is small, using radio buttons could be even more efficient and less error prone.

The input form for the *Deadwood* chapter of MAIRA Fig. 23.12 was, for example, re-arranged after a few years of use to better reflect such considerations:

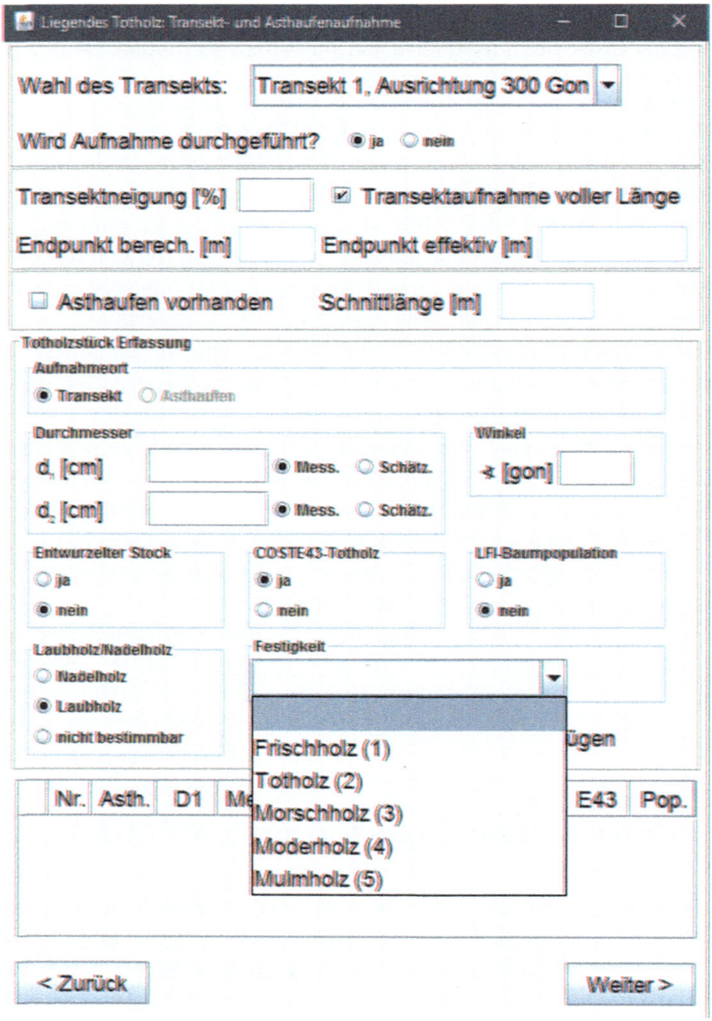

Fig. 23.12 The input form for MAIRA's *deadwood* chapter (in German)

- There is a *Yes/No* radio button at the top of the input form, specifying whether data is recorded for the currently selected transect. If the user selects No, the whole input form for the current transect is disabled and data recording is not possible. If the user selects *Yes,* the input form is enabled and data can be entered. On a subsequent No, previously entered data for the current transect is deleted, but only if the user explicitly confirms this deletion when prompted by a GUI message.
- The different properties of a piece of deadwood are presented to the user as radio buttons if two or three choices are possible (e.g. hardwood or softwood =

Laubholz/Nadelholz in Fig. 23.12) and as a choice list if there are more options (e.g. deadwood solidity = Festigkeit in Fig. 23.12.

23.7.4 Evolution of MAIRA

During the 9 years of NFI4, MAIRA was subject to an on-going development process in order to keep up with changing requirements over time. While most of the attributes assessed were left unchanged, a few new attributes were introduced (Sect. 23.6.2).

Furthermore, the recording process required optimisation as well, which meant additional changes to the software. Feedback from customers led to the introduction of various enhancements and improvements in the annual software development cycle to help increase data quality and consistency. Iterations of the following steps were carried out:

- Obtain feedback from the field staff and data analysis team.
- Incorporate relevant issues from this feedback into the software enhancements.
- Collect higher quality data with better recording efficiency.

These steps were not strictly formalised, but took place as an integral part of the software development cycle.

As few changes to methods and the software as possible were made during a single inventory to assure consistency and comparability over time, as well as throughout the entire NFI series. However, in some cases it was necessary to adjust the algorithms and methods to improve data quality. For example, the method used to choose trees for the tariff measurement (Sect. 2.3.4.5) was modified during NFI4 and was one of the few larger conceptual changes that took place.

23.7.5 General Data Quality Considerations

The main data quality concern in MAIRA is to avoid the recording of erroneous data. We are aware that not all errors can be prevented by technical means. However, we make efforts to prevent systematic errors, as some of the examples in the previous sections show, so that the data obtained during each field season is largely ready to use at the end of the session. Nevertheless, when a field team is aware of a mistake, the raw data is corrected during a short time window directly after the field season. Subsequent systematic corrections of raw data in the database are not part of our standardised process unlike in other systems where database checks and corrections of assessed data are standard (Italian NFI 2009; Open Foris 2016).

The weighted importance of attributes is a further basic principle in MAIRA's data quality philosophy. Specifically, the attributes that are critical for subsequent evaluations and results, such as single tree data, undergo an extremely rigorous and complex quality check. Less important attributes, e.g. rockfall characteristics, are checked properly but not as extensively.

For attributes where an unbiased measurement of the current value is essential (e.g. dbh), MAIRA makes the field staff carry out a new measurement without being influenced by previous inventory values. This new measured value is compared with a calculated estimate. Substantial deviations may mean the user, in severe cases, will have to repeat the measurement.

23.8 Conclusions

The two field applications used in NFI4, MAIRA and SILVIS, are connected by a common and clearly defined data flow. A centralised database contains all the raw data collected in the four inventories, NFI1 – NFI4. Data originating from this database serves as predetermined data for each field survey with both MAIRA and SILVIS. The data flow is supported by several scripts, but could be automated further in the future.

Our main software design principle is the *separation of concerns* using the Model-View-Controller architecture pattern. The Java Swing framework did not allow a complete separation and we therefore extended Swing for our own purposes. However, for the upcoming NFI, we will use a new Java version offering the JavaFX framework, which supports this main design principle.

In the process of developing software for the NFI, we gained valuable experience from monitoring the interaction of the software with the field staff, as well as data quality and data consistency that could be useful in other contexts:

- Use sophisticated business logic to guide and assist field staff throughout the data-recording steps so that additional interaction (e.g. GUI messages) only arise in contexts where errors are likely. Too many messages reduce accuracy and interrupt the workflow. Moreover, data can only be assessed efficiently if the field staff are satisfied with the field applications.
- Use specific GUI components designed for unambiguous data recording to take into account the fact that a field tablet, pen and virtual keyboard are used as input devices in the NFI. Using a virtual keyboard is, however, awkward because it occupies space on the screen, leaving less space for the input forms. Paradoxically, using the pen under the light conditions prevailing in a forest environment requires large GUI components and thus careful planning of the layout.
- Aim to create a "single source of truth" for metadata, such as for the definitions of value range and data format. Ideally, the definitions used in the field manual, MAIRA and the database metadata should share a common metadata source to achieve the highest possible consistency.
- Wherever possible, consistently complete tasks in a sequential manner. Although a predefined software engineering process was used in NFI4, consisting of annual software development cycles and different project phases, much of the work was

still done in parallel. In particular, the Manual of the Field Survey should be completed before the development work starts as it specifies the software to be used.
- Consider using the Manual as a "single source of truth" because it is used as for the specification. This would mean taking into account more technical aspects and structuring and standardising the Manual more strictly.

We consider the effort required to develop applications in-house at WSL has been worthwhile because it allows us to implement the specific needs of the NFI more directly and efficiently. As a result, fewer compromises need to be made to fulfil user requirements. Further, easy communication within the same institute between customers and software engineers fosters effective implementation. In particular, the business logic, with its numerous dependencies, can be precisely mapped into our software. Compared to commercially available and configurable products, our software is also much more comprehensive. All the effort involved in developing our software has led to better data quality and data availability. At the end of a field season, analyses can be started immediately and reports can be produced without delay.

References

Codd EF (1971) Further normalization of the data base relational model. IBM Research Report, San Jose, RJ909

Elmir A, Elmir B, Bounabat B (2013) Towards an assessment-oriented model for external information system quality characterization. IJCSI Int J Comput Sci Issues 10(4):2

Field-Map: Field-Map Software and Hardware Catalogue. http://www.fieldmap.cz/download/FM_Catalogue_en.pdf, downloaded 16.02.2017

Fridman J et al (2014) Adapting national forest inventories to changing requirements – the case of the Swedish national forest inventory at the turn of the 20th century. Silva Fenn 48(3):1095

Gamma E, Helm R, Johnson R, Vlissides J (2009) Design patterns: elements of reusable object-oriented software. Addison-Wesley, Boston. 42nd printing

ISO/IEC 25010:2011. Systems and software engineering – Systems and software Quality Requirements and Evaluation (SQuaRE) – system and software quality models

Italian NFI, Gasparini P, Bertani R, De Natale F, Di Cosmo L, Pompei E (2009) Quality control procedures in the Italian national forest inventory. J Environ Monit 11:761–768. https://doi.org/10.1039/b818164k. https://www.ncbi.nlm.nih.gov/pubmed/19557227

Laplante PA (2007) What every engineer should know about software 1560 engineering, vol 40, 1st edn. Taylor & Francis, Boca Raton

Open Foris (2016) Openforis collect, user's manual, Ver. November 2016. http://www.openforis.org/fileadmin/docs/collect/OF_Collect_MANUAL_October2016_WEB.pdf

Swiss Confederation (2016) HERMES 5. Projektmanagementmethode für alle Projekte. http://www.hermes.admin.ch/onlinepublikation/index.xhtml

Traub B, Meile R, Speich S, Rösler E (2017) The data storage and analysis system of the Swiss National Forest Inventory. Comput Electron Agric 132:97–107. https://doi.org/10.1016/j.compag.2016.11.016

Part IX
Coordination of the Swiss National Forest Inventory Data Analysis System

Chapter 24
Coordination of the Swiss National Forest Inventory Data Analysis System

Meinrad Abegg

Abstract The Swiss National Forest Inventory (NFI) is one of the most important environmental monitoring programs in Switzerland. The core objective is to provide different interest groups with data and information on Swiss forests. This information can be tailored to customers' needs by those producing the data and reports in the NFI program. To support the producers in this, the program leaders initiated the NAtional Forest Inventory Data Analysis System (NAFIDAS). Since customers may have different needs at different times, the system has to be constantly ready and regularly issue data for reports and analyses that comply with very high quality standards. In addition, changes and updates in the data and the algorithms must be trackable at all times.

24.1 Introduction

The Swiss National Forest Inventory (NFI) is one of the most important environmental monitoring programs in Switzerland. The core objective is to provide different interest groups, referred to here as customers, with data and information on Swiss forests (Fig. 24.1). This information can be tailored to customers' needs by those producing the data and reports in the NFI program (the producers in Fig. 24.1). To support the producers in this, the program leaders initiated the NAtional Forest Inventory Data Analysis System (NAFIDAS), described in Chap. 20. Since customers may have different needs at different times, the system has to be constantly ready and regularly issue data for reports and analyses that comply with very high quality standards. In addition, changes and updates in the data and the algorithms must be trackable at all times.

The NAFIDAS system consists of four parts: the web interface, database, data analysis application and data derivations. The producers (NFI staff) and the

M. Abegg (✉)
Scientific Service NFI, Swiss Federal Institute for Forest, Snow and Landscape Research WSL, Birmensdorf, Switzerland
e-mail: meinrad.abegg@wsl.ch

C. Fischer, B. Traub (eds.), *Swiss National Forest Inventory – Methods and Models of the Fourth Assessment*, Managing Forest Ecosystems 35,
https://doi.org/10.1007/978-3-030-19293-8_24

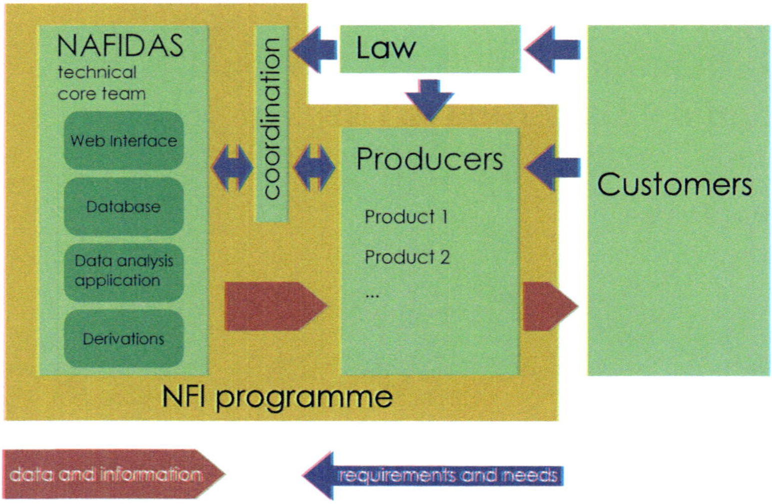

Fig. 24.1 The NAFIDAS system in relation to NFI coordination, producers and customers, as well as to Swiss law. Red arrows: flow of NFI data and information from the NAFIDAS system to the customers. Blue arrows: flow of requirements from the customers to the NAFIDAS system

customers interact with these as shown in Fig. 24.1 Some of the customers' needs are considered so important that they have been written into national laws. The complexity of the NAFIDAS system led the heads of the NFI program to initiate a single point of contact for the coordination of the NAFIDAS system to ensure:

1. Information is exchanged between different members of the NAFIDAS team within the NFI program.
2. The human resources needed to maintain the system, implement new data and further develop the system are used as efficiently as possible according to what has most priority.
3. Solutions can be tailored to suit a generically designed system.
4. The documentation of the decision processes is long term and easily accessible so that older decisions can be referred to in decision-making. The documentation should also be sufficiently complete to ensure the coordination can be handed over to another person at any time without losing any relevant information.

The focus of this chapter is on describing the NAFIDAS core team, their customers and producers and how their interactions are coordinated, as well the detailed process involved in meeting the objectives listed above.

24.2 NFI Customers and Their Information Needs

The customers interested in NFI data and information need a wide range of data about the forests in Switzerland in specific forms. The data may range in form from raw data measured in the field to statistically processed information summarised in result tables, with information on specific forest properties. It is always accompanied by additional information on the data, e.g. a short description of the variables when published on the NFI website or an edited text when published in a printed report.

The customers for this NFI data and information are:

- **Political administration institutions**: these may or may not have a core interest in forestry on a national or a regional level. They are mostly interested in having data on Swiss forests to support their decisions and their policy making, and usually rely on reports, where information is preselected and aggregated in result tables with additional texts on their interpretation. However, some administration departments need very specific information. If this is not available in pre-calculated result tables on the NFI website, it can be specifically calculated on demand. For a very quick orientation on the state of the Swiss forest, certain stakeholders in the administration require highly aggregated information in a form that allows them to capture the state and the development of the Swiss forest at a glance.
- **International organisations**: the most important organisations that depend on NFI data for their reporting are: the UN's Food and Agriculture Organisation (FAO), the Ministerial Conference on the Protection of Forests in Europe (Forest Europe) and the United Nations Framework Convention on Climate Change (UNFCCC) (FOREST EUROPE 2015; MacDicken et al. 2015; Pan et al. 2011). They rely greatly on NFI data when reporting on the development of forests in Europe and worldwide or on the national carbon dioxide balance. These organisations usually have standardised forms to complete for such reports, and mostly require tailored data derivations to obtain statistical estimates.
- **Research and educational institutions**: these institutions are interested in having up-to-date information about Swiss forests to pass on to students, especially if they are from the forestry sector. NFI data can also help answer various forest-related research questions. The data and information these institutions need range from raw data to written reports on the state of Swiss forests.
- **Media**: forest issues are regularly reported on in newspapers and on the radio and television. NFI is often contacted after a natural event affecting the forest or if new research findings or reports on forests are published. But sometimes the media reports something of general interest for the Swiss public about the forest. Information is given to the media according to their specific information needs or they are referred to result tables on the NFI website.
- **Consulting engineers**: consultants may need comparable data on Swiss forests for, e.g. producing a management plan. For this, they can use the reports and result tables available on the NFI website or request specific evaluations.

- **People involved in forestry**: forest owners and entrepreneurs, for example, who need comparable data on particular forests, which they can mostly obtain from the reports and result tables available on the NFI website.
- **Interested laypeople**: they usually want very general information about the forest. They may refer to the result tables on the NFI website or need appropriate access to general and interesting information on Switzerland's forests.

24.3 Legal Requirements

Monitoring the forest is so important in Switzerland that some requirements are specified in the following national laws (Sect. 1.3).

- **Federal Act on Forest** (Bundesgesetz über den Wald, WaG): which specifies that conducting an NFI is mandatory, and states in Article 34 that "the state and the cantons guarantee the information of the administration and the public about ... the state of the forest as well as the forest and wood economy" (SR 921.0 1991).
- **Ordinance on Geoinformation** (Geoinformationsverordnung, GeoIV): which specifies that the data from NFI has to be stored in such a way that its availability and quality level are guaranteed, and exported in "appropriate" data formats and stored safely (SR 510.620 2008).

24.4 Products and Producers

Several NFI products have been developed to satisfy the needs and requirements of NFI customers in terms of data, information and their availability. Each of these products is assigned to one person or 'producer' Fig. 24.1 who is responsible for it and for fulfilling the requirements of particular target customers. The form of these products in the NFI varies (Sect. 20.2.2) but the most important are:

- **NFI Report:** The NFI produces a report in the form of a book with its latest results roughly once every 10 years (Brändli 2010). The book summarises the most important findings on the state and the development of the Swiss forest. The data is mostly in the form of result tables providing answers to specific questions, such as: what is the growing stock of the main forest tree species in different regions? Statistical information and comments on the methods and on how to interpret the results are provided.
- **Online publication:** every 4–5 years a collection of result tables and corresponding maps are put on the NFI Project website (Abegg et al. 2014). The available information has been continually increased thanks to major technical improvements, and retrieving data has been improved with better search tools. At the moment 57,000 pre-calculated result tables are available in each of the four languages, German, French, Italian and English (www.lfi.ch).

Each result table is accompanied by additional information on the variables used and the statistical interpretation of the results.

- **NFI cockpit:** this was created to give a quick impression of the state and development of the Swiss forest (Brändli and Brändli 2015). The cockpit summarises graphically the main NFI findings in highly aggregated form according to the main NFI indicators (www.lfi.ch). This data collection uses pre-calculated figures from the NAFIDAS system, which can be accessed automatically using NFI's web interface.
- **International reports:** forest-related reports from, e.g. FAO, Forest Europe or UNFCCC (Forest Europe 2015; MacDicken et al. 2015; Pan et al. 2011), usually need very specific information to complete their questionnaires, which are normally the same for all countries. Particular forest properties may be defined slightly differently than in the NFI, which means special data derivations, statistical calculations and extrapolations may be needed. Thorough data documentation is essential to be able to reconstruct the data at a later date.
- **Further reports:** these include the Forest Report (Rigling and Schaffer 2015), a general report on Swiss forests issued roughly every 10 years, which is based to a large extent on NFI data, as well as other sources of information. Sometimes national or international organisations request NFI to complete a questionnaire to obtain the information they require. For all these reports, providing reliable estimates and detailed information about the content and calculations is essential.
- **Special evaluations:** a special evaluation of NFI data may be requested if specific information is needed that is not available in existing NFI reports or on the website. Its form can range from just a single figure to complex evaluations in the form of a report. Sometimes this service must therefore be paid for.
- **Raw and derived data:** representative data on forests is important for many researchers. Such data is freely available on request for research purposes assuming an agreement on data use can be reached. It is normally delivered in raw format, mostly without statistical editing, depending on the specific needs. In addition to this data, detailed documentation on the variables is supplied as well.
- **Outreach articles:** NFI publishes outreach articles for forest practitioners in forest-related magazines. These articles describe interesting findings in the most recent NFI evaluations. To ensure their storylines are interesting and relevant, they often need quick and reliable evaluations of the latest NFI data. Some types of data, e.g. regional maps displaying information, are usually very welcome. Here again, providing documentation about the data is very important.
- **Outreach information for laypeople:** the NFI website (www.lfi.ch) allows special access for laypeople with information about, e.g. forest trees in Switzerland. This requires having well-structured data so that maps can be quickly calculated in real time and a selection of interesting statistics provided about the Swiss forest.

24.5 Requirements and Needs

The people responsible for a product (the producers in Fig. 24.1) find out about or anticipate most of NFI customers' corresponding needs and requirements. Some of these, however, can be directly deduced from Swiss law. A third group of needs and requirements in the NAFIDAS system come directly from the customers themselves when they give feedback on a product directly to an NFI team member or in an organised user workshop.

Customers' needs and requirements may concern one or more parts of the NAFIDAS system (Chap. 20): the raw database, the derivation database and the web application. The most important requirements in the development and the maintenance of the NAFIDAS system concern the:

- **Interaction design**: both the external and the internal web interface of the NAFIDAS system are subject to specific user needs. This is mostly related to ensuring the required information can be found quickly and intuitively. External users are primarily interested in the result tables and explanations of the calculations involved. Internal users (Project members) also need to have easy and intuitive access to the data and documentation, but at the same time they need to be able to produce complex high quality report tables as efficiently as possible.
- **Content**: the specific data about the Swiss forest required for the two main NFI products, the NFI report and the web-publication, as well as for other national and international reports, must be up-to-date. External users' needs vary depending on their interests. Internal users also often need specific data (derivations) for research purposes or as a basis for the NFI reports.
- **Documentation**: to ensure NFI reporting is of high quality, the documentation in the NAFIDAS system must be on many levels. Each variable, whether raw or derived, must be explained. Access to the derivation script, which is usually in SQL, help users understand the variable in more detail. For data to be improved, e.g. by adapting a data derivation, the interdependencies between the variables must be known. Since the data may be changed at any time, several versions of the data and its metadata are needed, e.g. to reproduce the content of an earlier data delivery and identify possible deviations from the current data. The content of already finished products, e.g. the result tables calculated for an NFI report, must be reproducible in the future as well. These result tables must be accompanied by additional information, such as the variables needed or the statistical estimation procedure applied. This documentation should be available on a long-term basis.
- **Data structure**: for a generic system like NAFIDAS, a clear data structure is very important. The data analysis application is specifically designed to use a certain structure, which allows a generic evaluation of the data (Sects. 20.2 and 20.4). Consistent and clear ways of deriving data to represent populations and domains (Sects. 20.3.1 and 2.5.3) simplify the data supplies for researchers and allow thorough documentation to be produced with very little effort.

- **Quality assurance**: not only should the results, calculated with the available data, always be correct, but the available data should also be well-documented so that its quality can be easily checked.
- **Efficiency**: the system should ensure that producing and supplying the data and the corresponding documentation is very straightforward and quick to implement.

24.6 The NAFIDAS Team

The team hold regular meetings to coordinate the various needs and user requirements with the following roles represented, sometimes with the same participant having several roles:

- **Point of contact**: this person organises the meetings, collects requirements and ideas, identifies future needs and possible bottlenecks in the system, estimates the work load, keeps track of the deadlines and, most importantly, guarantees the information flow to all the relevant internal stakeholders in the system. This person must be in regular contact with the technicians as well as the users of the system and the 'producers'.
- **Technical core team**: these people are responsible for the technical implementation of the four parts of the NAFIDAS system: the data-analysis application, database infrastructure, database content and web interface. One person is mainly responsible for one of the four parts each, including for the corresponding work to be done.
- **Producers**: the people responsible for the most important NFI products ensure the needs associated with these products are addressed.
- **Project or group leaders**: these people are responsible for the employees and can decide on the allocation of human resources within the NAFIDAS system.
- **Experts**: individuals with special expertise and competence regarding certain subjects (data-derivation modules), who can report on related progress or problems.
- **Extended NAFIDAS team**: people who do not take part in meetings but who are kept up-to-date and sent the minutes of the meetings.

24.7 Coordination Process

Regular coordination meetings are held to ensure the NAFIDAS coordination objectives are met (Sect. 24.1) with four main agenda items at each meeting. Items (2), (3) and (4) are part of an extra process, which is described further in the following section.

1. **Status exchange**: updates on the status of the larger, more important tasks are exchanged, and any problems and issues discussed, as required in objective (1) information exchange (Sect. 24.1).
2. **Prioritisation of requirements**: the point of contact lists any requirements for the NAFIDAS system that have arisen, which are then discussed, prioritised and, where necessary, further working steps are planned.
3. **Adoption of solution proposals**: proposed solutions are either rejected or accepted and added to the to-do list so that deadlines can be set and their progress tracked.
4. **To-do list**: The to-do list is discussed, responsibilities for tasks assigned and deadlines for new high-priority tasks set. Any failures to meet deadlines are discussed and new deadlines set.

24.8 Identification of Requirements and Solution Finding

Five steps are involved in moving from the identification of a problem to its solution:

- **Step 1: Identification of requirements**: particular requirements may arise from personal demands on the system, made by either normal team members or a project leader. They may also arise while drafting a report, during a user workshop or during the project evaluation, which take place once every few years. These needs are collected by the coordinator, who conducts the first preselection and decides whether the particular task concerns only the NAFIDAS system or is something methodological to be discussed outside NAFIDAS. Tasks, for which the need and priority are clear, and the work required to implement them is small, are put directly on the to-do list so that a deadline can be set and a person can be assigned the responsibility. More complex tasks, e.g. those involving several experts and multiple working steps, are discussed at the meeting in step 2 (coordination). At this point, such requirements mostly still lack a detailed specification.
- **Step 2: Coordination**: At the NAFIDAS coordination meetings, all complex requirements identified in step 1 are discussed so that the need for further clarification can be identified and its priority discussed. These requirements and their priority are often controversial and challenged regardless of who proposed them, e.g. a project leader, technician or user at a workshop. If the participants agree that the requirement should be implemented and has high priority at the current stage of the project, a deadline for the pre-evaluation is set and the group and leader responsible for the task selected.
- **Step 3: Problem evaluation and elaboration of possible solutions**: An expert group meets once or several times to discuss the problem or requirement and clearly describe it. The experts are the technical specialists who may have to implement the new feature and the people who represent the "need side" of the requirement because they are, e.g. responsible for a NFI product. Possible solutions are discussed from different perspectives until a consensus can be

reached on what is the most efficient solution or the one that leads to the best quality output. Since this part of the solution finding is crucial for the quality of the system, the experts discussing the issues should 'fight hard' for their point of view, while keeping in mind the overall objectives. Sometimes these discussions result in no technical changes to the system being made, but just a few operational procedures adapted. This part of the process guarantees the achievement of objectives (2) and (3) (Sect. 24.1). One or several solutions are discussed during these meetings, and the cost for each solution in terms of e.g. working time for implementation and maintenance of the system is estimated.

- **Step 4: Presentation and decisions**: based on the solutions outlined in step 3 and the estimated cost, participants at the meeting decide on the implementation and the schedule for implementation. This point in the process provides the last chance for participants to express caveats and initiate a re-evaluation of the issue. Once it has been decided to implement a solution, it is divided up into single tasks which are set on the to-do list with deadlines.
- **Step 5: Implementation**: the tasks are implemented in the system according to the schedule of the to-do list.

24.9 Documentation

Minutes are taken at all meetings (objective (4) in Sect. 24.1), to summarise the discussion and decisions taken. Decisions are highlighted so that they can be identified quickly. The detailed proposals for more complex problems are stored in addition to the minutes. The to-do list lists all to-dos including those already done and the deadlines and the date of their completion. These files are stored in a file system on a server to which all NFI employees have access.

24.10 Time and Effort Required

The coordination meeting takes place every 2 months, with normally ten people attending the roughly 2-h meeting. Although it takes only a few hours to prepare the main meeting, i.e. draw up the agenda, and prepare the to-do list and minutes, more time is needed for prior discussions of NAFIDAS issues. These help to identify needs for action and enable the person at the point of contact to keep track of how the NAFIDAS system is working.

Usually two to five people take part in the expert meetings, depending on the subject, which take place in-between the main coordination meetings. They are repeated until the problem has been discussed thoroughly and solutions are ready for presentation, usually once but sometimes even four times or more, depending on the complexity of the problem and the need for particular experts to evaluate them further. Preparing the expert meetings may involve drawing up an agenda, but sometimes experiments, thorough data analyses or even simulations have to be conducted to provide a solid basis for adequate decisions.

24.11 Discussion and Conclusion

The design of the NAFIDAS coordination described here seems to be effective for the organisation of NAFIDAS. The team members are generally very satisfied with the system, and feedback from customers has been positive as well, with the punctual delivery of data and information to the customers being particularly appreciated. New ideas occur and new features, some of which are very complex, are constantly arising and the wish list grows longer. This shows how it will be particularly important in future for the coordination of the NAFIDAS system to be well organised.

In a program with the size and responsibilities of the NFI, or of NAFIDAS, which is one of its core parts, it is very important for certain quality standards and the most critical processes to be clearly defined. At the same time, it is essential for the NAFIDAS team to be able to develop new ideas freely and to feel responsible for the parts of the system assigned to them. Additionally, implementing any new ideas has to be feasible with the available human resources. To find the right balance between performing the defined processes, keeping the workload down while having considerable individual freedom is one of the main challenges for the coordination of NAFIDAS.

In summary, the coordination of NAFDAS seems to be mostly very effective. The discussions are sometimes tough to lead, but the results tend to be very satisfactory.

References

Abegg M, Brändli UB, Cioldi F, Fischer C, Herold-Bonardi A, Huber M, Keller M, Meile R, Rösler E, Speich S, Traub B (2014) Fourth national forest inventory – result tables and maps on the internet for the NFI 2009–2013 (NFI4b). [Published online 06.11.2014] Available from World Wide Web. http://www.lfi.ch/resultate/. Swiss Federal Research Institute WSL, Birmensdorf

Brändli UB (Red.) (2010) Schweizerisches Landesforstinventar. Ergebnisse der dritten Erhebung 2004–2006. Eidgenössische Forschungsanstalt für Wald, Schnee und Landschaft WSL, Birmensdorf; Bundesamt für Umwelt, BAFU, Bern. 312

Brändli UB, Brändli R (2015) Swiss NFI cockpit – forest indicator trends. Poster. [Published online 12.11.2015] Available on the Internet. http://www.lfi.ch/publikationen/publ/NFIcockpit-en.php. Swiss Federal Research Institute WSL, Birmensdorf

Forest Europe (2015) State of Europe's Forests 2015

MacDicken, K.; Jonsson, O.; Piña, L.; Marklund, L.; Maulo, S.; Contessa, V.; Adikari, Y.; Garzuglia, M.; Lindquist, E.; Reams, G.; D'Annunzio, R. Global Forest Resources Assessment (2015) How are the world's forests changing? Food and Agriculture Organization of the United Nations (FAO), Roma, 2016

Pan Y, Birdsey RA, Fang J, Houghton R, Kauppi PE, Kurz WA, Phillips OL, Shvidenko A, Lewis SL, Canadell JG, Ciais P, Jackson RB, Pacala SW, McGuire AD, Piao S, Rautiainen A, Sitch S, Hayes DA (2011) Large and persistent carbon sink in the world's forests. Science 333:988–993

Rigling A, Schaffer HP (eds) (2015) Forest report 2015. Condition and use of Swiss forests. Federal Office for the Environment, Bern. Swiss Federal Institute WSL, Birmensdorf. 144 pages

SR 510.620 (2008) Verordnung über Geoinformation (Geoinformationsverordnung, GeoIV) vom 21. Mai 2008. AS 2008 2809

SR 921.0 (1991) Bundesgesetz über den Wald (Waldgesetz, WaG) vom 4. Oktober 1991. AS 1992 2521

Glossary

Meinrad Abegg

Term	Description
Accessibility boundary	Boundary that separates accessible from non-accessible area.
Accuracy and precision	These are terms from statistical sampling and refer to how close a value comes to the (always unknown) 'true value'. Accuracy refers here to systematic deviations from the 'true value', and precision to the variability of sample estimates around the 'true value'.
Aerial photo interpretation	Derivation of information about an object of interest from aerial photos.
Afforestation	Planting or seeding of trees with the objective to convert a non-forest area (unstocked or too loosely stocked) into a forest area.
Airborne digital sensor (ads)	Digital camera sensor for aerial imaging, e.g. from Leica, based on push-broom technology. The resulting image data are long image stripes that are recorded simultaneously from different viewing angles.
Analysis table	Specially structured table in the derivation database that contains derived data.
Areal damage	Areal damage in forest stands. In the NFI, this refers to sample plots where trees have died or been severely damaged on at least 10% of the assessed area since the previous measurement.
Assortment	Wood product of defined dimensions and/or quality according to agreements in the wood trade. In the NFI, potentially usable assortments are calculated according to modelled stem dimensions.
Attribute	Characteristic or property of an object, such as a single tree or a sample plot. Synonyms are *variable* and *characteristic*.
Axle load	The axle load of a vehicle is the ratio of the total mass (vehicle and payload) allocated to one axle. The axle load is displayed in tons (t).

(continued)

M. Abegg (✉)
Scientific Service NFI, Swiss Federal Research Institute WSL, Birmensdorf, Switzerland
e-mail: meinrad.abegg@wsl.ch

C. Fischer, B. Traub (eds.), *Swiss National Forest Inventory – Methods and Models of the Fourth Assessment*, Managing Forest Ecosystems 35,
https://doi.org/10.1007/978-3-030-19293-8

Term	Description
Azonal site	An azonal site represents the entirety of environmental influences that impact on living beings at a certain location. On zonal sites, vegetation adapted to the general climate of the particular zone or area can establish. If the conditions are extreme, the range of species shifts, and the site is then declared azonal.
Bias	Systematic error in an estimate.
Bifurcation	Tree with a bifurcated stem. In the NFI, bifurcations up to a height of 9 m are recorded.
Biodiversity function	Forest function with the objective of preserving or promoting biodiversity.
Bole wood	Stem wood >7 cm in diameter without bark or a stump.
Business-as-usual scenario	Timber harvesting scenario based on the forest characteristics and forest policy implemented up until the end of 2009.
Carrying capacity	Load for which a road was constructed.
Change	Change in a unit of interest on a population or an area.
Classification unit	Attribute of an area or population of interest that divides them into different units or groups. In the data analysis software (DAA), any derived variable can be chosen as a classification unit.
Coarse roots	Roots of woody plants with a diameter >5 mm.
Colour infrared image	Aerial image with band composite near-infrared, red and green.
Components of change	Components of the change in a population of trees, such as growth, felling and mortality.
Continuous forest inventory	Forest inventory in which a representative sub-grid (as in the Swiss NFI) or a sub-unit of the investigation area is assessed each year.
Control survey	An inspection done as part of the ongoing Quality Control program. The sample plots are selected according to topic and are visited directly after the field measurements by an instruction team (cold check), who then focus on particular attributes of interest or remeasure the whole plot.
Conversion	Changing a silvicultural system by selective logging of a stand using the present stocking, e.g. converting coppice stands to high forest or layered stands to all-aged stands (plenter conversion).
Crown coverage	Fraction of the sample plot area covered by trees >3 m in height and shrubs regardless of height.
Cutting interval	Time between two cuttings in a plenter forest.
Data analysis application (DAA)	Software package in NAFIDAS for the statistical analysis of forest inventory data.
Data quality objectives (DQO)	Combination of measurement quality objectives (MQO) and data quality limits (DQL).
DBH threshold	Minimum diameter at breast height of trees for calliper measurement. In the NFI, the threshold is 12 cm.
Derivation	Inventory-based calculation instruction for a variable.
Derivation database	Database with a structure adapted to inventory analysis. All data in the derivation database is generated by calculations, transformations or imputations, and is permanently stored.

(continued)

Term	Description
Diameter at breast height (dbh)	Diameter of a stem at a height of 1.3 m above ground.
Domain	Syn. area domain: Area-related entity where information is obtained in a forest inventory, for example the area of a forest or of the country.
	Type *reference domain*: Area on which information is provided such as accessible forest without shrub forest (german: Auswertungseinheit).
	Type *area domain of interest*, syn. Reference unit in Chap. 20 (german: Aussageeinheit): A geographically defined extent with clearly demarcated borders and known area, such as the whole of Switzerland, a canton or any administrative district. In the statistical analysis of the NFI, these are used as post strata.
Dominant diameter (ddom)	Diameter of the 100 thickest trees per hectare.
Established regeneration	All deciduous and coniferous tree species (no shrub species) >1.3 m above ground and a dbh <12 cm.
Estimate	Value as result of an estimator calculation.
Estimation	In probability sampling, estimation is the process of finding an estimate (or an approximation) of the unknown population parameter of interest.
Estimator	Rule for calculating an estimate of a population parameter based on observed data.
Expansion factor	Factor to calculate the local density (per hectare) of a target variable.
Field application	Software to capture field data or data from the NFI interview survey of the forest services.
Field attribute	Attribute measured on elements of the population on the sample plot or concerning the sample-plot centre for area-related information.
Field manual	The manual describes in detail the field observations that the field teams need to make, and includes definitions of all attributes and terms, as well as a complete description of the measurement process.
Field survey	Any type of measurement or assessment conducted on the field plots during a forest inventory.
Final felling	Silvicultural intervention to remove the remaining shelter of a stand.
Fine roots	Roots of woody plants <5 mm in diameter.
First phase sample	First phase (large sample with easy to capture information) of a two-phase sampling forest inventory.
Forest	Ecosystem continuously covered with trees. The Swiss NFI defines areas as forest if they are at least 20% covered by trees >3 m in height and are at least 25–50 m in width (depending on the tree cover).
Forest boundary line	Boundary between stocked and unstocked area, used for the NFI forest decision.

(continued)

Term	Description
Forest cover map	GIS-based map that distinguishes 'forest' from 'non-forest' area according to the consistent Swiss NFI definition.
Forest definition	The forest definition describes the requirements for measuring and interpreting the land cover and land use of forests. The criteria in the NFI are tree height, crown coverage, extent and land use.
Forest development plan	Planning instrument of the cantonal forest services that describes the forest area and defines forest-related objectives and development.
Forest edge	Boundary area between forest and other elements of the landscape. The forest edge may consist of a shelter belt, shrub belt and/or herbaceous fringe.
Forest function	Societal demands currently or potentially fulfilled by the forest. The main forest functions are: providing protection against natural hazards, wood production, protection of drinking water and welfare (nature protection, recreation).
Forest interspace	Shortest distance between the forest boundary lines through the sample-plot centre when the sample-plot centre is located on the unstocked side of the forest boundary line.
Forest inventory	Periodical assessment of tree and stand attributes as a basis for planning on the enterprise, regional or national level.
Forest management plan	Planning instrument for forestry enterprises with a validity period of 10–20 years. It consists of a description of the forest, a plan for future silvicultural interventions and a budget for future felling.
Forest road survey	Survey of the local forest services to acquire information on the state and the development of the forest roads.
Forest truck road	Roads >2.5 m in width capable of carrying vehicles with an axle load of 10 t at the time of the assessment.
Forest width	Shortest distance between forest boundary lines through the sample-plot centre when the sample-plot centre is located on the stocked side of the forest boundary line.
Global forest resources assessment (FRA)	Process in the framework of the Food and Agriculture Organization of the United Nations (FAO) to report on the state of forests worldwide.
Global navigation satellite system (GNSS)	System for a worldwide exact positioning based on satellite signals.
Greenhouse gas inventory (GHGI)	Inventory of turnover of climate-influencing gases in a country.
Growing stock	Wood of the stem including bark (stemwood) of living and standing trees and shrubs with a dbh >12 cm.
Herbaceous fringe	Unmanaged or only slightly managed buffer area without woody species between a forest edge and cultivated land.
High forest	Forest type with trees mostly established from seeds.
Impact assessment	Instrument to assess the impact of an activity or measure. In the Swiss NFI, the analysis and assessment of the target group oriented impact of NFI information and products.
Increment	Increase in the wood volume of trees. In the NFI, the increment consists of the stemwood increment of all surviving and

(continued)

Term	Description
	ingrowing trees and the predicted stemwood increment of the felled or mortality trees for half of the measuring interval.
Ingrowth	Trees in the inner (of two) concentric sample plots that developed a dbh of 12 cm between two measuring points in time.
Input form	Digital form to record data with different input components, such as text fields or dropdown lists, which is typically organised and displayed in windows.
Interpretation area	Area of 50 m × 50 m with the sample-plot centre at the intersection of the diagonals.
Intervention	Silvicultural treatment in a forest stand, such as tending, thinning, or regeneration felling.
Inventory	In the forestry context, the process of assessing forest characteristics for a defined area. The inventory is used to provide estimates for certain areas or populations of interest within a certain time period.
	In NAFIDAS the inventory defines the selection of data with a uniform semantic for a specific time frame.
Inventory cycle	Timeframe in a forest inventory in which a complete sampling grid following a (principally) uniform method is assessed.
Land cover	Land cover is the physical material on the surface of the earth.
Large branches	Branch wood with a diameter ≥7 cm measured over the bark.
Line intersect sampling	Procedure for selecting the sampling elements according to where they intersect with a line (transect).
Litter	Non-woody tree parts and woody tree parts <7 cm lying on the ground.
Losses	Trees that have disappeared between two consecutive inventories, e.g. due to forest management, avalanches, landslides or forest fires.
Lying deadwood	Dead trees or tree parts lying on the ground. In the Swiss NFI, this can be assessed either in the tally tree assessment, where only trees with a dbh >12 cm are recorded, or along a transect line, where all lying deadwood >7 cm is recorded.
MAIRA	Software to capture field data in the Swiss NFI.
Measurement quality objectives/MQO	Expected tolerance level to be met when assessing attributes, such as +/− 1 cm for dbh measurements.
Merchantable wood	Aboveground woody parts of trees >7 cm measured over bark.
Mortality	In the Swiss NFI, trees that died naturally between two consecutive measurements because of e.g. wind or insects, or have disappeared e.g. in an avalanche, but have not been harvested.
NAFIDAS	The National Forest Inventory Data Analysis System of the Swiss NFI is an application for storing and analysing NFI data on the state and change of the Swiss forest on regional and national levels.
Net change	Change in a population parameter of interest between two measuring points in time.
NFI	National forest inventory. In the context of this report "NFI" refers to the Swiss NFI called 'LFI'. The terms NFI1, NFI2, NFI3, NFI4 are related to the inventory cycle or the data analysis respectively reporting period.

(continued)

Term	Description
Non-forest area	Area that does not fulfil the forest definition according to the Swiss NFI.
Nongrowth	Trees outside the inner circle and inside the outer circle of the concentric sample plots that grew over the dbh threshold of 36 cm between two measuring points in time (or simulation decades).
Observation	Value of an attribute of a subject, based on measurement or visual assessment.
Observer agreement	Degree of agreement among observers with the same level of training under comparable conditions. This is particularly important for evaluations of the reproducibility of attribute assessment.
One-phase estimation	An estimation procedure in the Swiss NFI based on the sample of terrestrial plots alone.
Panel (annual)	Sample plots to be visited during the field season within 1 year. Representative subsample of the entire NFI sample of an inventory cycle.
Parameter	Metadata to control the NAFIDAS DAA application such as inventory ID or target variable.
Population	Main unit for which estimates are obtained in an inventory that are not area-related. In the NFI, a population could be trees with a dbh $\geq$12 cm, lying deadwood $\geq$7 cm in diameter or young trees $\geq$10 cm in height.
Post stratification	Statistical method to reduce sampling error. The post-strata are sub-groups or sub-units of the investigated population or area. The strata are constructed only at the time of analysis according to given inventory data.
Predetermined data	Data originating from previous inventories used to control the ongoing assessment in NAFIDAS and the field application.
Prediction	In a statistical sense, a statistical approach generate values of the response variable on not yet observed elements.
Pre-haulage	Transport of wood from the end-point of skidding to the selling location of timber (central timber yard or train station) on roads with limited access for trucks.
Production region	Regions of Switzerland with relatively homogeneous growth and wood production conditions. In the Swiss NFI, the country is classified into the regions Jura, Plateau, Pre-Alps, Alps and Southern Alps.
Productive functions	Forest function with the objective of promoting wood production.
Protection forest	Forest providing protection against natural hazards.
Protective or protection function(s)	Forest function with the objective of using the forest to provide protection against natural hazards.
Quality assurance (QA)	The objective is to provide a framework to assure the production of complete, accurate, and unbiased forest information of known quality. This includes having highly automated processes, state-of-the-art technology as well as hard- and software, a comprehensive field manual as well as instructions and training, plausibility checks, manual cross-checks and taking into account any feedback notes from the field surveys. Risk management, an ongoing process, is one of the primary tasks of QA measures.

(continued)

Term	Description
Quality control (QC)	Quality control, such as repeating the aerial image interpretation, the repeat and control survey, and plausibility checks.
Raster point	25 equally distributed points overlaid over the interpretation area (50 m × 50 m) used to determine land cover, to measure tree height and to calculate the crown coverage for the NFI forest decision.
Raw data	Data collected in the field with the field application MAIRA and uploaded to the central database. Currently, several spatial vector datasets are also stored in the raw database.
Recco	Telemetric technology with passive reflectors (chip), developed to locate people buried in avalanches.
Reduction line	Line at the transition between 'forest' and 'non-forest' areas to enable correct estimations of a population within the forest area.
Reference point	Point with x-, y- and z- coordinates determined in the stereo aerial image to locate the centre of the sample plot in the field.
Reference stand	Stand enclosing the centre of the terrestrial sample plot.
Regeneration	In the Swiss NFI, it describes all juvenile woody plants ≥10 cm in height up to a dbh <11.9 cm. The assessment of the regeneration focuses mostly on juvenile trees.
Regulation of mixture	Tending intervention at the young growth and thicket stage to regulate species mixtures.
Repeat survey	Complete re-assessment of a subsample of the field plots by a QA team (blind check), conducted in the same field season.
Result table	The main objective of a result table is to display the estimate and the standard error of a target unit. The layout is determined according to the classification units and other NAFIDAS parameters selected by the user.
Risk management	Process for continually identifying and assessing processes, assessment techniques, measurement devices as well as soft- and hardware with a high risk. Risk management consists of the three main components: (1) risk assessment, (2) risk treatment and (3) risk acceptance, including risk communication.
	Risk management in NAFIDAS means that processes and software modules with a high-risk priority can be identified and prioritized at any time during development, especially when changes are made to the system.
Rotation period	Period between regeneration and final felling of an even-aged stand.
Roundwood	Wood without bark or a stump that is sortable according to the Swiss trading customs regulations established by the Swiss Forest Owners Association.
Salvage logging	Unplanned logging as a result of abiotic damage, e.g. snow or storms, or biotic damage, e.g. bark beetles.
Sample inclusion probability	Probability of selecting an object of interest in a sample survey.

(continued)

Term	Description
Sample plot	Randomly or systematically selected part of the forest area on which tree, stand or plot attributes are acquired.
Sampling frame	The area from which the samples are randomly drawn. In the Swiss NFI, samples are drawn from a systematic grid with a random starting point.
Sampling grid	Spatial distribution of sample points.
Sampling units	In the Swiss NFI, the (infinite) possible points (potential sample plot centres) in a sampling frame.
Shelterwood felling	An even-aged silvicultural system in which, in order to provide a source of seed and/or protection for regeneration, the overstorey is removed in two or more successive shelterwood cuttings, the first of which is usually the seed cutting (although it may be preceded by a preparatory cutting) and the last is the final cutting. Any intervening cuttings are termed 'removal' or 'secondary' cuttings.
Shrub	Woody plant species with branching of the stem at the ground level, which does not reach 5 m in height.
Shrub belt	Fringe of woody plant species with a dbh <12 cm in front of the shelter belt at the forest edge.
Shrub forest	Forest area where more than two thirds is covered with shrub species. Shrub forests are mostly covered with green alder (*Alnus viridis*) and knee pine (*Pinus mugo prostrata*), but also with hazel or similar stockings.
Small branches	Branch wood with a diameter <7 cm measured over the bark.
Software development cycle	A process or framework to define and control tasks in software development at each step in the software development process.
Stage of development	Stage in the development of a stand defined according to the size of the dominant tree (diameter or height). In the Swiss NFI the following stages of development are defined based on the dominant diameter: young growth/thicket (<12 cm), polet timber (12–30 cm), young timber (31–40 cm), medium timber (41–50 cm) and old timber (>50 cm).
Stand	Group of trees that are clearly differentiated from their surroundings in terms of tree species composition, stand age and/or structure. In the Swiss NFI, a stand area is $\geq$5 ares.
Stand border	Boundary separating tree groups that differ in terms of tree species composition, age and/or structure.
Stand stability	Resilience of a stand against disrupting influences. It is derived in the Swiss NFI from the assessment of mechanical stability under abiotic and biotic stresses over the next 10–20 years.
Standard error	The variation of the mean of an estimate in a sample.
State of Europe's Forests (SoEF)	A process in the framework of the Pan-European Ministerial Conference on the Protection of Forests in Europe.

(continued)

Term	Description
Stem	Main axis of the tree from the ground to the top of the crown.
Stem	Main woody axis of tree.
Stemwood	Wood aboveground from the base to the top of the stem (without the branches) measured over bark.
Strata	1. Grouping of the sample into homogeneous subgroups in order to reduce the variability of an estimate within strata. 2. Domains (area).
Stump	The aboveground base part of the stem that would remain after cutting the tree with a height <1.3 m aboveground.
Stump inventory	The part of the forest inventory on the assessment of stumps.
Subplot	In Swiss NFI, a circular plot with a centre shifted away from the sample plot used in regeneration assessment.
Surviving trees	Sample trees recorded as living trees in two consecutive measurement cycles.
Sustainable forest management	Management and use of forests that ensures to maintain their biological diversity, productivity, regeneration capabilities, vitality and ability to fulfil relevant current and future ecological, economic and social functions without affecting other ecosystems (as defined in Forest Europe).
Tally tree	Tree that is part of the sample of an inventory. In the Swiss NFI, tally trees have a dbh ≥12 cm and are located on a sample plot within the forest area according to the Swiss NFI forest definition.
Target variable	Measure of a population or area characteristic (e.g. basal are per hectare).
Tariff tree	Sample tree subjected to additional measurements to calculate a volume model (tariff) and to correct a possible bias in the volume estimate. It is part of a subsample of the NFI sample trees.
Thinning	Silvicultural intervention to directly or indirectly favour choice trees (elite trees) and to improve the forest structure, stability and quality of the remaining trees.
Timber harvesting	The main timber harvesting processes are felling and hauling (transport to road).
Timber harvesting (technique)	Organisational and technical work flow in timber harvesting, from the felling of a tree to hauling it a way.
Total mean increment (TMI)	Potential average amount of wood produced within 50 years since the stand was established in an even-aged high forest. It is a measure of the performance of a site, independent of the current stocking.
Tree	Plant with an upright woody stem >5 m in height under normal growth conditions.
Two-phase (double) sampling	Forest inventory based on a large sample (first phase) with information that is easy to obtain and a sub-sample (second phase) with attributes that are more difficult to obtain. This procedure supports an efficient and precise estimate of the variables of interest.

(continued)

Term	Description
Two-stage estimation	Process of producing unbiased estimates of tree volume. In the Swiss NFI, tree height and upper stem diameter are measured on a random sub-sample of the standing, living trees, and the single stem volume of these trees is predicted with a very precise volume function.
	With two-stage estimations of growing stock and biomass, the estimates are based on: (a) single tree volumes predicted with tariff functions on all standing living trees (first stage), and (b) differences between the single-stem volume predicted with the volume function and the single-stem volume predicted with the tariff function on a random subset of standing living trees (second stage).
	Two-stage estimations minimise the risk of bias in the estimates arising from the (local) deviations of the tariff functions from volume functions.
Two-stage sampling	In the Swiss NFI applied for the selection of tariff trees. It defines a selection procedure to gain a subsample of tally trees, based on a selection probability proportional to tree size.
Type of analysis	Defines the type of analysis in NAFIDAS according to whether the focus is on the state, change components or gross change.
Variable	Unambiguously defined meaning (attribute) concerning a population or an area. A variable classifies the population or the area of interest according to certain criteria. Examples are dbh (tree-related) or forest ownership (area-related). In NAFIDAS a variable defines a database structure and semantic specification.
Vegetation height model	A vegetation height model is a raster model of the height of aboveground vegetation. It is calculated from the difference between a surface model and a terrain model and does not contain any anthropogenic objects with height such as houses.
Vegetation period	Period where substantial growth of woody species is assumed (April to August).
Young trees	In the NFI, trees >10 cm in height with a dbh <12 cm.

Index

A

Aboveground biomass, 46, 243
Aerial images, 4, 9, 10, 12, 68, 95–97, 118, 126, 129, 130, 149, 150, 163, 340, 342, 370, 377, 381
Analysis tables, 315–318, 323, 333
Annual panels, 40, 54, 63–66, 71–73, 75–80, 148, 149, 187, 331, 347, 349, 359–361
Architecture, 106, 278, 300, 309–313, 320, 377, 382, 387–390, 392–394, 400
Assortments, 189, 231–237, 259, 261, 286
Azonal site, 21, 181, 418

B

Basal area, 46, 53, 85, 86, 206, 222, 223, 226, 227, 250, 251, 261, 265, 268, 271–273, 276, 277, 287, 301–303, 306, 308, 309, 319, 321–324, 326, 348
Bias, 12, 40, 71, 212, 213, 218, 341, 352, 361, 362, 364
Biodiversity function(s), 5, 97
Biomass, 12, 40, 41, 46, 50, 62, 69, 71, 85, 86, 126, 136, 139, 206, 243–247, 277, 282, 292, 293
Blind check, 13, 342, 347
Boles, 232, 237, 261
Business as usual (BAU) scenario, 288, 289, 291–294

C

Cantons, 4, 5, 7, 8, 10, 14, 18, 41, 67, 82, 139, 149, 151, 154, 156, 251, 306, 309, 312, 314, 334, 344, 369, 408
Carrying capacity, 45, 154, 370
Change components, 41, 75, 79, 80, 82–89, 302, 309, 325, 327
Classification units (CUs), 301, 309, 321–324, 326, 330, 333, 334
Cold check, 347
Concrete, 118, 191, 193, 347, 389
Continuous inventory, 5, 9, 146, 153, 390, 391, 393
Continuous survey system, 11
Control surveys, 152, 341, 346–348, 363, 364, 387

D

Database, 13, 18, 51, 101–108, 150, 161, 253, 258, 266, 288, 300, 302, 304–320, 328–331, 333, 345, 348
Data catalogue, 8, 9, 200, 250, 306, 312
Data collection, 5, 6, 8–14, 40, 50, 57, 62–66, 68, 72, 73, 101, 106, 108, 109, 145, 146, 150, 160, 198, 199, 201, 331, 369, 370, 373, 377–401
Dead organic matter (DOM), 281–283, 286, 311

C. Fischer, B. Traub (eds.), *Swiss National Forest Inventory – Methods and Models of the Fourth Assessment*, Managing Forest Ecosystems 35,
https://doi.org/10.1007/978-3-030-19293-8

Deadwood, 12, 40, 41, 44, 56, 57, 74, 113, 161, 172, 244, 245, 247, 281–283, 286, 292–294, 386, 392, 397, 398
Degree of cover/coverage (DC), 121, 122, 165, 171, 252
Derivation, 13, 57, 68, 69, 71, 85–89, 102, 208, 276, 300–302, 304–309, 314–321, 328–333, 380, 405, 407, 409, 410
Derived attribute, 145, 159, 315
Derived data, 159, 195, 302, 305, 315, 318, 330, 395, 409
Diameter at breast height (dbh), 11, 39, 44, 81, 136, 168–171, 205, 208, 212, 216, 218, 223, 234, 245, 246, 250, 261, 268, 271, 274, 277, 283, 301–303, 315, 348, 350, 351, 353–355, 360–362, 396, 397, 400

E

Established regeneration, 171, 348

F

Felling, 82, 190, 258, 259, 266, 270–273, 276, 288
Field applications, 161, 165, 169, 313, 377–382, 390–393, 400
Field manual, 13, 161, 165, 189, 312, 340, 341, 344–346, 364, 396, 400
Field season, 147, 149–151, 153, 155, 156, 163, 198, 342, 345–347, 378, 382–384, 387, 390, 391, 395, 399, 401
Field survey manual, 146, 159, 161
Field surveys, 7, 10–12, 95, 96, 145–157, 159, 161, 187, 188, 195, 197, 199, 200, 250, 302, 328, 339–364, 384, 400, 401
First-phase sampling, 11, 68, 87, 320, 328, 336
Foliage, 243, 245–247, 281, 282
Forest administration, 4, 7, 9, 12, 18
Forest areas, 4, 8, 10, 13, 40, 41, 43, 44, 48, 62, 69, 95, 96, 108, 113, 116, 129, 139, 288, 290, 323, 324, 326
Forest boundary line (FBL), 11, 49, 96, 107, 108, 111–119, 121–123, 147, 165, 342, 343
Forest cover map, 10, 11, 17, 45, 68, 125, 126, 129, 137, 195
Forest development plans, 13, 188
Forest districts, 13, 67, 138, 139, 148, 154, 156, 187, 188, 191, 193, 202, 236, 320, 370, 377, 378
Forest edges, 12, 45, 108, 111–113, 130, 132, 145, 161, 162, 166, 168, 190, 192, 195, 200, 250, 251, 386
Forest establishment, 189
Forest functions, 4, 8, 9, 13, 96, 145, 189, 251
Forest inventory, v, 4, 7, 14, 45–48, 59, 74, 75, 78, 79, 81, 82, 129, 136, 300, 386
Forest monitoring, 339, 340
Forest planning, 5, 13, 151, 156
Forest resources, 8, 9, 18, 139, 205–223, 281
Forest roads, 4, 8, 13, 18, 43, 45, 154, 156, 161, 163, 188, 190–193, 199, 201, 202, 367–373
Forest road survey, 10, 45, 153, 154, 156, 188–193, 195, 202, 302, 369, 370
Forests, 3, 39, 95, 102, 107, 111, 147, 159, 187, 197, 208, 235, 244, 249, 257, 265, 281, 285, 300, 339, 367, 377, 405
Forest service, 8, 12–14, 17, 45, 145, 151, 153, 156, 173, 187–195, 202, 236, 237, 257, 328, 344, 368–370
Forest truck roads, 45, 190–195, 367, 370, 372
Forest use, 96, 112, 121, 123, 153, 173, 189

G

Gravel, 191, 193
Greenhouse gas inventory (GHGI), 5, 6, 14, 16, 292, 294, 331
Growing stock, 10, 41, 50, 62, 67–69, 71, 72, 74, 125, 136–139, 246, 259–262, 266, 272, 275–277, 285–292, 408
Growth model, 84, 206, 222, 265, 268, 277

H

Harvesting costs, 4, 9, 190, 257–260, 286, 287, 291, 292, 294
Harvesting productivity model (HeProMo), 257–262, 286
Hauling route, 12
Herbaceous fringe, 168
Hot checks, 346

I

Impact assessment, 14
Increment, 9, 74, 83–86, 147, 206, 216, 222–223, 226–229, 234, 253, 268, 274, 277, 287–289, 292, 302
Ingrowth, 81–85, 87, 123, 274, 277, 282, 288, 290, 350

Internal web application, 302, 304, 305, 309, 311, 319, 331
Interpretation area, 11–13, 116–119, 135, 172–174, 386
Interventions, 147, 154, 156, 168, 173, 257, 271–273, 275, 276, 288, 291, 294, 382
Interview survey, 10, 11, 45, 145–157, 187–195, 197, 201–202, 252, 257, 260, 328, 377, 378
Inventories, 4, 40, 96, 102, 159, 187, 197, 206, 234, 243, 250, 257, 271, 282, 287, 340, 368, 377, 405

K
Kappa, 141, 347, 350–352, 354, 357–362

L
Land cover, 12, 96, 116–118, 124, 131, 342
Land uses, 4, 42, 96, 121, 122, 126, 129, 292, 342
Line intersect sampling, 12, 44, 56, 172, 244
Litter, 247, 281–283, 286, 292–294
Logging, 147, 173, 271, 273, 276, 390
Lying deadwood, 12, 40, 44, 48, 56, 57, 59, 61, 69, 161, 172, 200, 244, 245, 247, 281, 386

M
MAIRA, 146, 151, 160, 165, 169, 198, 200, 313, 341, 345, 377–380, 382–388, 390, 393–400
Management scenario simulation model (MASSIMO), 5, 223, 258, 259, 266–278, 286–288, 293, 294, 315, 318
Marking points, 162, 163, 378
Measurement quality objectives (MQO), 351
Metadata, 101, 102, 105, 107, 302, 305, 307–309, 312, 315, 316, 318–320, 328–331, 400, 410
Microhabitats, 9

N
National forest inventory (NFI), 3, 40, 95, 101, 107, 119, 145, 159, 187, 197, 205, 231, 243, 249, 257, 265, 281, 285, 300, 339, 368, 377, 405
National forest inventory data analysis system (NAFIDAS), 13, 189, 223, 231, 232, 258, 300–332, 380, 405–414
Naturalness, 189
Needs assessments, 7
Net change, 57, 71, 74–78, 85
Network of forest road, 8, 13
Non-forest, 9, 12, 62, 63, 69, 96, 99, 108, 111, 112, 114, 116, 119–123, 130, 145, 150, 151, 161, 165, 167, 195, 199, 200, 320, 326, 334, 342, 343, 346, 347, 385
Nongrowth, 82–85, 87, 88, 222, 223, 282

O
Other wooded land (OWL), 10, 11, 13

P
Parameter set, 304–306, 308, 311, 318
Piggyback transport, 192, 193
Plausibility tests, 13, 378
Post-stratification, 10
Potential vegetation, 12
Pre-determined data, 151, 380–382, 390, 396, 397, 400
Productive functions, 4
Protection forests, 6, 17, 118, 126, 251–252, 266, 270, 272, 276, 290, 292, 293
Protective functions, 4, 5, 97, 173, 249, 251, 252, 260, 272, 273, 288, 293, 294
Pseudoturnover (PT), 352, 353, 359–362
Public web application, 303, 306–308, 310, 311

Q
Quality assurance (QA), 319, 329, 331, 332, 339–341, 347, 351, 363, 370, 371, 373, 411
Quality control (QC), 103, 201, 319, 340–341, 352, 363, 372

R
Raster points, 11, 12, 107, 108, 116–123, 342
Raw data, 13, 17, 64, 68, 101, 102, 145, 302, 306, 308, 314, 317–319, 328, 331, 377–401, 407
Raw database, 13, 300, 302, 314, 380, 410
RECCO, 147, 162

Reduction line, 50, 62, 166
Reference domains, 76–78, 301, 307, 309, 321–323, 325–327, 330, 333
Reference points, 12, 96, 108, 163, 165, 378, 381
Reference stand, 165, 168, 171, 173, 174, 348, 349
Reference unit, 251, 309, 320, 322, 333, 334
Regeneration, 171–172, 200, 274–275, 289, 348
Remote sensing, 7, 10, 17, 18, 41, 99, 101–104, 129, 132, 136, 145, 251, 339
Remote sensing products, 10, 18, 99
Renaturalised road, 192
Repeat survey, 148, 152, 157, 198, 199, 340, 341, 347–351, 361, 363, 364, 387
Requirements, 410–411
Risk assessment, 341
Risk management, 319, 341

S
Sample-plot, 10, 45, 95, 101, 106, 116, 130, 145, 161, 198, 216, 234, 257, 266, 381
Sample plot radius, 48, 55
Sample-plot centres (SPC), 6, 11, 12, 45, 96, 108, 113, 116–119, 121–123, 147, 161–163, 165, 167–169, 171–173, 199, 200
Sampling design, 39–89, 323
Sampling grid, 9, 18, 64, 68, 95, 127, 302, 304, 309, 320, 321, 325, 330, 333
Shelter belt, 168, 250
Shrub belt, 168, 250
Shrub forest, 10, 41–43, 62, 76, 77, 115, 150, 151, 153, 159, 165, 187, 199, 200, 262, 294, 301, 309, 322, 325, 342, 349, 384, 385
Shrubs, 10–12, 17, 40–44, 111, 115, 118, 119, 128, 153, 159, 165, 167, 168, 170, 171, 173, 199, 250, 348
Silviculture, 13
SILVIS, 156, 188, 377–380, 382, 388, 390, 400
Software development cycle, 390–392, 399, 400
Stage of development, 173, 250
Stand, 12, 40, 43, 50, 51, 74, 111, 126, 137, 161, 170, 173, 174, 189, 200, 216, 223, 234, 250–253, 257, 260, 265, 266, 268, 270–273, 276, 277, 287, 341, 346, 348, 349, 353, 357–361, 385
Stand age, 223, 268, 272, 273, 348, 353–355, 360, 361
Stand assessment, 161, 173, 200
Stand borders, 168, 349, 355, 361
Stand stability, 161, 174, 200, 252, 346, 348, 354, 357–361
State inventory, 271, 277, 302
Stereo-image interpretation, 10, 63, 101, 108, 132, 342
Stocking boundary, 162, 168
Stumps, 9, 44, 113, 114, 169, 170, 206, 216, 223, 234, 235, 244, 250
Sub-grids, 5, 10, 12, 146, 148, 321
Subplots, 58, 61, 70, 78, 171, 200, 274, 277, 288
Subpopulations, 49, 58, 69, 70, 80–86, 322–324, 326, 327
Sub-sample, 11, 50, 51, 68, 71, 81, 82, 87, 206, 212, 347
Survey method, 7, 8
Survey team, 6, 148, 151–153, 160, 198, 344
Surviving trees, 268, 282, 287
Sustainable forest management (SFM), 4, 9, 16, 250

T
Tally trees, 11, 43, 44, 48–53, 55, 56, 71, 74, 81, 83, 87, 88, 348
Target variables, 13, 40, 41, 45–48, 51, 58, 60, 62, 66, 68–70, 72–80, 82–85, 247, 302, 304, 305, 309, 318, 321–328, 330, 333, 334
Tariff trees, 12, 50–56, 71, 85–90, 169, 170, 200, 206, 216, 222, 223, 234, 235, 348, 353, 355, 386
Taxation stretch, 11, 12
Terrestrial inventory, 145–157
Terrestrial survey, 11, 13, 130, 146, 150, 154, 156, 159, 161, 197
Timber harvesting potentials, 258, 266, 275, 285–292, 294
Timber harvesting (technique), 153, 189, 190, 258
Timber yield, 5
Time study, 197–202
Training, 13, 18, 96, 103, 132, 133, 145, 149, 151, 152, 156, 198, 199, 201, 340–342, 344–347, 361, 364, 368, 369, 373
Transect lines, 12, 56, 57, 61, 172
Transects, 11, 12, 45, 56, 57, 61, 172, 398
Tree volumes, 12, 170, 206, 218, 222, 223, 249, 268

Two-phase sampling, 68, 69, 126, 320, 323, 335

U
Unbiased, 14, 57, 59–61, 66–68, 85, 89, 206, 213, 218, 320, 340, 396, 397, 400
Upper stem diameter, 11, 56, 170, 348
Utilization, 372

V
Variables, 10, 40, 96, 106, 126, 159, 188, 206, 234, 250, 260, 268, 283, 300, 339, 379, 407
Vegetation height model (VHM), 10, 17, 125–133, 137, 140, 320
Vertical structure, 168, 173
Volumes, 4, 8, 40, 46, 50–52, 57, 70, 71, 82, 85, 86, 88, 126, 147, 205, 206, 208, 212–214, 216, 218, 221–224, 228, 229, 234, 237, 243, 244, 246, 258, 260, 261, 268, 274, 277, 287, 288, 293, 348

W
Wood supply, 17
Wood volume, 8, 10, 205, 244
Woody plant species, 12, 161

Y
YASSO07, 281–283

9783030192921